Mushrooms *of* West Virginia *and the* Central Appalachians

To Jackson,
a budding
mycologist!
All my best,
Auntie Tamara

Mushrooms *of* West Virginia *and the* Central Appalachians

William C. Roody

THE UNIVERSITY PRESS OF KENTUCKY

Publication of this volume was made possible in part by a grant
from the National Endowment for the Humanities.

Published by The University Press of Kentucky
Scholarly publisher for the Commonwealth, serving Bellarmine University,
Berea College, Centre College of Kentucky, Eastern Kentucky University,
The Filson Historical Society, Georgetown College, Kentucky Historical Society,
Kentucky State University, Morehead State University, Murray State University,
Northern Kentucky University, Transylvania University, University of Kentucky,
University of Louisville, and Western Kentucky University.

Editorial and Sales Offices: The University Press of Kentucky
663 South Limestone Street, Lexington, Kentucky 40508-4008
www.kentuckypress.com

13 12 11 10 09   4 5 6 7 8

Frontispiece: Typical mushroom woods—dark, wet, mysterious, and full of promise.
Monongahelia National Forest.

Library of Congress Cataloging-in-Publication Data

Roody, William C.
Mushrooms of West Virginia and the Central Appalachians / William C. Roody
p. cm.
Includes bibliographical references (p.)
ISBN 0-8131-2262-7 (cloth: alk. paper)
ISBN 0-8131-9039-8 (paper: alk. paper)
1. Mushrooms—West Virginia—Identification. 2. Mushrooms—
Appalachian Region—Identification. 3. Mushrooms—West Virginia—Pictorial works.
4. Mushrooms—Appalachian Region—Pictorial works. I. Title.
QK617.R64 2002
579.6'09754—dc21
ISBN13: 978-0-8131-9039-6 (pbk.: alk. paper)

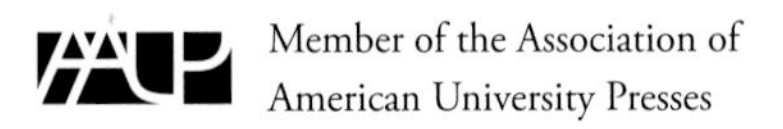
Member of the Association of
American University Presses

Printed in Korea by Pacifica Communications

Disclaimer

This book is intended to serve as a guide for identification of wild mushrooms. Although edibility is discussed, this is not a guide to the safe consumption of wild mushrooms. Many more species exist than can be presented in a general guidebook such as this, and the user is sure to encounter some that are similar to those featured here, including poisonous species. The information contained in this book regarding the edibility of wild mushrooms reflects current knowledge based on traditional use as well as the considerable experience of the author. While every effort has been made to alert the user to the dangers associated with eating wild mushrooms, should the user choose to eat wild mushrooms, he or she does so strictly at his or her own risk. Neither the author nor the publisher can accept responsibility for misidentifications made by the user, nor allergic reactions from consuming species that are generally regarded as edible. Anyone who wishes to gather wild mushrooms for eating should consult other books that specifically address this subject.

This book is dedicated to my dear mother,

Vera L. Bennett Roody

Contents

Preface

Wild mushrooms can be found in all terrestrial habitats, but the greatest variety and abundance occur in association with trees, either as saprobic decomposers of wood and forest litter, as parasites, or as symbiotic partners. Because West Virginia is a forested state that normally receives plentiful rainfall, the mycobiota is very rich. Pastures, parklands, and other grassland habitats nourish many additional species that contribute to the amazing diversity of fungi that are present throughout the Appalachians. Although an inventory of mushrooms in the state is far from complete, several hundred species have already been recorded. For the mushroom hunter, this bounty provides sheer delight and considerable challenge. Every outdoor excursion offers the chance of finding a previously unencountered mushroom. The real possibility even exists of discovering a species that is completely new to science.

It would require multiple volumes to describe and illustrate all of the known mushrooms from an area as large and biologically diverse as West Virginia. However, this guide will serve as a substantial introduction to the mushrooms of the state and of the broader region encompassing the central Appalachians. Nearly 400 species are described and illustrated, and many more are discussed in the text. The featured mushrooms represent a balance between those that are common and conspicuous within the region and some that are less well known. Priority has been given to the important edible and poisonous species. The photographs were taken by the author and with few exceptions show the mushrooms in their natural habitats.

The method for identifying species is simplified for the guide's easier use by those without previous experience or familiarity with wild mushrooms. A glossary of specialized mycological terms is provided.

The author holds a firm belief that through the pleasures of wild mushrooms one may discover a new appreciation for the beauty, ecology, and tangible rewards of the natural world.

Acknowledgments

I have many people to thank for their assistance and support in the preparation of this book. First and foremost, I am enormously grateful to Donna Mitchell, who has contributed in so many ways that it is impossible to recount them all. Hank Mashburn and Walt Sturgeon have been particularly helpful in sharing their knowledge of West Virginia mushrooms. Ernst Both, Raymond Fatto, Donna Mitchell, Walt Sturgeon, Tom Volk, and Don Yeltman reviewed parts of the manuscript and made many valuable suggestions. Sam Norris skillfully produced the line drawings in the pictorial key to groups. Additional mycological friends and colleagues who have assisted with identifications, or upon whom I have imposed for their wise counsel regarding the taxonomy and distribution of mushrooms, include Alan and Arleen Bessette, Glenn Freeman, Robert Hunsucker, Emily Johnson, Jay Justice, Gary Lincoff, Owen McConnell, Joe Miller, Orson K. Miller, Susan Mitchell, Steven L. Stephenson, Steve Trudell, and Rod Tulloss.

I am pleased to express my gratitude to the Wildlife Diversity Program at the West Virginia Division of Natural Resources for financial support. In particular I want to thank Donald Phares, Assistant Chief of Special Projects, and especially Kathleen Leo, Project Leader–Nongame Wildlife, whose personal encouragement provided my incentive for translating my love of mushrooms into this book. Brian McDonald, coordinator for the Natural Heritage Program, has also given his full support to this project and other efforts to document West Virginia fungi. Paul J. Harmon provided invaluable assistance with technical matters in the preparation of the manuscript. I also want to thank West Virginia DNR staff members Tom Allen, Jeff Hajenga, Howard Jones, Walt Kordek, Sam Norris, Donald Phares, Craig Stihler, Jack Wallace, Jennifer Wykle, and Jim Vanderhorst, who supplied me with mushrooms for study and have made valuable collections of voucher specimens for the fungus herbarium. Tina Hall and Dean Walton of The Nature Conservancy also contributed collections of fungi during their assignment to the Elkins, West Virginia, office.

To these individuals and to the many other mushroom friends who share the joys of mushrooming, I am profoundly grateful.

Introduction

What Are Mushrooms?

The most satisfactory answer to this question is that a mushroom is the fruiting body of a fungus, designed to produce and disseminate reproductive spores. Mushrooms have evolved a great variety of forms and dispersal mechanisms in order to accomplish this, but the basic function remains the same. The main body of a fungus is usually hidden from view beneath the ground or within various substrates. It is composed of a network of microscopic thread-like filaments (hyphae), which individually are many times thinner than human hair and are invisible to the naked eye. A mass of interwoven hyphae is collectively called mycelium and can often be seen as a cottony or web-like "mold." It is the mycelium that carries on the feeding and growing processes of a fungus. Since fungi cannot manufacture their own energy through photosynthesis as green plants do, they must absorb carbon and other nutrients from other living or dead organic matter. The mycelium secretes digestive enzymes that break down pre-existing organic matter. The resulting dissolved nutrients are then absorbed by the mycelium. If a food source remains available, the mycelium may live for many years, periodically producing mushrooms or other fruitbodies to insure the dispersal of spores to new locations.

Because the fruitbodies have distinct features, they can be used to identify different species of fungi. The process is complicated by the fact that most mushrooms are temporary, ephemeral structures that appear sporadically, depending on environmental conditions. Their presence or absence is greatly influenced by the vagaries of temperature and rainfall, as well as by the limits imposed by nutritional requirements. Since fleshy mushrooms are 95 percent water, ample rainfall is especially important. In dry years a fungus may not produce fruitbodies.

The spores produced by fungi are similar in function to the seeds of plants. They are carried to new locations by wind and rain, or by insects and other animals. When a viable spore comes to rest on a suitable substrate, it germinates and sends forth a primary hypha. If this hypha comes in contact with a useable food source before its own meager reserves are exhausted, it begins to grow and branch, eventually forming a colony of mycelia. A single mushroom can produce and disperse an astronomical number of spores, but very few survive to establish a new mycelium. Except for rare species, which should not be collected, picking a mushroom is not harmful to the fungus, so as long as the mycelium and substrate on which it grows are not excessively disturbed.

Classification and Scientific Names of Mushrooms

The scientific names for mushrooms reflect a system of classification based on natural relationships. The names are formed in Latin and consist of two parts, which are usually italicized when written. The first part of the name (the first letter of which is always capitalized) is the genus, and it encompasses any number of mushrooms that have a restricted set of characteristics in common. The second part of the name (never capitalized) is the species. The species name indicates uniqueness from other closely related members of a genus. In some cases the classification is further refined to accommodate variations within a species. These may be designated as varieties and are indicated as "var." after the species name (for example, *Amanita muscaria* var. *formosa*). When formally written, author citations are included with the Latin name. The author citations, which are not italicized, serve as abbreviated bibliographic references that are mainly useful to those who want to consult the original work in which a species was first described, or to track its nomenclatural history.

The categories of Family and Order represent a higher level of relationship with broader affinities. Until recently, the classification of mushrooms was based on morphological features of the fruitbody and spores. Modern methods of classification (systematics) are increasingly based on molecular studies. New techniques, especially DNA sequencing, have shattered many traditional concepts of relationship, resulting in the current state of taxonomic flux. Consequently, when a new mushroom guidebook is published, it often includes some unfamiliar names and relationships for known mushrooms. This often causes confusion and frustration for the user who learned a different name for a mushroom from an older book. A standardized list of common names for mushrooms would greatly alleviate some of these problems and may be forthcoming in the not-too-distant future. Fortunately, the species epithet in scientific names seldom changes, even if a particular taxon is transferred to a different genus or family. In addition to the ongoing restructuring of systematics, there is sometimes honest disagreement among investigators regarding the proper placement of species. For the most part, the classification and nomenclature used in this book follow *Ainsworth and Bisby's Dictionary of the Fungi*, 9th ed. (2001).

At times the renaming of mushrooms may seem hopelessly confusing, but it is well to remember that the mushrooms themselves do not change. Using the most recent scientific names is useful for purposes of communication among mycologists and understanding phylogenetic relationships, but it has little to do with actually recognizing mushrooms in the field.

Eating Wild Mushrooms

For many people, the desire to gather wild mushrooms for eating is a primary motivation to learn about mushrooms. This is indeed one of the rewarding aspects of wild mushrooms. However, this activity must be approached with an element of caution since some mushrooms contain dangerous toxins. Collecting wild mushrooms for food is not an inherently risky business for prudent individuals. On the contrary, it is a wholesome activity that has been enjoyed for centuries in many cultures. In fact mushroom hunting approaches a national pastime in parts of Europe, Asia, and Africa. Our American cultural heritage, on the other hand, comes to us in large part from the British, who are notoriously reserved when it comes to eating wild mushrooms. As a result many Americans have acquired a collective aversion to eating wild mushrooms that has only recently begun to dissipate. In some parts of the United States, especially on the west coast, gathering and eating wild mushrooms is fast becoming a popular activity. In the Appalachians, there is a long tradition of eating a few species such as morels and bradleys (*Lactarius volemus*), but gathering a broader variety of wild mushrooms for food is still a relatively esoteric pursuit, and little competition exists for many good edibles. For those who wish to indulge in eating wild mushrooms, it would be beneficial to keep the following rules in mind.

1. **Never eat any mushroom unless you are absolutely certain of its identity *and* know that it is an established edible species.**

2. **Collect only fresh specimens from uncontaminated sites.** Many cases of "mushroom poisoning" are due to consuming overripe specimens that harbor harmful bacteria. Always refrigerate mushrooms as soon as possible after collecting. Also, bear in mind that mushrooms can take up toxic heavy metals and chemicals from contaminated ground.

3. **Cook all mushrooms well and eat only a small amount the first time you try a new species.** Cooking destroys certain toxins that are present in some mushrooms such as morels and honey mushrooms. Cooking also breaks down the cell walls, making mushrooms more digestible and nourishing. Since allergic reactions are not uncommon with mushrooms, eating a small amount to test your tolerance for a previously untried species is always good practice.

Poisonous Mushrooms

A relatively small percentage of wild mushrooms are toxic, but a few are potentially life threatening if ingested. It behooves anyone who is intent on eating wild mushrooms to first become familiar with the harmful species. Though only a handful of wild mushrooms are dangerously poisonous, it takes just one mistake to send the careless collector to the hospital . . . or worse. Special care should be taken to recognize *Amanita virosa* (p.62) and its close relatives. Members of this group are responsible for most mushroom-related fatalities in North America. Several species of Amanita are poisonous, and the edibility of many others in the genus is unknown. Even though a few such as *Amanita rubescens* and *Amanita fulva* are considered to be good edibles when well-cooked, it is strongly recommended to avoid all members of the genus Amanita for food, especially if you are a beginner.

The Deadly Galerina, *Galerina autumnalis* (p.33), is another common mushroom that contains potentially lethal toxins. There are many similar "little brown mushrooms" with which it can be confused, and none should be eaten. Some people have mistaken the Deadly Galerina for the edible Honey Mushroom, *Armillaria mellea* (p.30).

Other common mushrooms that are frequently responsible for moderate to severe gastrointestinal-type poisoning include *Chlorophyllum molybdites* (p.64), *Omphalotus illudens* (p.146), and *Boletus subvelutipes* (p.292). In addition to the wholesale rejection of the genus *Amanita,* one should avoid eating mushrooms belonging to *Entoloma, Inocybe, Cortinarius,* and *Hebeloma,* since these genera contain a high percentage of known toxic species. One should also avoid small species of *Lepiota,* some of which are dangerously poisonous. The toxicity of the mushrooms in this guide is noted in the species descriptions, but for more thorough information on this topic, see the recommended reading list (p.505).

How to Collect Mushrooms

The proper method for collecting mushrooms depends on one's purpose for doing so. If you are collecting unknown mushrooms to be identified, it is important to pick the entire fruitbody, including the base of the stalk. Whenever possible, try to obtain specimens of various ages in order to see the full range of developmental features. Make field notes to record the substrate, habitat, associated plants, colors, and odors of fresh specimens, as well as any staining or bruising that results from handling them. Collections that are destined for an herbarium should be kept in separate containers. Waxed paper bags are ideal for this, or specimens can simply be rolled up in a makeshift cylinder of wax paper with the ends twisted. Film canisters or small divided fishing tackle boxes are handy for transporting small or delicate specimens. Fleshy mushrooms are highly perishable. On returning from the field, refrigerate the mushrooms until they can be "worked on," except for any specimens to be used for spore prints.

When gathering known edible mushrooms for purposes of eating, choose only fresh specimens from uncontaminated sites. Do not pick along busy roadsides, in areas of industrial waste, or on lawns or golf courses where herbicides or other chemicals are routinely applied. It is best to "field clean" edibles by cutting away the stalk base and brushing off any loose soil or debris from the cap. A small soft-bristled paint brush works well for this. It is much more difficult to remove grit and debris from mushrooms once they have mingled with others in a basket. You may also want to cut away any blemishes or flesh damaged by insects. The extra care taken at the time of collection will pay big dividends later when preparing mushrooms for cooking.

It is generally best to transport mushrooms in a rigid container. A woven basket is the carrier of choice since it protects the specimens and allows air circulation. Mushroom baskets come in a wide variety of sizes and styles. Any basket will do, but many mushroomers take great pride in their choice and will search far and wide to find the perfect one for them. Although it is common practice to collect morels in a plastic bag, this is not recommended. Mushrooms kept in non-breathable plastic will spoil quickly, and they will almost certainly be ruined if left in a vehicle parked in the sun.

Spore Prints

Individual mushroom spores are too small to be seen with the naked eye. Although they differ in size, shape, and ornamentation, a microscope is required to examine these features. Spores also differ in color, and in combination with morphological features of the fruitbody, the color of a spore deposit can be very helpful in determining genera. This is especially true for gilled mushrooms and boletes, but is also important with some other groups. In some cases the spore color can distinguish safe edible mushrooms from poisonous look-alikes.

When spores are viewed individually under a microscope, many are colorless (hyaline) or indistinctly colored. However, when observed in a mass deposit, the color intensifies and becomes discernible.

On a gilled mushroom, the spores are formed at the tips of club-like structures (basidia) that are located on the gill face. Once mature, the spores are ejected into the space between the gills, and then fall into the air currents beneath the cap to be carried away to new locations. If the air is still, or if the surrounding air currents are restricted, the spores will drop vertically and accumulate directly below the cap. When mushrooms grow in overlapping clusters, natural spore deposits from one fruitbody are often formed on the caps of those below. A more reliable method for determining spore color is to make a spore print. With gilled mushrooms, boletes, and stalked tooth fungi, this is done by cutting the cap from the stalk and placing it face down on a sheet of white paper (or on clear glass such as a microscope slide), then covering it with a bowl or similar object to inhibit air currents (see photo). Depending on the species and freshness of the fruitbody, it may take anywhere from a half-hour to several hours before an accumulation of spores is sufficient to accurately determine the color. Using plain white paper, or transparent glass that can then be placed on a white background, is recommended so that subtle differences in color can be observed. This is especially important for light-colored spores. Sometimes a spore print is unobtainable because the specimens are immature, sterile, or have been adversely affected by temperature or humidity. With boletes and polypores that have small pore openings, it is important to orient the cap so that the tubes are as vertical as possible. This will allow the spores to fall free and not collect on the narrow tube walls.

When saving specimens for a herbarium, it is often desirable to keep them intact. To obtain a spore print without removing the cap from the stalk, place the stalk through a hole in the center of a white file card (just wide enough to accommodate the stalk), so that the fertile underside of the cap rests on the paper. Then position the card with the inserted mushroom over the opening of a vessel that will accommodate the stalk length, and place the entire assembly somewhere that is sheltered from air currents.

How to Use This Book

Starting with the pictured key to major groups on page 10, turn to the section that best fits the type of mushroom you are identifying. Review the introductory comments to the section, then proceed to the photo illustrations and search for a match. Choose the closest match and then refer to the written descriptions. The mushroom you are attempting to identify may or may not be illustrated, but the descriptions often include a discussion of similar species. A correct identification is made when both the illustration *and* the written description correspond with your collection. Some flexibility when interpreting characters is necessary, but don't *force* a species to fit a description. Remember that there are many more mushrooms in nature than are featured in this book.

For larger groups, such as gilled mushrooms and boletes, dichotomous field keys to subsections are provided in order to reduce the number of possible matches into smaller, more manageable units. When using these keys, it is important to follow the numerical progression from the beginning. With practice and familiarity with the groups, it may be possible to bypass the keys and go directly to the appropriate subsection. For convenience, the species accounts are arranged alphabetically within each section.

Many non-illustrated and not fully described species are discussed in the text under "comments." These remarks are only intended to provide clues or suggestions regarding possible look-alikes. Do not base identifications solely on the brief comments. For those species mentioned but not illustrated elsewhere in this guide, refer to the additional sources (see recommended books) to search for more complete descriptions and illustrations. Having multiple references will improve your success in identifying less common mushrooms.

Also keep in mind that the size of some mushrooms can vary considerably. The measurements given in the descriptions represent the normal size range, but larger or smaller specimens are sometimes encountered.

Pictured Guide

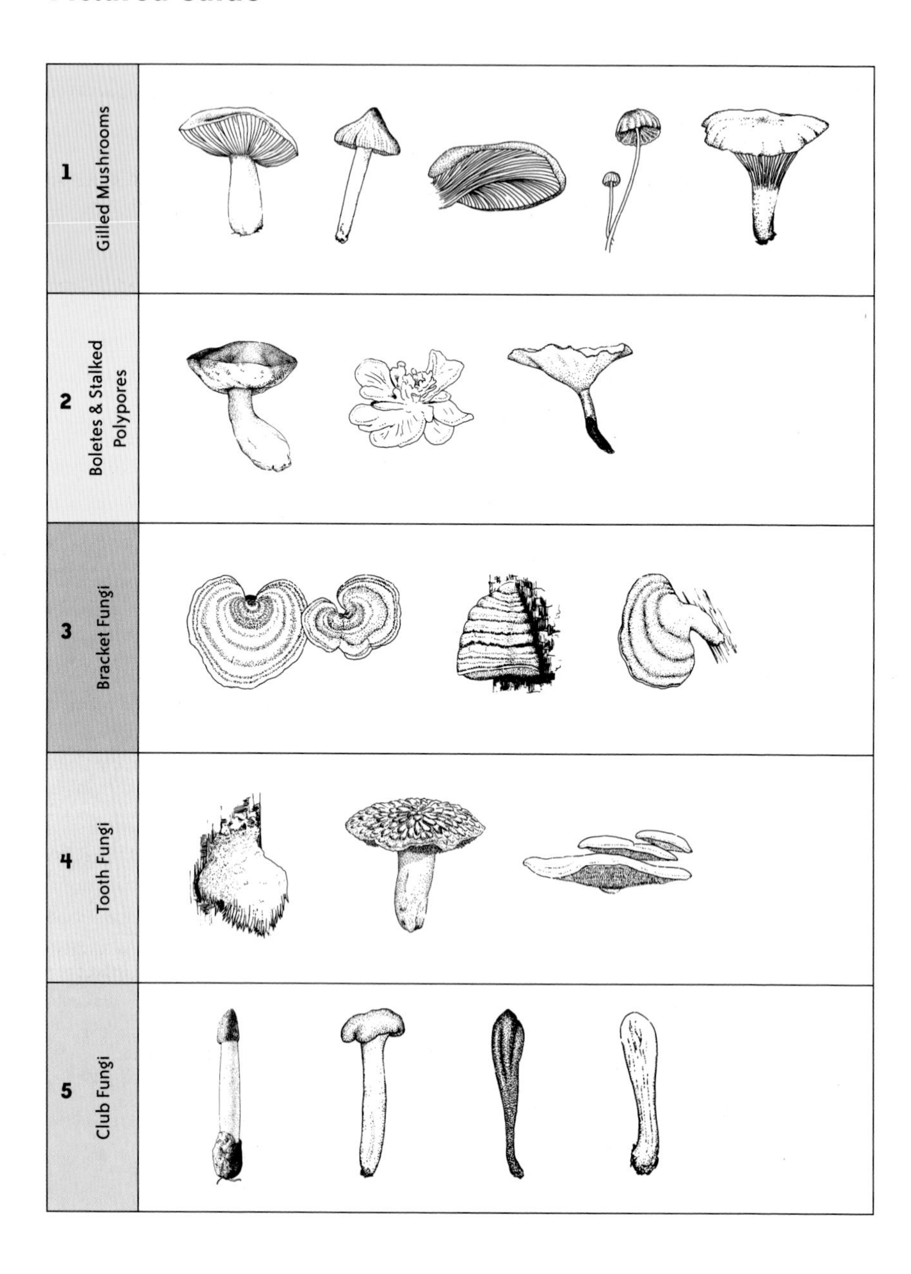

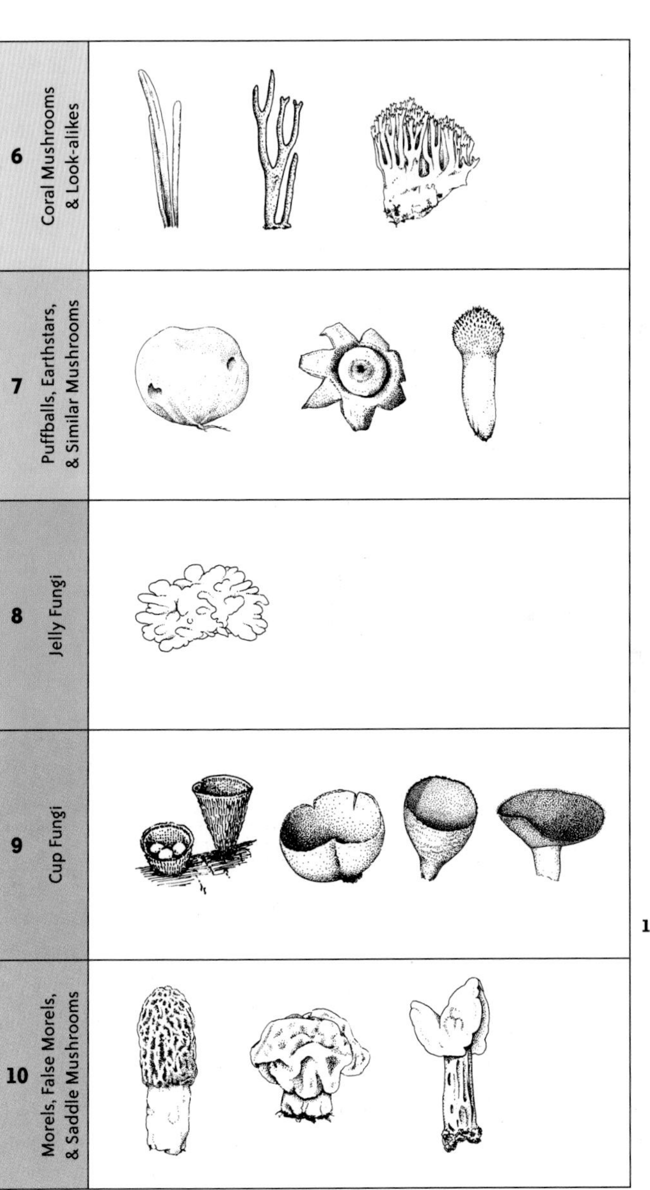

1 Gilled Mushrooms

2 Boletes & Stalked Polypores

3 Bracket Fungi

4 Tooth Fungi

5 Club Fungi

6 Coral Mushrooms & Look-Alikes

7 Puffballs, Earthstars, & Similar Mushrooms

8 Jelly Fungi

9 Cup Fungi

10 Morels, False Morels, & Saddle Mushrooms

11 Mycoparasites

Rozites caperatus, p.75

Gilled Mushrooms

Section 1

The gilled mushrooms constitute the majority of mushrooms included in this book. This is also the easiest group to categorize since it represents the form that most of us think of when applying the term "mushroom." Until fairly recently gilled mushrooms in the genus *Agaricus* were the only widely available mushrooms in supermarkets. The store-bought "button mushrooms" are immature *Agaricus bisporus.* Gilled mushrooms in the wild encompass a wide range of shapes, sizes, and colors. Many are more or less umbrella-shaped with a cap and stalk, others are shelf-like and stalkless, but all have radiating vertical plates (gills) on the underside of the cap. The reproductive spores are produced on the faces of these gills. Whether the gills are free from the stalk (not touching) or variously attached to the stalk is often important for identifying species. Cutting a mushroom in half lengthwise through the center of the cap will provide the best view for determining the type of gill attachment. Other features such as color, spacing, thickness, width, and the presence of interveining between the gills may also be useful. Since the number of gilled mushrooms represented in this book is comparatively large, the following dichotomous key will break them down into smaller groups that have similar features.

Key to the Groups of Gilled Mushrooms

Note: It is important to follow the key in numerical progression

1. Stalk lateral or lacking.....**Section 1-A** (p.16)
1. Stalk central or eccentric.....2.

2. Stalk with a conspicuous membranous or cottony ring.....3.
2. Stalk without a ring (but may have an appressed fibrillose ring zone).....4.

3. Fruitbody growing on wood.....**Section 1-B** (p.29)
3. Fruitbody growing on the ground or other non-woody substrates.....**Section 1-C** (p.44)

4. Gills descending the stalk (decurrent), or if broadly attached, exuding latex (a white, clear, or colored fluid) when broken.....5.
4. Gills free from the stalk, or variously attached but not decurrent, and not exuding latex when broken (a few Russula spp. in Sec. 1-J have subdecurrent gills).....7.

5. Cut or broken flesh and gills exuding a clear, white, or colored latex.....*Lactarius* **Section 1-D** (p.81)
5. Cut or broken flesh and gills not exuding latex.....6.

6. Fruitbody small with a stalk ¼ in. (6 mm) or less thick at the apex.....**Section 1-E** (p.111)
6. Fruitbody medium to large with a stalk typically more than ¼ in. (6 mm) thick at the apex....**Section 1-F** (p.125)

7. Fruitbody small with a stalk that is typically ¼ in. (6 mm) or less thick at the apex.....8.
7. Fruitbody medium to large with a stalk that is typically more than ¼ in. (6 mm) thick at the apex.....10.

8. Fruitbody growing on wood.....**Section 1-G** (p.152)
8. Fruitbody growing on soil or on other non-woody substrate.....9.

9. Spore deposit white to pale pinkish buff.....**Section 1-H(a)** (p.166)
9. Spore deposit pink, brown, or blackish.....**Section 1-H(b)** (p.189)

10. Fruitbody growing on wood.....**Section 1-I** (p.205)
10. Fruitbody growing on soil or other non-woody substrate.....11.

11. Flesh and stalk brittle, typically breaking easily and cleanly....*Russula* **Section 1-J** (p.215)
11. Flesh and stalk fibrous and not especially brittle.....12.

12. Spore deposit white to cream or pinkish buff.....**Section 1-K(a)** (p.235)
12. Spore deposit pink, brown, or blackish.....**Section 1-K(b)** (p.266)

Gilled Mushrooms with a Lateral Stalk, or Stalk Lacking

The mushrooms in this section are saprobes or parasites that grow on wood or, less often, on soil. Most are laterally attached to the substrate, often occurring in overlapping shelf-like clusters. The spore color of species with overlapping caps is sometimes evident in the field when the spores from one fruitbody accumulate on the caps of those beneath.

Latin name: *Crepidotus malachius* (Berk. & M.A. Curtis) Sacc.
Common name: Soft-skinned Crepidotus
Order: Agaricales
Family: Cortinariaceae
Width of cap: 1–2 in. (2.5–5 cm)
Cap: Kidney or fan-shaped; white to grayish, hygrophanous; surface smooth with dense white hairs at the point of attachment, margin striate; **flesh** white, odor mild, taste mild or bitter.
Gills: Crowded, radiating from a basal point of attachment, broad; whitish at first, becoming pinkish then rusty brown as the spores mature.
Stalk: Rudimentary or lacking; white-tomentose when present.
Spore print: Rusty brown to dark brown.
Spores: 5–7.5 μm, globose, smooth to finely punctate, pale brown.
Occurrence: In groups or overlapping clusters on logs and stumps of broad-leaved trees; saprobic; early summer-fall; occasional.
Edibility: Not edible.

Comments: Clusters of these mushrooms might be mistaken for small Oyster Mushrooms, *Pleurotus ostreatus* (p.25), or Angel Wings, *Pleurocybella porrigens* (not illustrated), which are similar in form but produce a white to pale lilac spore deposit. *Crepidotus applanatus* (not illustrated) is very similar but has narrow gills and smaller spores (4–5.5 μm). The genus name *Crepidotus* means "slipper-shaped." The name *malachius* comes from the Greek word malac, meaning "soft" or "tender."

Latin name: *Crepidotus mollis* (Schaeff.) Fr.
Synonyms: *Crepidotus fulvotomentosus* Peck; *Crepidotus calolepis* Fr.
Common names: Soft Slipper; Jelly Crep
Order: Agaricales
Family: Cortinariaceae
Width of cap: ½–2½ in. (1.3–6.5 cm)
Cap: Convex, kidney-shaped to fan-shaped, or semicircular, often thickened at the point of attachment; ochre-brown to orange-brown or tan, becoming whitish to grayish brown on drying; surface smooth or with scattered small scales; **flesh** thin, white, watery, upper layer gelatinous, odor not distinctive, taste mild or sometimes bitter.
Gills: Close, radiating from the point of attachment; pallid at first, becoming grayish brown to cinnamon in age.
Stalk: Rudimentary or lacking.
Spore print: Brown.
Spores: 7–9 x 4.5–6 μm, broadly elliptical, smooth.
Occurrence: Solitary to clustered on logs, fallen limbs, stumps, or standing deadwood of broad-leaved trees, rarely on conifer wood; saprobic; summer-fall; occasional.
Edibility: Not edible.

Comments: The gelatinous upper layer of the cap is elastic and can be stretched slightly at the margin. *Crepidotus crocophyllus* (not illustrated) is similar but has yellow to ochre-orange gills. Other species of *Crepidotus* differ microscopically. *Panellus stipticus* (p.23) is smaller and has a distinct stubby stalk, bitter tasting flesh, and white spores. Species of *Pleurotus* and *Hohenbuehelia* have white or pale lilac spores. *Mollis* means "soft."

Latin name: *Crepidotus versutus* Peck
Common name: None
Order: Agaricales
Family: Cortinariaceae
Width of cap: ½–1 in.(1–2.5 cm)
Cap: Fan to kidney-shaped, sometimes nearly circular, margin incurved; white to grayish; surface dry, downy to hairy; **flesh** thin, white, odor and taste not distinctive.
Gills: Radiating from the point of attachment, moderately well-spaced; white at first, becoming rusty brown in age.
Stalk: Rudimentary or lacking.
Spore print: Cinnamon-brown.
Spores: 7–11 x 4.5–6 μm, elliptical, smooth to minutely punctate, yellowish to yellowish brown.
Occurrence: Scattered in small groups or overlapping clusters on fallen branches, on the underside of logs, or other decaying wood of broad-leaved trees, infrequently reported on bare soil; saprobic; summer-fall; common.
Edibility: Unknown.

Comments: When growing on vertical substrates, the caps are laterally attached and shelf-like. When growing on the underside of a horizontal substrate, the fruitbodies are attached at the upper surface of the cap. *Crepidotus herbarum* (not illustrated) is similar but has smaller spores (6–7.5 x 3–4 μm) and a pale yellowish brown spore deposit. *Schizophyllum commune* (not illustrated) has a densely hairy white to grayish cap with gill-like folds underneath that are split lengthwise on the edge. The name *versutus* is derived from the Latin word versut meaning "clever," perhaps referring to the ability of the fruitbody to adapt its form depending on its position on a substrate.

Latin name: *Hohenbuehelia geogenia* (DC) Singer
Common name: Shoehorn Oyster Mushroom
Order: Agaricales
Family: Pleurotaceae
Width of cap: 1–4 in. (2.5–10 cm)
Height of fruitbody: 1½–4 in. (4–10 cm)
Cap: Spathulate to shoehorn-shaped, or deeply funnel-shaped with an opening on one side; fawn brown to umber brown; surface smooth, gelatinous (especially noticeable when moist); **flesh** soft and flexible, odor and taste farinaceous.
Gills: Close, decurrent; white to ochraceous.
Stalk: Lateral, short, solid; whitish to grayish.
Spore print: White
Spores: 5–6 x 4–5 μm, broadly ovoid, smooth, hyaline, non-amyloid.
Occurrence: In small groups or clusters on the ground, often arising from buried wood; saprobic; fall; occasional.
Edibility: Edible.

Comments: *Hohenbuehelia petaloides* (not illustrated), which has a whitish to grayish brown cap with a whitish bloom when young, is usually found on stumps and decaying logs, but also grows on buried wood and wood mulch. Similar species of *Pleurotus* lack a gelatinous layer in the cap tissue and grow conspicuously on wood. *Geogenia* means "arising from the earth."

Latin name: *Lentinellus ursinus* (Fr.) Kühner
Synonym: *Lentinus ursinus* Fr.
Common name: Bear Lentinus
Order: Russulales
Family: Auriscalpiaceae
Width of cap: 1–4 in. (2.5–10 cm)
Cap: Convex to flat, fan to kidney-shaped, or semicircular, margin often lobed, incurved at first; reddish brown, paler toward the margin; surface more or less covered with dark brown hairs that are especially dense at the point of attachment to the substrate, margin sometimes nearly glabrous; **flesh** sharply acrid tasting (sometimes slowly), or bitter.
Gills: Radiating from the point of attachment to the substrate; close; whitish to pinkish brown or tan; edges coarsely serrated.
Stalk: Lacking.
Spore print: White.
Spores: 3–4.5 x 2–3.5 μm, subglobose, minutely spiny, hyaline, amyloid.
Occurrence: Solitary or more typically in groups, often in overlapping clusters on decaying wood of broadleaf trees, less often on conifer wood; saprobic; summer-fall; occasional.
Edibility: Not edible.

Comments: The similar *Lentinellus vulpinus* (not illustrated) has a much paler pinkish white cap with radial rib-like veins, and whitish hairs at the point of attachment. It too has serrated gill edges and strongly acrid tasting flesh. *Ursinus* means "bear-like," in reference to the hairy brown cap.

Latin name: *Panellus serotinus* (Schrad.) J.G. Kühn
Synonym: *Pleurotus serotinus* (Schrad.) Gillet
Common name: Late Fall Oyster
Order: Agaricales
Family: Tricholomataceae
Width of cap: 1–4 in. (2.5–10 cm)
Cap: Convex, semicircular or fan-shaped to kidney-shaped in outline, margin incurved; color variable, olive-yellow becoming olive-brown to greenish brown, sometimes with violaceous tints; surface smooth, glabrous to slightly velvety, viscid when young or when moist; **flesh** firm, thick, white, odor not distinctive, taste mild to bitter.
Gills: Broadly attached to decurrent, close; whitish to yellowish, or orangish yellow.
Stalk: Short, thick, lateral, or sometimes absent; yellowish; downy or scaly.
Spore print: Pale yellow.
Spores: 4.5–6 x 1.5–2 μm, sausage-shaped, smooth, hyaline, amyloid.
Occurrence: Solitary or more often in overlapping clusters on standing deadwood or fallen trunks, stumps and branches of broad-leaved and conifer trees; saprobic; fall-early winter; common.
Edibility: Edible.

Comments: The Late Fall Oyster is easily recognized by its color, stature, and especially its appearance late in the year. It is not highly regarded as an edible but can be gathered when few other fleshy mushrooms are available. *Serotinus* means "late fruiting."

Latin name: *Panellus stipticus* (Bull.) P. Karst.
Common names: Luminescent Panellus; Bitter Oyster
Order: Agaricales
Family: Tricholomataceae
Width of cap: ½–1¼ in. (1.3–3 cm)
Cap: Convex with an incurved margin at first, semicircular to kidney-shaped or shell-shaped in outline; dingy white to pale brown, orangish buff or tan; surface dry, scurfy to minutely scaly; **flesh** yellowish pink, tough, odor not distinctive, taste bitter or astringent.
Gills: Broadly attached or subdecurrent with numerous crossveins; close to crowded; colored more or less like the cap.
Stalk: ¼–⅝ in. (0.5–1.5 cm) long when present; thick, eccentric to lateral; whitish to yellowish brown; pruinose to scurfy, or hairy at the base.
Spore print: White.
Spores: 3–6 x 2–3 μm, sausage-shaped, smooth, hyaline, amyloid.
Occurrence: In groups on logs, stumps, and fallen branches of broad-leaved trees; saprobic; can be found year round but is most frequently encountered in the fall; common.
Edibility: Not edible, and by some accounts possibly poisonous.

Comments: This mushroom will glow in the dark when fresh or sometimes when revived in water after drying. *Crepidotus mollis* (p.18) and other small *Crepidotus* species are similar, but they have brown spores and their fruitbodies are not luminescent. *Schizophyllum commune* (not illustrated) has a densely hairy white to grayish cap and longitudinally split gill-like folds on the underside. The specific epithet, which is sometimes spelled *stypticus,* means "astringent" or "styptic." The fruitbodies are reputed to have been used in folk medicine to staunch bleeding.

Latin name: *Phyllotopsis nidulans* (Pers.) Singer
Synonyms: *Pleurotus nidulans* (Pers.) P. Kumm.
Common name: Orange Mock Oyster
Order: Agaricales
Family: Tricholomataceae
Width of cap: 1–3 in. (2.5–7.5 cm)
Cap: Convex to flat with an inrolled margin at first, fan-shaped to shell-shaped in outline; orange to yellow-orange, or sometimes mostly yellow when young, hygrophanous, becoming paler on drying; surface densely covered with short, stiff hairs; **flesh** orangish, tough, typically with a strong odor of spoiled cabbage, but occasionally odorless, taste unpleasant.
Gills: Radiating from the point of attachment, moderately close; orange to yellow-orange or paler.
Stalk: Lacking, although the caps may narrow to a stalk-like base.
Spore print: Pale pink.
Spores: 6–8 x 3–4 μm, sausage-shaped, smooth, hyaline, non-amyloid.
Occurrence: Solitary or more often in overlapping clusters on decaying wood of broad-leaved trees, also reported to occur on conifer wood; saprobic; summer-fall; occasional.
Edibility: Not edible.

Comments: This beautiful and distinctive mushroom is easily recognized when fresh by the brilliant orange color and the unpleasant odor of most collections. *Paxillus panuoides* (not illustrated), which grows only on conifer wood, is similar but has an ochre-brown cap with yellow to pale yellow-orange forked gills, and yellowish buff spores. *Nidulans* means "nesting."

Latin name: *Pleurotus ostreatus* (Jacq.) Quél.
Synonym: *Pleurotus sapidus* (Schulzer ex Kalchbr.) Sacc. (according to some authors)
Common name: Oyster Mushroom
Order: Agaricales
Family: Pleurotaceae
Width of cap: 2–8 in. (5–20 cm)
Cap: Convex with an incurved margin at first, often becoming flattened with an upturned wavy margin in age, fan-shaped, oyster shell-shaped, or semicircular in outline; surface glabrous, smooth, satiny when dry; color variable, white to cream, grayish to brown, sometimes with lilac tones; **flesh** tough at the point of attachment, otherwise tender, odor pleasant, anise-like, taste mild.
Gills: Radiating from point of attachment, often decurrent, moderately close to crowded; cream at first, becoming yellowish in age or in drying.
Stalk: Short, eccentric to lateral, or more or less central when growing on top of a horizontal substrate, sometimes absent; whitish, base often fuzzy.
Spore print: White to pale lilac.
Spores: 7–12 x 3.5–5 μm, elliptical, smooth, hyaline, non-amyloid.
Occurrence: Gregarious, usually in overlapping clusters, sometimes solitary, on living or dead standing trunks, logs, and stumps of broad-leaved trees, less frequently on hemlock or pine; parasitic and saprobic; spring-early winter; common.
Edibility: Edible and very good when young and fresh, discard the tough basal portion. Small black beetles and other insects that often inhabit the gills can be evicted by tapping on the caps.

Continued on next page

Comments: This variable mushroom is actually a complex of closely related species or forms with overlapping features. Generally, the early spring and summer forms are white to cream-colored, while fall and early winter collections tend to be more grayish or grayish brown. Some mycologists consider the paler early forms to be a separate species (*Pleurotus pulmonarius*). For culinary purposes it makes little difference, since both are good edibles that can often be harvested in large quantities. The late season forms are less prone to insect infestation. Oyster mushrooms are easily cultivated and commercially produced varieties are now available in many supermarkets. The Angel Wing, *Pleurocybella porrigens* (not illustrated), which is smaller and thinner than the Oyster Mushroom, grows in late summer and fall on decaying conifer wood. The Elm Oyster, *Hypsizygus tessulatus* (not illustrated), has broadly attached or notched gills, a more prominent eccentric stalk, and has "water spots" on its whitish cap. It grows singly or in groups of two or three on living broad-leaved trees, especially elm. Species of *Crepidotus* have brown spores. *Lentinellus* species have hairy caps and serrated gill edges. See also the descriptions and comments under *Panellus serotinus* (p.22) and *Hohenbuehelia geogenia* (p.20). The name *Pleurotus* means "side ear," referring to the form and growth pattern. *Ostreatus* means "oyster-like."

Latin name: *Plicaturopsis crispa* (Pers.) D. Reid
Synonym: *Trogia crispa* (Pers.) Fr.
Common name: Crimped Gill
Order: Polyporales
Family: Atheliaceae
Width of cap: ⅜–1 in. (1–2.5 cm)
Cap: Fan-shaped or shell-shaped with incurved, lobed margin; upper surface tomentose, concentrically zoned with yellow-orange to ochre or reddish brown bands, paler on the margin; **flesh** thin, flexible, becoming hard and brittle when dry, odor and taste not distinctive.
Gills: Underside of the cap with moderately well-spaced, branching gill-like or vein-like folds or wrinkles; whitish to pale ochre.
Stalk: A rudimentary extension of the cap, or lacking.
Spore print: White.
Spores: 3–4 x 1–1.5 µm, cylindrical to sausage-shaped, usually containing 2 oil drops, smooth, hyaline.
Occurrence: Gregarious in overlapping clusters on dead branches and trunks of broad-leaved trees, especially beech and birch; saprobic; year-round; occasional.
Edibility: Not edible.

Comments: The wrinkled or folded fertile surface makes this mushroom easy to recognize. Multiple branching caps are not uncommon. The fruitbodies shrivel up when dry and revive in wet weather. Other similar wood-decaying "turkey tails" have true gills or pores, or are completely smooth on the underside. *Crispa* means "having a waved or curled margin."

Latin name: *Tectella patellaris* (Fr.) Murrill
Synonym: *Panus opercualtus* Berk. & M.A. Curtis
Common name: Veiled Panus
Order: Agaricales
Family: Tricholomataceae
Width of cap: ⅜–¾ in. (1–2 cm)
Cap: Convex and pendulous, kidney-shaped, shell-shaped, cup-like, or nearly round in outline, margin inrolled, often with attached remnants of a membranous partial veil; brown to reddish brown; **flesh** tough, odor and taste not distinctive.
Gills: Radiating from a short stub-like stalk or point of attachment; close to moderately well-spaced; colored like the cap or paler; at first concealed by a thin whitish buff membranous veil.
Stalk: Often lacking; when present, eccentric and very short, or expressed as a short lateral extension of the cap.
Spore print: White.
Spores: 3–4 x 1–1.5 μm, sausage-shaped, smooth, hyaline, amyloid.
Occurrence: In groups or clusters on fallen twigs and small branches of broad-leaved trees; saprobic; summer to early winter; occasional.
Edibility: Unknown. Too small to be of culinary interest.

Comments: This small, inconspicuous mushroom is easily overlooked and may be more common than is reported. Its pendulous form and presence of a partial veil on young specimens are distinctive features. *Patellaris* means "small dish-shaped."

Gilled Mushrooms with a Ring on the Stalk, Growing on Wood

Normally the mushrooms grouped here are conspicuously growing on wood. However they can sometimes appear to be growing on soil or humus if the woody substrate is extensively decomposed or buried. The stalk ring is formed from a partial veil that extends from the cap margin to the stalk on young specimens. As the cap expands, the veil typically separates from the cap margin but remains attached to the stalk as a well-defined membranous collar or skirt-like ring. Species that merely have a zone of appressed fibrils on the stalk are placed in other sections.

Latin name: *Armillaria mellea* (Vahl) P. Kumm.
Synonym: *Armillariella mellea* (Vahl) P. Karst.
Common names: Honey Mushroom; Bootlace Fungus
Order: Agaricales
Family: Tricholomataceae
Width of cap: 1½–4 in. (4–10 cm)
Length of stalk: 2–6 in. (5–15 cm)
Cap: Convex at first, becoming broadly convex to nearly flat; pale yellow-brown to honey colored with olivaceous tones, darker in the center; surface viscid when wet, nearly glabrous or with fine yellowish brown scales, especially concentrated on the disc; **flesh** white, thick at the center, odor and taste not distinctive.
Gills: Broadly attached or somewhat decurrent; white, becoming pinkish buff and spotted brownish with age.
Stalk: More or less equal but often tapered and sometimes compressed near the base; whitish at first, soon becoming olivaceous or brownish; fibrous and tough; surface floccose to scurfy; partial veil thick and cottony, leaving a membranous white ring on the upper stalk, ring often has a yellow margin.
Spore print: White to cream.
Spores: 7–9 x 5.5–7 μm, broadly elliptical, smooth, hyaline, non-amyloid.
Occurrence: In caespitose clusters on decaying wood of broad-leaved trees, or on the ground from buried wood, especially around old stumps, less often growing on conifer wood; summer-fall, occasionally appearing in spring; common.
Edibility: Poisonous raw, but edible and very good when *well cooked.* Reports of gastrointestinal disturbances from consuming this mushroom are probably due to undercooking. Most mycophagists consider the Honey Mushroom to be safe and delicious. In some years it can be harvested in large quantities. Great care should be taken not to mistake the Deadly Galerina, *Galerina autumnalis* (p.33) for the Honey Mushroom. Although both species occur on logs and stumps, and share some features, they can be easily and decisively distinguished by spore color. The Deadly Galerina has brown spores. Honey Mushroom spores are white.

Comments: What was once considered to be a single species in North America having various forms and ecological roles is now known to be a complex of up to ten closely related species. The determination of species based solely on field characters is difficult. At least three species are known to occur in West Virginia, including the easily recognized Ringless Honey Mushroom, *Armillaria tabescens* (not illustrated), which is very similar to *Armillaria mellea* except that it does not have a ring on the stalk. *Armillaria gallica* (not illustrated), which has a pinkish brown to reddish brown scaly cap, a web-like partial veil, and a swollen club-shaped stalk, normally grows on the ground but can sometimes be found on stumps and logs. Compare with *Galerina autumnalis* (p.33) and *Pholiota mutabilis* (p.39).

The Honey Mushroom is one of the few gilled mushrooms that can be a serious forest pathogen. It has the ability to live as a saprobe on decaying logs and stumps, and from there it can spread by underground mycelial cords (rhizomorphs) to the root systems of living trees and other plants, which may then be killed by the fungus. The flattened black rhizomorphs resemble bootlaces and can often be found beneath the sloughing off bark of dead trees. Wood that is permeated with the mycelium of the Honey Mushroom becomes luminescent and is called "foxfire." *Mellea* means "honey colored."

Latin name: *Cystoderma granosum* (Morg.) A.H. Sm. & Singer
Synonym: *Armillaria granosa* (Morg.) Kauffm.
Common name: None
Order: Agaricales
Family: Tricholomataceae
Width of cap: 1–3½ in. (2.5–9 cm)
Length of the stalk: 2–3 in. (5–7.5 cm)
Cap: Convex; dull orange to ocraceous orange; surface dry, granular, sometimes wrinkled; **flesh** white, odor and taste not distinctive.
Gills: Adnate to slightly adnexed, crowded; white to pale tan.
Stalk: More or less equal or tapering upward from a swollen base; surface granular from the base up to a flaring ring on the upper stalk, smooth above the ring; colored like the cap below the ring, whitish to pale brown above.
Spore print: White.
Spores: 4–5 x 3 μm, elliptical, smooth, hyaline, amyloid.
Occurrence: Solitary or more often in small groups, at times in caespitose clusters on decaying wood of broad-leaved trees, especially beech, birch, and maple; saprobic; late summer-fall; occasional.
Edibility: Unknown.

Comments: *Phaeomarasmius erinaceellus* (p.37) is reddish brown, smaller, and has brown spores. *Cystoderma amianthinum* (not illustrated) is smaller and grows in scattered groups on the ground. *Cystoderma terrei* (not illustrated) grows on the ground in conifer woods and has an orange to brick-red cap, and a scaly stalk. *Granosum* means "grainy."

Latin name: *Galerina autumnalis* (Peck) A.H. Sm. & Singer
Synonym: *Pholiota autumnalis* Peck
Common name: Deadly Galerina
Order: Agaricales
Family: Cortinariaceae
Width of cap: 1–2½ in. (2.5–6.5 cm)
Length of stalk: 1–3 in. (2.5–8 cm)
Cap: Hemispheric at first, becoming convex to nearly flat, sometimes with a slight umbo; yellow-brown to orange-brown or dark brown, paler with age; surface glabrous, viscid when moist, shiny, margin smooth to obscurely translucent striate; **flesh** thin, pale brown, odor not distinctive.
Gills: Adnate to subdecurrent, but may detach from the stalk with age, close; pale yellowish brown to orange-brown.
Stalk: Equal, smooth but with lengthwise fibrillose streaking; brown becoming darker toward the base, with whitish mycelium at the point of attachment; partial veil leaving a thin, fibrillose or membranous ring on the upper stalk; ring white, but usually appearing brown from a deposit of spores on the upper side; sometimes the ring will deteriorate and not be readily apparent.
Spore print: Rusty brown.
Spores: 8.5–10.5 x 5–6.5 μm, elliptical, smooth or minutely roughened.
Occurrence: Scattered in small groups or clusters, occasionally solitary, on decaying wood of broad-leaved trees (often moss-covered logs and stumps), sometimes growing from buried wood and appearing terrestrial; saprobic; early spring-early winter; common.
Edibility: Deadly poisonous; contains dangerous amatoxins that destroy the liver.

Continued on next page

Comments: This is one of many small brown mushrooms that have a ring on the stalk and grow in groups on decaying wood. One must take special care not to confuse the Deadly Galerina with such edibles as the Honey Mushroom, *Armillaria mellea* (p.30), or the Changeable Pholiota, *Pholiota mutabilis* (p.39). Recent DNA analysis has indicated that *Galerina autumnalis* is synonymous with Galerina marginata, a species that was originally described from Europe. Because *Galerina marginata* has nomenclatural precedence, North American collections may soon be known by that name. The genus name *Galerina* means "helmet-shaped." The specific epithet *autumnalis* means "growing in the fall." Although this mushroom can be found almost anytime of the year, it is most often encountered in the cooler weather of spring and fall.

Latin name: *Gymnopilus luteus* (Peck) Hesler
Synonym: *Pholiota lutea* Peck
Common name: Yellow Gymnopilus
Order: Agaricales
Family: Cortinariaceae
Width of cap: 2–4 in. (5–10 cm)
Length of stalk: 1½–3½ in. (4–9 cm)
Cap: Convex with an incurved margin at first; buff-yellow to yellow-orange or ochraceous orange; surface dry, smooth, silky or downy, at times slightly scaly at the center; **flesh** pale yellow, odor sweet and spicy, taste bitter.
Gills: Adnexed to sinuate, close; pale yellow, becoming rusty red.
Stalk: Equal or enlarged downward; colored like the cap, staining rusty yellow when handled; partial veil leaving a thin, fibrillose to membranous ring on the upper stalk that with age may be reduced to a narrow but prominent annular zone.
Spore print: Rusty brown.
Spores: 6–9 x 4–5 µm, elliptical, surface finely roughened, dextrinoid.
Occurrence: Scattered or in clusters on decaying logs, stumps, and standing trunks of trees; saprobic; summer-fall; occasional.
Edibility: Not edible, hallucinogenic.

Comments: *Gymnopilus spectabilis* (not illustrated) is similar but typically has an orange-brown, silky to fibrillose or slightly scaly cap, and adnate to slightly decurrent gills. Its flesh also has a spicy odor and intensely bitter taste. Both of these species cause hallucinations and possible nausea if eaten. *Luteus* means "yellow."

Latin name: *Leucopholiota decorosa* (Peck) O.K. Miller, T.J. Volk, & Bessette
Synonym: *Armillaria decorosa* (Peck) A.H. Sm. & Walters
Common name: Decorated Pholiota
Order: Agaricales
Family: Tricholomataceae
Width of cap: 1–2½ in. (2.5–6.5 cm)
Length of stalk: 1½–2½ in. (4–6.5 cm)
Cap: Broadly conical to convex with an incurved margin at first; surface densely covered with long, pointed, and recurved rusty brown scales; **flesh** white, odor not distinctive, taste mild or bitter.
Gills: Adnexed, close; white; gill edges uneven.
Stalk: Equal or swollen at the base, often curved; colored like the cap and densely scaly from the base up to a flaring, shaggy ring zone, smooth and white above.
Spore print: White.
Spores: 4–6 x 3.5–4 μm, broadly elliptical, smooth, hyaline, amyloid.
Occurrence: Solitary or in small groups on decaying logs, stumps, and fallen branches of broad-leaved trees; saprobic; fall; occasional.
Edibility: Unknown.

Comments: In some years this beautiful and distinctive mushroom is not uncommon in the high elevation northern hardwood forests of Appalachia. It normally fruits fairly late in the season, often on sugar maple. The genus name *Leucopholiota* means "white Pholiota," in reference to the color of the gills and spores. *Pholiota squarrosoides* (p.40) is similar but it has a viscid cap cuticle beneath its scaly cap and produces brown spores. Other species of *Pholiota* also have brown spores. Some *Cystoderma* species are also similar, but their caps are granulose or finely scaly rather than covered with long, shaggy scales. Compare with *Phaeomarasmius erinaceellus* (p.37), which is smaller and has a dark reddish brown scaly cap. *Decorosa* means "elegant" or "handsome."

Latin name: *Phaeomarasmius erinaceellus* (Peck) Singer
Synonym: *Pholiota erinaceella* (Peck) Peck
Common names: Powder-scale Pholiota; Hedgehog Pholiota
Order: Agaricales
Family: Cortinariaceae
Width of cap: ⅜–1½ in. (1–4 cm)
Length of stalk: 1–2 in. (2.5–5 cm)
Cap: Hemispheric to convex at first, becoming broadly convex to nearly flat; surface dry, densely covered with dark reddish brown erect scales or granules, fading to yellow-brown with age; margin often ragged with remnants of a partial veil; **flesh** yellowish, odor not distinctive, taste bitter or metallic.
Gills: Adnexed, crowded; whitish at first, becoming ochraceous to tawny; gill edges crenulate.
Stalk: Equal, densely covered from the base up with scales that terminate in a slightly flaring fibrous ring near the apex; colored like the cap or paler on the scaly portion, whitish and lacking scales above the ring.
Spore print: Cinnamon-brown.
Spores: 6–8 x 4–4.5 μm, elliptical, smooth, pale brownish.
Occurrence: Solitary or in small groups on decaying logs, stumps, and fallen limbs of broad-leaved trees; saprobic; early summer-fall; occasional.
Edibility: Unknown.

Comments: *Gymnopilus sapineus* (not illustrated), which grows on conifer wood, is similar but lacks scales on the stalk and has a yellowish orange, minutely scaly cap, and rusty spotted gills. Compare with *Cystoderma granosum* (p.32), and *Leucopholiota decorosa* (p.36). *Erinaceellus* means "like a small hedgehog."

Latin name: *Pholiota aurivella* (Batsch) Fr.
Common name: Golden Pholiota
Order: Agaricales
Family: Strophariaceae
Width of cap: 1½–5 in. (4–12.5 cm)
Length of stalk: 2–3 in. (5–7.5 cm)
Cap: Convex to broadly convex, sometimes with a broad umbo; yellow to yellowish orange or yellow-brown ground color, at least partially covered with flattened and recurving rusty brown scales; surface viscid to glutinous when wet; margin incurved at first, often ragged with fragments of a partial veil; **flesh** yellow, odor and taste not distinctive.
Gills: Adnate to adnexed, close to crowded; pale yellowish at first, becoming rusty brown to orange-brown.
Stalk: Equal or tapering upward, central or eccentric; colored like the cap but usually paler, whitish above the ring; surface scaly from the base upward to a scant, fibrous ring that often disappears with age, or is reduced to an inconspicuous annular zone near the apex.
Spore print: Brown
Spores: 7–10 x 4.5–6 μm, elliptical with an apical germ pore, smooth, brownish, weakly dextrinoid.
Occurrence: In clusters or sometimes solitary on decaying logs, trunks, and stumps of broad-leaved and conifer trees, also on trunks of living trees, often emanating from wounds several feet up; parasitic and saprobic; fall to early winter, less frequently in spring; common.
Edibility: Not recommended. Edible according to some accounts, but there are reports of gastrointestinal disturbances from consuming this or other closely related species.

Comments: The cap scales on the Golden Pholiota can wash off in the rain and then may somewhat resemble the Honey Mushroom, *Armillaria mellea* (p.30), which also grows in clusters on wood, but produces white spores. Compare also with *Pholiota squarrosoides* (p.40). *Aurivella* means "golden fleece."

Latin name: *Pholiota mutabilis* (Schaeff.) P. Kumm.
Synonyms: *Galerina mutabilis* (Schaeff.) P.D. Orton; *Kuehneromyces mutabilis* (Schaeff.) Singer & A.H. Sm.
Common names: Changeable Pholiota; Two-toned Pholiota
Order: Agaricales
Family: Strophariaceae
Width of cap: ¾–2½ in. (2–6.5 cm)
Length of stalk: 1–2½ in. (2.5–6.5 cm)
Cap: Convex to flattened with a broad umbo; hygrophanous, yellowish brown to cinnamon-brown, becoming yellow to ochre toward the center upon drying; surface smooth, glabrous, or with scattered whitish fibrils toward the margin; margin obscurely striate; **flesh** thin, whitish, odor and taste not distinctive, or slightly fruity.
Gills: Adnate to short decurrent, close; whitish to pale yellow at first, becoming cinnamon-brown.
Stalk: Slender, often curved, equal; more or less colored like the cap, becoming darker brown toward the base; partial veil leaving a thin, fibrillose to membranous ring on the upper stalk, or often, ring soon disappearing but leaving a distinct annular zone; surface scaly below the ring, smooth or vertically lined above.
Spore print: Cinnamon-brown.
Spores: 5.5–7.5 x 3.5–5 μm, oval, truncated at apex, smooth.
Occurrence: Often in dense clusters on stumps, logs, and near the base of dead trees, especially birch and beech and other broad-leaved trees, but also occurring less frequently on conifers; saprobic; summer-fall; occasional.
Edibility: Edible but best avoided (see comments below).

Comments: The two-toned caps and the downward curved scales on the stalk below the ring are good field characters. Although the tender caps of this mushroom are edible, it is easily confused with a number of similar small, brown lignicolus species, including the highly toxic *Galerina autumnalis* (p.33). *Pholiota malicola* (not illustrated), which often fruits fairly late in the season, has a smooth yellowish to orange-yellow cap, and does not have scales on the stalk. *Mutabilis* means "changing," in reference to the hygrophanous nature of the cap.

Latin name: *Pholiota squarrosoides* (Peck) Sacc.
Common name: Scaly Pholiota
Order: Agaricales
Family: Strophariaceae
Width of cap: 1½ x 4½ in. (4–11.5 cm)
Length of stalk: 2–5 in. (5–12.5 cm)
Cap: Broadly conical to convex, becoming nearly flat with age, with or without a broad umbo, margin incurved when young; ground color whitish to pale ochraceous, sometimes developing olivaceous tones with age, especially when wet; surface covered with conical, tawny scales that are more concentrated toward the center, cuticle viscid beneath the scales when wet, sticky when dry; **flesh** thick, whitish, odor mild to somewhat fragrant, taste not distinctive.
Gills: Adnate to adnexed, close to crowded; whitish at first, becoming rusty cinnamon as the spores mature.
Stalk: More or less equal; colored like the cap, surface covered with recurved, tawny scales from the base upward, ending and merging with a somewhat flaring, cottony, evanescent ring near the apex, white and smooth above the ring.
Spore print: Cinnamon-brown.
Spores: 4–6 x 2.5–3.5 μm, broadly elliptical, smooth.
Occurrence: Usually in caespitose clusters on decaying logs and stumps, or sometimes on trunks of living broad-leaved trees, especially birch, beech, and maple; parasitic and saprobic; summer-fall; common.
Edibility: Not recommended; edible for some people, but gastrointestinal disturbance from eating this species has been reported.

Comments: This attractive wood-decaying mushroom causes a "heartrot" of living trees. *Pholiota squarrosa* (not illustrated) is very similar but has a dry scaly cap, a greenish cast to the gills, and a garlic-like odor. Some mycologists consider these two taxa as merely varieties of the same species. Young specimens of the Scaly Pholiota might be confused with *Leucopholiota decorosa* (p.36), which grows singly or in small groups and has white spores. Compare also with *Pholiota aurivella* (p.38) and *Armillaria mellea* (p.30). *Squarrosoides* means "resembling *Pholiota squarrosa.*" *Squarrosa* means "roughened with scales."

Latin name: *Pholiota veris* A.H. Sm. & Hesler
Common name: None
Order: Agaricales
Family: Strophariaceae
Width of cap: ¾–2 in. (2–5 cm)
Length of stalk: 1½–3 in. (4–7.5 cm)
Cap: Convex to flat; yellowish brown to pale cinnamon, or pale straw colored when fresh, hygrophanous, drying to nearly white from the center outward; surface somewhat viscid, margin translucent striate; **flesh** thin, pale cinnamon, odor and taste not distinctive.
Gills: Adnate, close; whitish to pale tan at first, becoming pale cinnamon with age.
Stalk: Equal; colored like the cap or paler; surface nearly glabrous; partial veil leaving a thin membranous ring on the upper stalk.
Spore print: Brown.
Spores: 5.5–7.5 x 3.7–4.8 μm, elliptical to ovate, smooth, pale tan.
Occurrence: In small groups or clusters on decaying wood, wood mulch, or buried wood of broad-leaved trees; saprobic; early spring-summer; common.
Edibility: Unknown.

Comments: This is one of the first gilled mushrooms to appear in the spring. *Tubaria confragosa* (not illustrated) is similar but has a dry, reddish brown cap with a fibrous whitish "bloom" when young. *Agrocybe acericola* (not illustrated) has an ochre to cinnamon-brown cap with a smooth margin, and white cord-like strands (rhizomorphs) at the stalk base. Compare with the dangerously poisonous *Galerina autumnalis* (p.33). *Veris* means "of the spring."

Latin name: *Pleurotus dryinus* (Pers.) P. Kumm.
Synonym: *Pleurotus corticatus* (Fr.) Quél.
Common name: Veiled Oyster
Order: Agaricales
Family: Pleurotaceae
Width of cap: 1½–4 in. (4–10 cm)
Length of stalk: 1½–4 in. (4–10 cm)
Cap: Convex at first, becoming more or less flattened, margin inrolled, often ragged with partial veil remnants; white to cream or pale yellow; surface cottony to scaly in the center, especially evident in older specimens; **flesh** thick and tough, odor usually pleasant, somewhat anise-like, or sour in older specimens.
Gills: Decurrent, subdistant to moderately close; white to cream or yellowish with age.
Stalk: More or less equal or enlarged downward, central to eccentric, at times nearly lateral; whitish, velvety, with a thin cottony to membranous ring on the upper stalk that becomes inconspicuous or may disappear with age.
Spore print: White.
Spores: 9–12 x 3.5–5.5 µm, elliptic, smooth, hyaline, non-amyloid.
Occurrence: Solitary or in small clusters on logs, stumps, and trunks of broad-leaved trees, also on trunks of living trees; saprobic; summer-fall; occasional.
Edibility: Edible.

Comments: This is the only species of *Pleurotus* that has a partial veil. The somewhat similar *Pleurotus levis* (= *Panus strigosus,* not illustrated) is larger and has a densely hairy stalk that lacks a ring, and grows from wounds on living trees, often high on the trunk, as well as on deadwood. *Hypsizygus tessulatus* (not illustrated) has a whitish to grayish cap with darker "water spots," broadly attached gills, and a smooth, ringless stalk. The name *dryinus* refers to an association with oak, which is a common substrate for this mushroom.

Latin name: *Stropharia hornemannii* (Fr.) Lundell & Nannf.
Synonym: *Stropharia depilata* (Pers.) Sacc.
Common names: Hornemann's Stropharia; Lacerated Stropharia.
Order: Agaricales
Family: Strophariaceae
Width of cap: 2–5 in. (5–12.5 cm)
Length of stalk: 3–5 in. (8–12.5 cm)
Cap: Convex to broadly bell-shaped, becoming flattened with age, often with a broad umbo; color variable from cinnamon-brown or reddish brown, to dingy yellowish brown, often tinged violet, and olivaceous gray on the margin; surface smooth and viscid, with scattered whitish fragments of a partial veil at the margin, at least when young; **flesh** white to pale yellowish, odor and taste not distinctive.
Gills: Adnate, sometimes with slightly decurrent lines at the stalk apex, moderately close; whitish at first, becoming smoky purplish gray, but gill edges remaining white.
Stalk: Equal; whitish, covered with dense cottony scales from the base upward to a persistent flaring membranous ring, smooth above the ring, base with white cord-like rhizomorphs.
Spore print: Purple-brown.
Spores: 10–13 x 5–7 μm, elliptical with an apical germ pore, smooth.
Occurrence: Sometimes solitary but more often in groups on or around well-decayed logs and stumps of conifer wood, also reported to occur on wet mossy ground; saprobic; late summer-fall; uncommon.
Edibility: Not edible.

Comments: In West Virginia, this distinctive mushroom occurs in high elevation mixed woods late in the season. *Pholiota albocrenulata* (not illustrated), which is somewhat similar but grows on wood of broad-leaved trees (especially maple), has a deep reddish brown to vinaceous brown scaly cap, a reddish brown scaly stalk, and finely serrated gill edges. The specific epithet honors the Dutch botanist Jens Wilken Hornemann (1770–1841).

Gilled Mushrooms with a Membranous Ring on the Stalk, Growing on Soil or Humus

This section includes representatives from several genera, including many species in the important genus *Amanita*, which has both edible and dangerously poisonous members.

Keep in mind that a partial veil that forms a ring on the stalk may fall away and disappear in older specimens. Also, in some cases the ring may merge with scales or other material on the stalk and be less apparent. Mushrooms that have a thin band of appressed fibrils (from a collapsed web-like veil) rather than a more substantial membranous ring on the stalk are grouped elsewhere.

A few mushrooms included in this section can occur either on the ground or on decaying wood. Because the distinction between humus and the advanced state of decomposing wood is not always obvious, if there is doubt as to the substrate also review the wood-inhabiting species in Section 1-B.

Latin name: *Agaricus abruptibulbus* Peck
Common name: Abruptly-bulbous Agaricus
Order: Agaricales
Family: Agaricaceae
Width of cap: 2–6 in. (5–15 cm)
Length of the stalk: 2½–6 in. (6.5–15 cm)
Cap: Convex to broadly convex, becoming flattened; white to creamy white or tinged yellowish with age, staining reddish when waterlogged; surface smooth to silky-fibrillose, or slightly scaly with age, bruising yellow; margin sometimes ragged with remnants of the partial veil; **flesh** white, fragile, with a pleasant odor of anise or almond, taste mild.
Gills: Free from the stalk, crowded; whitish to grayish pink when young, becoming pinkish brown then chocolate-brown as the spores mature; covered at first with a whitish membranous veil that typically has yellowish or pale brownish cottony patches on the underside.
Stalk: Slender, nearly equal with an abrupt bulb at the base; white, bruising yellow, especially at the base; smooth to slightly fibrillose-silky; partial veil leaving a whitish skirt-like ring on the upper stalk.
Spore print: Chocolate-brown.
Spores: 6–8 x 4–5 μm, elliptical, smooth, pale brown.
Occurrence: Solitary or gregarious on the ground in woods, woodland margins, grassy groves, parklands, and landscaped areas; saprobic; summer-fall; common.
Edibility: Edible and good with caution (see comments below).
Continued on next page.

Comments: *Agaricus silvicola* (not illustrated), if indeed a separate species, is nearly identical. It has a swollen but not abruptly bulbous stalk base, and smaller spores (5–6.5 x 3.5–4.5 m). One must be careful not to confuse either of these species with the deadly *Amanita virosa* (p.62), which has a sac-like volva at the stalk base, white gills that do not change color, and white spores. *Agaricus xanthodermis* (not illustrated), which is also poisonous, has a white cap, often with a brownish center, bruises yellow, and has an unpleasant chemical odor that is frequently described as "ink-like." The edible *Agaricus arvensis* (not illustrated) is similar to *Agaricus abruptibulbus* with its white cap and stalk that bruise yellow, and flesh having a pleasant anise-like odor. However, *Agaricus arvenis* is generally a more robust mushroom that lacks an abrupt bulb at the stalk base, and it grows in grassy areas and pastures.

Latin name: *Agaricus campestris* L.
Common names: Meadow Mushroom; Pink Bottom
Order: Agaricales
Family: Agaricaceae
Width of cap: 1¼–3½ in. (3–9 cm)
Length of the stalk: 1–2 in. (2.5–5 cm)
Cap: Hemispherical to nearly globose at first, soon convex, becoming broadly convex to flattened with age; color variable, white to off white or grayish brown; surface smooth to silky-fibrillose, or somewhat scaly; **flesh** white to pale pinkish, odor and taste mushroomy or slightly of anise.
Gills: Free from the stalk; pink at first, becoming dark chocolate-brown at maturity.
Stalk: More or less equal, short, stout; whitish; surface smooth or slightly downy; partial veil leaving a fragile and sometimes indistinct membranous ring on the upper stalk that may disappear with age.
Spore print: Chocolate-brown.
Spores: 6.5–8 x 4.5–5 µm, elliptical, smooth, pale brown.
Occurrence: Usually gregarious on soil in lawns, fields, pastures, and other grassy places, often forming fairy rings or arcs; saprobic; summer-fall, but most abundant in cooler weather, occasionally appearing in spring; common.
Edibility: Edible and good.
Continued on next page

Comments: This close cousin of the cultivated Button Mushroom (*Agaricus bisporus*) is thought by many to have superior flavor to commercially produced varieties. The Meadow Mushroom can often be gathered in large quantities and a surplus is easily preserved by drying, freezing, or canning. It is one of the most popular edible wild mushrooms. There are several similar whitish mushrooms with free gills that grow in grasslands. The cap and stalk of the poisonous *Agaricus xanthodermis* (not illustrated) quickly stain bright yellow when bruised, and its flesh has an unpleasant chemical odor. The similar but more substantial Horse Mushroom, *Agaricus arvensis* (not illustrated) also stains yellow when bruised, but has a pleasant anise odor and is an excellent edible. *Agaricus pocillator* (not illustrated) and *Agaricus placomyces* (not illustrated) have whitish caps with brown or gray to blackish scales concentrated near the center. Edibility of these species is not established and both should be regarded as suspect. Like the Meadow Mushroom, the poisonous *Chlorophyllum molybdites* (p.64) can grow in fairy rings in lawns and pastures. It is a much larger mushroom that has buff to brown scales in the center of the cap, a moveable ring on the stalk, and produces greenish spores. *Leucoagaricus naucinus* (p.70) is an all white mushroom that produces white spores. It has a movable ring on the stalk and gills that remain white at all stages. Whenever collecting any white mushrooms for the table, take special care not to confuse them with the dangerously poisonous *Amanita virosa* (p.62). *Campestris* means "of the fields."

Latin name: *Amanita brunnescens* var. *brunnescens* G.F. Atk.
Common names: Cleft-foot Amanita; Brown Blusher
Order: Agaricales
Family: Amanitaceae
Width of cap: 2–5 in. (5–12.5 cm)
Length of the stalk: 2–6 in. (5–15 cm)
Cap: Ovoid at first, then convex to broadly convex, becoming flattened or shallowly depressed with a slight umbo; brown to grayish brown or olive-brown, darker in the center, often radially streaked; surface somewhat viscid, with scattered whitish cottony patches that may wash off or wear away with age; **flesh** white to brownish, odor of raw potatoes.
Gills: Narrowly free from the stalk, close; white.
Stalk: Gradually tapering upward from an enlarged, abruptly bulbous base that has wedge-like vertical clefts; whitish to tinged pinkish brown, typically with reddish brown stains, especially at the base of the stalk; surface smooth to scurfy; partial veil leaving a whitish, membranous, skirt-like ring on the upper stalk.
Spore print: White.
Spores: 7.5–10 x 7–8.5 μm, globose to subglobose, smooth, hyaline, amyloid.
Occurrence: Solitary or in small groups on the ground in oak woods and mixed broad-leaved and conifer woods; mycorrhizal; summer-fall; common.
Edibility: Not recommended; reports vary, but possibly poisonous.

Comments: This medium to large mushroom is recognized by the reddish brown stains that slowly develop on the fruitbody with age or when injured, and the potato-like odor (especially noticeable at the base of the stalk). *Amanita brunnescens* var. *pallida* (not illustrated) has a white or cream-colored cap and stalk but is otherwise nearly identical. Compare with *Amanita rubescens* (p.60), which has a rounded or turnip-shaped stalk base, and lacks a distinct odor. Compare also with *Amanita porphyria* (p.59). *Brunnescens* means "becoming brown."

Latin name: *Amanita cokeri* (E.-J. Gilbert & Kühner) E.-J. Gilbert
Common name: Coker's Amanita
Order: Agaricales
Family: Amanitaceae
Width of cap: 3–6 in. (7.5–15 cm)
Length of the stalk: 3–6 in. (7.5–15 cm)
Cap: Hemispheric to nearly globose at first, becoming convex to flattened with age; white to ivory; surface with soft, conical, white to brownish concentrically arranged warts (which may wash off in rainy weather), cuticle viscid beneath the warts; **flesh** white, odor and taste not distinctive.
Gills: Narrowly free from the stalk, close to crowded; white to cream.
Stalk: Tapering upward from a swollen, spindle-shaped (ventricose) rooting base that has concentric rows of recurved scales on its upper portion; colored like the cap, usually with brownish stains at the base; partial veil leaving a skirt-like membranous ring on the upper stalk; smooth above the ring, scurfy to somewhat scaly below.
Spore print: White.
Spores: 11–14 x 6.5–9.5 μm, elliptical, smooth, amyloid.
Occurrence: Solitary or in small groups in mixed woods, especially with oak and pine; mycorrhizal; summer-fall; common.
Edibility: Not edible.

Comments: Coker's Amanita is one of several large, whitish amanitas belonging to the subgenus Lepidella (meaning small scales). Many of these have strong odors that are variously described as smelling like chlorine, old ham, or smelly socks. Although it can be difficult to distinguish individual species in the field, the lepidellas can be recognized as a group by their medium to large size, warted caps with ragged margins, fragile partial veils that easily fall away, and strong odors. *Amanita abrupta* (not illustrated) is similar to Coker's Amanita but has an abruptly bulbous, non-rooting stalk base. *Amanita rhopalopus* (not illustrated) has a strong odor of old ham and an elongated rooting stalk base that lacks coarse scales. *Amanita polypyramis* (not illustrated) has powdery to cottony patches rather than conical warts on the cap, and a nearly smooth stalk that is often shaped like a bowling pin. The specific epithet *cokeri* honors American mycologist William Chambers Coker (1872–1953).

Latin name: *Amanita daucipes* (Mont.) Lloyd
Common names: Carrot-foot Amanita; Turnip-foot Amanita.
Order: Agaricales
Family: Amanitaceae
Width of cap: 2½–10 in. (6.5–25 cm)
Length of the stalk: above ground portion 3–6 in. (7.5–15 cm)
Cap: Convex, becoming broadly convex to nearly flat; creamy white to pale pinkish orange; surface dry, densely covered with pinkish to salmon minute spines or irregular-shaped cottony warts; margin ragged; **flesh** thick, whitish, odor sweetish, nauseous, like old ham.
Gills: Free to narrowly free from the stalk; crowded; white to pale cream.
Stalk: Slightly tapering upward from a large, swollen base; colored like the cap or more whitish, especially near the apex, surface scurfy to wooly; stalk base large, up to 5 in. (12.5 cm) thick, rooting, carrot-shaped to spindle-shaped, often with lengthwise splits and orangish to reddish brown stains; partial veil remains as a skirt-like ring on the upper stalk.
Spore print: White.
Spores: 9–11 x 5.5–6.5 μm, elliptical, smooth, amyloid.
Occurrence: Solitary or in scattered groups in oak woods, mixed woods, and on disturbed ground such as roadside banks; mycorrhizal; summer-fall; occasional to locally common.
Edibility: Unknown.

Comments: This is one of the easiest of the Lepidella amanitas to recognize. The salmon-pink tones on the cap and stalk, oversized rooting stalk base, and peculiar nauseous odor are distinctive. The fragile ring often falls off the stalk but leaves remnants on the ground near the base. *Daucipes* means "carrot foot."

Latin name: *Amanita jacksonii* Pomerleau
Common names: American Caesar's Mushroom; Eastern Caesar's Mushroom
Order: Agaricales
Family: Amanitaceae
Width of cap: 2–6 in. (5–15 cm)
Length of the stalk: 3–6 in. (7.5–15 cm)
Cap: Ovoid at first, becoming convex to broadly convex or nearly flat, usually with a distinct umbo; brilliant red to orange-red, becoming yellow near the margin when expanded; smooth, glabrous, or sometimes with a patch of white volval material on the disc; margin striate; **flesh** white to yellowish, odor and taste not distinctive.
Gills: Free from the stalk, crowded; pale yellow; covered at first with a pale yellowish orange partial veil.
Stalk: Slender, equal or tapering upward; cream to yellow overlaid with yellowish orange smears or scale-like patches; partial veil leaving a persistent membranous skirt-like ring near the apex; stalk base seated in a large, elongated, white, sac-like volva that is mostly buried in the soil.
Spore print: White.
Spores: 7.5–10 x 5.5–7 µm, elliptical, smooth, non amyloid.
Occurrence: Solitary or in groups on the ground in mixed woods, especially with oak and pine, sometimes growing in rings or arcs beneath trees; mycorrhizal; summer-fall; fairly common in some locations.
Edibility: Edible but not popular. The flavor is somewhat fishy.

Comments: The American Caesar's Mushroom is one of the most striking mushrooms in our woods. A very similar orange-brown species that has yet to be named is also common in the central Appalachians. It has a much smaller saccate volva, and lacks orangish patches on its stalk. R.E. Tulloss, who is studying the taxonomy of Amanita in North America, has given this species the temporary designation of "*Amanita* #16." Its edibility has not been established. The poisonous *Amanita parcivolvata* (p.239) has a powdery yellow stalk that lacks a partial veil ring and sac-like volva at the base. Compare also with *Amanita flavoconia* (p.54). In the past there has been much confusion regarding the nomenclature of the American Caesar's Mushroom. Until thoroughly studied this mushroom was thought to be the same as the European *Amanita caesarea.* However, the entity that occurs in eastern North America is not the same, and according to R.E. Tulloss the correct name for our species is *Amanita jacksonii,* named in honor of Canadian amateur mycologist and botanical illustrator Henry Alexander Carmichael Jackson (1877–1961).

Latin name: *Amanita flavoconia* G.F. Atk.
Common name: Yellow Patches
Order: Agaricales
Family: Amanitaceae
Width of cap: 1¼–3½ in. (3–9 cm)
Length of the stalk: 2–4 in. (5–10 cm)
Cap: Ovoid at first, becoming convex to nearly flat with age; orange to bright yellow-orange; covered at first with chrome-yellow warts that are easily rubbed or washed off, surface smooth and viscid beneath the warts; **flesh** white, odor not distinctive.
Gills: Barely free from the stalk, close; white or tinged yellow on the edges; covered at first with a yellowish partial veil.
Stalk: Equal or slightly tapered upward from a small rounded bulb at the base; white to yellowish orange; smooth to scurfy; base usually with chrome-yellow flakes of universal veil material loosely adhering to the bulb, or in the soil around the base; partial veil leaving a skirt-like ring on the upper stalk.
Spore print: White.
Spores: 7–9 x 5–8 μm, elliptical, smooth, hyaline, amyloid.
Occurrence: Solitary or in groups on the ground in broad-leaved and mixed woods, especially with hemlock, also in high elevation red spruce forests; mycorrhizal; summer-fall; common.
Edibility: Unknown, possibly poisonous.

Comments: This beautiful mushroom is common throughout the central Appalachians. The yellow patches on the cap and stalk base are fragments of a universal veil that encloses the young button stage. The veil material is easily washed off the cap by rainfall and some specimens are completely glabrous, although there are usually telltale bits of the yellow veil on the surrounding soil. *Amanita frostiana* (not illustrated) is nearly identical but has a striate cap margin, a rim on the basal bulb of the stalk, and non-amyloid spores. *Amanita flavorubescens* (p.55) is a more robust mushroom with pale yellow warts on the cap, and reddish brown stains on the stalk base. Some color forms of *Amanita muscaria* (p.56) may be similar but are typically much larger mushrooms with whitish to pale yellow warts on the cap, and concentric scaly bands on the stalk base. Compare also with *Amanita parcivolvata* (p.239) and *Amanita jacksonii* (p.52). *Flavoconia* means "yellowish" and "conical."

Latin name: *Amanita flavorubescens* G.F. Atk.
Synonym: *Amanita flavorubens* Berk. & Mont.
Common name: Yellow Blusher
Order: Agaricales
Family: Amanitaceae
Width of the cap: 1½–4½ in. (4–11.5 cm)
Height of the stalk: 2–5 in. (5–12.5 cm)
Cap: Convex, becoming flattened with age; golden yellow to yellowish bronze with scattered yellowish patches on top; surface smooth beneath the patches, with or without short striations on the margin; **flesh** white, slowly staining wine red when cut or broken, odor and taste not distinctive.
Gills: Free from the stalk, crowded, creamy white.
Stalk: More or less equal with a swollen to bulbous base; stuffed at first, then becoming hollow; white to yellowish, usually with reddish brown stains at the base; surface nearly smooth to scurfy; partial veil leaving a skirt-like white membranous ring that sometimes has a yellow fringe on the upper stalk; base often with loosely attached fragments or concentric zones of yellow universal veil material.
Spore print: White.
Spores: 8–10 x 5.5–7 μm, elliptical, smooth, hyaline, amyloid.
Occurrence: Solitary or in small groups on the ground in mixed broad-leaved and conifer woods, woodland margins, and parklands, especially beneath oak; mycorrhizal; early summer-fall; occasional to fairly common in some years.
Edibility: Unknown, possibly poisonous.

Comments: Compare with *Amanita rubescens* (p.60), which has pinkish gray warts on the cap and more extensive reddish staining on the stalk base. *Amanita flavoconia* (p.54) has a deep yellow-orange cap, and lacks reddish stains on the stalk. Compare also with *Amanita muscaria* var. *formosa* (p.56), which can sometimes be similar in color to the Yellow Blusher. *Flavorubescens* means "yellow becoming reddish."

Latin name: *Amanita muscaria* var. *formosa* Pers.
Common names: Fly Amanita; Yellow-orange Fly Agaric; Soma
Order: Agaricales
Family: Amanitaceae
Width of cap: 2–8 in. (5–20 cm)
Length of the stalk: 3–8 in. (7.5–20 cm)
Cap: Nearly globose at first, then convex to broadly convex, becoming flattened or depressed in the center with age; yellow to orange, or reddish orange, with the deepest color at the center and paler toward the margin, fading with age or exposure to sunlight; surface viscid when moist, with numerous whitish to buff cottony warts; margin vaguely lined, and often adorned with fragments of the universal veil; **flesh** white, odor and taste not distinctive.
Gills: Free, or barely attached to the stalk, crowded, white to cream, covered at first with a partial veil.
Stalk: Equal to slightly tapered upward from an enlarged basal bulb; white to pale cream, or pale yellow, usually discoloring when handled; surface fleecy to torn scaly; basal bulb globose to subglobose, with concentric scaly rings on the upper portion that may or may not terminate in a distinctly rimmed collar; partial veil leaving a skirt-like whitish to buff membranous ring clinging to the upper stalk.
Spore print: White.
Spores: 8.7–12.2 x 6.5–8.2 *μ*m, broadly elliptical, smooth, hyaline, non amyloid.
Occurrence: Solitary or more often gregarious on the ground in mixed woods, conifer plantations, or under trees in lawns and parklands, especially oak, pine, hemlock, and Norway spruce; sometimes growing in rings or arcs; mycorrhizal; late spring-late fall; common.
Edibility: Poisonous.

Comments: This large and showy mushroom is often abundant in summer and fall. It is the most common of three varieties of the Fly Amanita that occur in West Virginia, all of which are poisonous. *Amanita muscaria* var. *alba* (not illustrated) has a white to silvery white cap, but is otherwise nearly identical to variety *formosa. Amanita muscaria* var. *persicina* (not illustrated) is a southern North American species that has only been reported in West Virginia from the shale barrens in Pendleton county. It has a pale orange to peach-orange cap, and lacks prominent concentric rings at the stalk base. *Amanita pantherina* var. *velatipes* (p.58) is similar in size and stature to the Fly Amanita, but it has a cream to pale brown cap. Compare also with *Amanita flavoconia* (p.54), and *Amanita parcivolvata* (p.239). Although varieties of the Fly Amanita have a long history of ritual use as an inebriant when eaten, this is a potentially dangerous practice since the toxins involved vary in their concentrations, and ingesting them can cause nausea and disorientation, or worse. The name *muscaria* refers to "fly" and originated from the belief that the caps of this mushroom placed in a dish of milk will attract and stupify flies. *Formosa* means "beautiful."

Latin name: *Amanita pantherina* var. *velatipes* (G.F. Atk.) Jenkins
Synonym: *Amanita velatipes* G.F. Atk.
Common name: Panther Amanita
Order: Agaricales
Family: Amanitaceae
Width of cap: 3–8 in. (7.5–20 cm)
Length of the stalk: 3–8 in. (7.5–20 cm)
Cap: Ovate at first, becoming broadly bell-shaped to convex or nearly flat with age; yellowish cream to tan, or more brownish, especially in the center; surface smooth, viscid when moist, with white to grayish buff cottony warts that are more or less concentrically arranged, margin striate; **flesh** whitish, odor not distinctive.
Gills: Free from the stalk, crowded; white to creamy yellow.
Stalk: More or less equal or tapering upward from a bulbous base; white to cream; surface smooth to somewhat scurfy or scaly near the base; basal bulb large, globose or elongated, with or without concentric scaly bands below a collar-like rim; partial veil leaving a persistent, membranous whitish ring on middle to lower stalk.
Spore print: White.
Spores: 8–13 x 6–8 µm, broadly elliptical, smooth, hyaline, non-amyloid.
Occurrence: Usually gregarious on the ground under conifer trees in woods, parklands, and lawns, especially with Norway spruce, also under broad-leaved trees, sometimes growing in rings or arcs; mycorrhizal; summer-fall; common.
Edibility: Poisonous.

Comments: The cap warts are easily washed off in wet weather. *Amanita pantherina* var. *pantherina* (not illustrated) is typically smaller and has a dark brown cap, but is otherwise similar. The name *pantherina* means "panther-like," in reference to the spotted cap. *Velatipes* means "veil foot," referring to the rimmed, boot-like volva.

Latin name: *Amanita porphyria* (Alb. & Schwein.) Secr.
Common names: Purple-brown Amanita; Gray-veil Amanita
Order: Agaricales
Family: Amanitaceae
Width of cap: 1½–3½ in. (4–9 cm)
Length of the stalk: 2–4 in. (5–10 cm)
Cap: Bell-shaped to convex or broadly convex, becoming flattened with age, with or without an umbo; grayish brown with subtle purplish tones; surface smooth, silky when dry, with or without randomly distributed grayish, cottony patches of universal veil; margin smooth or obscurely striate; **flesh** whitish, odor of raw potatoes.
Gills: Free from the stalk or barely attached, crowded; white to pale grayish.
Stalk: Slender, tapering slightly upward from a substantial roundish basal bulb that is typically flattened or shallowly depressed on top; colored like the cap or paler; surface smooth, often with vertical striations on the upper portion; partial veil leaving a thin, gray, membranous ring about midway or higher on the stalk.
Spore print: White.
Spores: 7.2–10.3 µm, globose, smooth, hyaline, amyloid.
Occurrence: Solitary or in scattered groups on the ground in conifer woods; mycorrhizal; fall; occasional to common locally.
Edibility: Poisonous.

Comments: This small to medium, rather drab mushroom is not uncommon late in the season in high elevation red spruce woods and in plantations of Norway spruce. The fruitbodies typically have an odor of raw potatoes (most noticeable at the base of the stalk). Compare with *Amanita brunnescens* var. *brunnescens* (p.49), which has an abruptly bulbous stalk base with wedge-shaped clefts and reddish brown stains. *Porphyria* means "purplish."

Latin name: *Amanita rubescens* (Pers.) Gray
Common name: Blusher
Order: Agaricales
Family: Amanitaceae
Width of cap: 2½–5½ in. (6.5–14 cm)
Length of the stalk: 3–7 in. (7.5–18 cm)
Cap: Broadly bell-shaped at first, becoming convex to flattened; color variable, yellowish cream to pinkish brown, tan, bronze, or reddish brown, paler toward the margin, covered with buff to pinkish gray, cottony warts; surface viscid, margin unlined; **flesh** thick, white, slowly staining reddish when cut, odor and taste not distinctive.
Gills: Free, or barely touching the stalk, close; white to pale pinkish, developing reddish brown stains with age, or where damaged by insects; at first covered with a thin membranous veil.
Stalk: Nearly equal or tapering upward from a swollen, turnip-shaped base; white or with a pinkish flush, typically stained reddish brown, especially near the base; surface smooth to somewhat scurfy; basal bulb generally smooth but sometimes with a few fragments of universal veil on the upper part, or occasionally with a slight rim; partial veil leaving a skirt-like, whitish ring on the upper stalk, upper surface of ring usually striate.
Spore print: White.
Spores: 7.5–10.5 x 7–7 μm, elliptical, smooth, hyaline, amyloid.
Occurrence: Usually gregarious under broad-leaved and conifer trees, especially oak and pine; mycorrhizal; summer-fall; common, often abundant.
Edibility: Edible with caution, must be cooked (see comments below).

Comments: The cap warts (universal veil) of this and similar species may be washed off in rainy weather. Generally speaking, mushrooms in the genus *Amanita* should be avoided for eating because of potentially serious consequences from misidentifications. Some poisonous species such as *Amanita pantherina* (p.58) and *Amanita muscaria* (p.56) could be mistaken for *Amanita rubescens.* These are variable species with overlapping color forms. Other amanitas have unknown edibility. For experienced collectors, however, the Blusher is a good edible mushroom when well-cooked. The similar *Amanita flavorubescens* (p.55) has a golden yellow cap with yellowish warts. It also develops reddish brown stains, but the staining is usually less pronounced and often restricted to the lower portion of the stalk. Compare with *Amanita brunnescens* var. *brunnescens* (p.49), which also has brown to reddish brown stains, especially on the stalk base. *Amanita rubescens* var. *alba* (not illustrated) is a paler version of the Blusher. It is mostly white or creamy white except for the characteristic reddish brown stains. The Blusher is frequently parasitized by *Hypomyces hyalinus* (p.495), a white mold that transforms infected fruitbodies into ghostly misshapened whitish clubs. Parasitized and non-parasitized fruitbodies often occur in close proximity. *Rubescens* means "becoming reddish."

Latin name: *Amanita virosa* (Fr.) Bertillon
Common names: Destroying Angel; Death Angel
Order: Agaricales
Family: Amanitaceae
Width of cap: 2–5 in. (5–12.5 cm)
Length of the stalk: 2½–6 in. (6.5–15 cm)
Cap: Ovate to broadly bell-shaped when young, becoming convex to flattened; white or sometimes with a slight yellowish tinge on the disc; surface glabrous, viscid when moist, margin smooth; **flesh** white, odor not distinctive.
Gills: Free, or barely reaching the stalk, close; white; covered at first by a membranous white partial veil.
Stalk: Tapering upward from a bulbous base that is sheathed with a substantial membranous sac-like volva; white; surface smooth to cottony-scaly; partial veil leaving a skirt-like membranous ring on the upper stalk; volva white, often buried in the soil except for the top margin that may be loose and free, or clinging to the stalk.
Spore print: White.
Spores: 8.6–11.7 x 7–9 μm, globose or subglobose, smooth, hyaline, amyloid.
Occurrence: Solitary or in scattered groups on the ground in various broad-leaved and conifer woods, especially with oak, also beneath trees in lawns and parks; mycorrhizal; early summer-fall; common.
Edibility: Dangerously poisonous. Most mushroom fatalities in North America are caused from eating this mushroom, or from its close relative *Amanita phalloides* (not illustrated), which has not yet been found in West Virginia. The toxins in these species destroy the liver and kidneys.

Comments: All parts of this mushroom are pure white. There are a few equally dangerous closely related species that are difficult to distinguish based on field characters. *Amanita bisporigera* (not illustrated) is generally smaller in stature, and appears earlier in the season. It produces only 2 spores on a basidium whereas the Destroying Angel has 4 spores per basidium. When KOH is applied to the cap of either of these species, it will produce a bright yellow staining reaction. *Amanita verna* (not illustrated) is nearly identical to *Amanita virosa* but is more slender in stature, appears earlier in the spring, and the cap does not turn yellow with the application of KOH. *Amanita alba* (p.236) has a striate cap margin, and lacks a ring on its stalk. As with several species of *Amanita,* the immature fruitbodies of *Amanita virosa* are first enclosed in a thick, white membrane (universal veil) that ruptures at the top as the developing cap and stalk break through and elongate, leaving a sac-like cup (volva) at the stalk base. This volva is sometimes referred to as the "death cup." The globose to egg-shaped immature fruitbody could be mistaken for a harmless puffball. However, a lengthwise cut through the center will reveal the structure of an embryonic gilled mushroom within, whereas the interior of a puffball is homogeneous. This feature can be seen in the photo of *Amanita jacksonii* (p.52). *Virosa* means "poisonous."

Latin name: *Chlorophyllum molybdites* (G. Mey.) Massee
Synonym: *Lepiota morganii* (Peck) Sacc.
Common names: Green-gilled Lepiota; Green-spored Parasol
Order: Agaricales
Family: Agaricaceae
Width of cap: 4–8 in. (10–20 cm)
Length of the stalk: 4–9 in. (10–23 cm)
Cap: Subglobose to short cylindrical or oval before the cap opens, then becoming broadly bell-shaped to convex and eventually nearly flat; whitish with a buff-brown to pale brown center that breaks up into scales as the cap expands, exposing white underlying tissue, young buttons may be entirely buff-brown; **flesh** thick on the disc, becoming thin toward the margin, white, odor not distinctive.
Gills: Free and remote from the stalk, close, broad; white at first, becoming dull olive-green to grayish green as the spores mature.
Stalk: Slender, equal or slightly tapering upward from a swollen base; white at first, becoming pale brownish with age or when handled; surface smooth; partial veil leaving a thick, ragged, collar-like movable ring on the upper stalk.
Spore print: Dull grayish green.
Spores: 9–13 x 6–9 μm, subovoid to elliptical with an apical germ pore, thick-walled, smooth, hyaline, dextrinoid.
Occurrence: Solitary or more often gregarious, often in fairy rings or arcs, on the ground in lawns, pastures, roadsides, or other grassy places; saprobic; summer-fall; occasional to fairly common in some locations.
Edibility: Poisonous.

Comments: This large and beautiful mushroom is easily mistaken for some edible species that share the grassland habitat, and consequently it is a frequent cause of mushroom poisoning. The problem of misidentification is compounded because the characteristic green color of the gills is not apparent in young specimens. The Shaggy Parasol, *Macrolepiota rhacodes* (not illustrated), which is edible, has a darker brown, more scaly cap, flesh that stains orange or brownish when cut (most noticeable in the stalk base), and white spores. The edible and delicious Parasol Mushroom, *Macrolepiota procera* (p.72) has white gills that remain so, and white spores. *Leucoagaricus naucinus* (p.70) is smaller than the Green-gilled Lepiota and has a pure white, smooth cap. *Chlorophyllum* means "green leaves," referring to the color of the mature gills. The specific epithet *molybdites* derives from the Greek word root "molybd," which means lead, also in relation to the mature gill color. Some mycologists have placed this species in the genus *Macrolepiota.*

Latin name: *Lepiota cepaestipes* (J. Sowerby) Sacc.
Synonym: *Leucocoprinus cepaestipes* (J. Sowerby) Pat.
Common name: Onion-stalk Lepiota.
Order: Agaricales
Family: Agaricaceae
Width of the cap: ¾–2½ in. (2–6.5 cm)
Height of the stalk: 1½–4½ in. (4–11.5 cm)
Cap: Ovoid at first, becoming bell-shaped to broadly convex with an umbo; whitish with a pinkish tan to grayish brown disc; surface granular to scaly, smooth on disc; margin striate, sometimes splitting; **flesh** thin, fragile, white, odor not distinctive, taste not distinctive or bitter.
Gills: Free from the stalk, crowded; white, becoming dingy white with age.
Stalk: Usually tapering upward from a swollen base, hollow; white to pinkish tinged, often bruising pale yellowish; surface glabrous; partial veil leaving a white, membranous ring on the upper stalk.
Spore print: White.
Spores: 8–10 x 6–7 μm, broadly elliptical with a germ pore, smooth, weakly dextrinoid.
Occurrence: Usually gregarious, often in clusters on wood mulch, straw, organic compost, and rich soil; saprobic; summer-fall; fairly common.
Edibility: Not edible, possibly poisonous.

Comments: This delicate mushroom will sometimes grow in potted plants. *Leucocoprinus birnbaumi* (not illustrated) is similar but has a yellow cap and stalk. *Lepiota americana* (not illustrated) is larger, has reddish brown scales on the cap, and flesh that stains reddish when exposed. *Cepaestipes* means "onion stem," alluding to the stalk shape.

Latin name: *Lepiota clypeolaria* (Bull.) Quél.
Common name: Shaggy-stalked Lepiota
Order: Agaricales
Family: Agaricaceae
Width of cap: 1–2½ in. (2.5–6.5 cm)
Length of the stalk: 1½–3½ in. (4–9 cm)
Cap: Broadly bell-shaped to convex, with an umbo; surface broken up into numerous ocraceous to brown, or reddish brown scales that are concentrated toward the disc, whitish beneath the scales; margin ragged with remnants of a partial veil; **flesh** thin, white, odor slightly unpleasant or not distinctive.
Gills: Free from the stalk, close; whitish.
Stalk: More or less equal, somewhat brittle; surface yellowish brown, covered with shaggy white, cottony fibers or scales from the base up to a poorly defined fragile ring that is often integrated with the floccose material on stalk; stalk smooth above the ring zone.
Spore print: White.
Spores: 12–16 x 5.5–6.5 μm, spindle-shaped, smooth, dextrinoid.
Occurrence: Solitary or in small groups in conifer and broad-leaved woods; saprobic; summer-fall; occasional.
Edibility: Poisonous.

Continued on next page

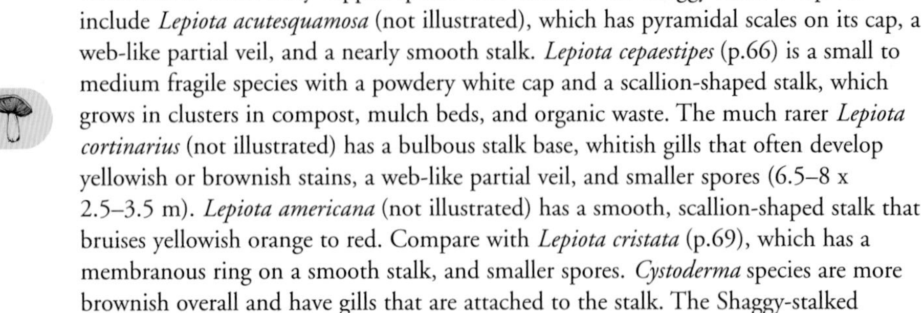

Comments: Other scaly-capped species that resemble the Shaggy-stalked Lepiota include *Lepiota acutesquamosa* (not illustrated), which has pyramidal scales on its cap, a web-like partial veil, and a nearly smooth stalk. *Lepiota cepaestipes* (p.66) is a small to medium fragile species with a powdery white cap and a scallion-shaped stalk, which grows in clusters in compost, mulch beds, and organic waste. The much rarer *Lepiota cortinarius* (not illustrated) has a bulbous stalk base, whitish gills that often develop yellowish or brownish stains, a web-like partial veil, and smaller spores (6.5–8 x 2.5–3.5 m). *Lepiota americana* (not illustrated) has a smooth, scallion-shaped stalk that bruises yellowish orange to red. Compare with *Lepiota cristata* (p.69), which has a membranous ring on a smooth stalk, and smaller spores. *Cystoderma* species are more brownish overall and have gills that are attached to the stalk. The Shaggy-stalked Lepiota is variably colored and may represent a complex of closely related species or varieties. Because some small lepiotas contain deadly toxins, all should be strictly avoided for eating. The specific epithet *clypeolaria* is derived from the Latin word clypeus, meaning "shield."

Latin name: *Lepiota cristata* (Alb. & Schwein.) P. Kumm.
Common names: Crested Lepiota; Stinking Parasol
Order: Agaricales
Family: Agaricaceae
Width of cap: ½–1½ in. (1.3–4 cm)
Length of the stalk: 1¼–2½ in. (3–6.5 cm)
Cap: Broadly bell-shaped to convex, becoming nearly flat with an umbo; orange-brown to reddish brown at first and remaining so on the disc, elsewhere becoming whitish as the cuticle breaks up into more or less concentrically arranged, flattened, brownish scales or granules, revealing the surface beneath, scales more sparse toward the margin; **flesh** thin, white to pale pinkish, odor pungent, like rubber tires, unpleasant.
Gills: Free from the stalk, close to crowded; white.
Stalk: Slender, equal, hollow; whitish or tinged pinkish; surface smooth to silky; partial veil leaving a delicate, membranous ring on the stalk that soon falls away.
Spore print: White.
Spores: 6–8 x 3–4.5 µm, bullet-shaped, smooth, hyaline, dextrinoid.
Occurrence: Usually in scattered groups or clusters on the ground in lawns, grassy places, woodland borders, open woods, or along paths; saprobic; summer-fall; common.
Edibility: Not recommended. Possibly poisonous and easily confused with similar species that are decidedly dangerous.

Comments: This small mushroom is distinguished by the pungent, chemical odor, which is a bit like burned rubber. Compare with *Lepiota clypeolaria* (p.67), and see comments under that species for notes on other look-alikes. No small lepiotas should be considered for eating since some of them contain deadly amatoxins. *Cristata* means "crested," referring to the cap ornamentation.

Latin name: *Leucoagaricus naucinus* (Fr.) Singer
Synonym: *Lepiota naucina* (Fr.) Quél.; *Lepiota naucinoides* Peck
Common names: Smooth Parasol; Smooth Lepiota
Order: Agaricales
Family: Agaricaceae
Width of the cap: 2–4 in. (5–10 cm)
Height of the stalk: 2–4½ in. (5–11.5 cm)
Cap: Nearly globose at first, expanding and becoming convex or flattened, at times with a slight umbo; chalk-white to creamy white, developing pale buff to pale grayish brown tones with age, especially at the center; surface dry, smooth to minutely roughened, or becoming finely cracked; **flesh** thick, white, odor and taste not distinctive.
Gills: Free from the stalk; close; white or tinged pinkish with age.
Stalk: Equal, typically with a bulbous base, becoming hollow with age; white, often developing pale brownish stains when handled; surface smooth; partial veil leaving a collar-like movable ring on the upper stalk.
Spore print: White.
Spores: 7–9 x 5–6 μm, broadly ovoid with an apical pore, smooth, hyaline, dextrinoid.
Occurrence: Solitary or in groups on the ground in lawns, pastures and other grassy places, less often in woods and pine plantations; saprobic; summer-fall; occasional.
Edibility: Not recommended (see comments below).

Comments: Although sometimes rated as edible, there are numerous accounts of individuals becoming ill after consuming this mushroom. Those who would experiment eating the Smooth Parasol should be careful not to confuse it with the potentially deadly *Amanita virosa* (p.62), which has a membranous skirt-like ring on the stalk and a cup-like sac (volva) at the stalk base. Otherwise these two all-white species are very similar. In the young "button stage," the poisonous *Chlorophyllum molybdites* (p.64) can also mimic the Smooth Parasol. Its cap can be nearly white to pale brown and smooth at first, but develops conspicuous scales as it expands. There are several white species of *Agaricus* that grow in grassland, but their mature gills and spores are chocolate-brown. The name *naucinus* is derived from the Latin word naucum, meaning "trivial."

Latin name: *Macrolepiota procera* (Scop.) Singer
Synonym: *Lepiota procera* (Scop.) Gray
Common name: Parasol Mushroom
Order: Agaricales
Family: Agaricaceae
Width of cap: 3–8 in. (7.5–20 cm)
Length of the stalk: 4–10 in. (10–25 cm)
Cap: Nearly round to oval at first, becoming bell-shaped to broadly convex or flattened, usually with an umbo; surface at first pale brown to darker brown, dry and felt-like before the cuticle breaks up into coarse, flattened scales as the cap expands, revealing a white to pale tan ground color beneath; the concentric scales are more sparsely distributed outward from the disc, which remains solid brown; margin usually ragged; **flesh** white, soft, odor and taste not distinctive.
Gills: Free from the stalk, close to crowded, broad; creamy white, or pale brown with age.
Stalk: Slender, cylindrical, slightly tapering upward from a swollen base, stuffed or hollow; colored like the cap or paler; surface more or less patterned with scaly bands; flesh tough, fibrous; partial veil leaving a thick, membranous, fringed ring on the upper stalk, which becomes loose and movable on mature specimens.
Spore print: White.
Spores: 15–20 x 10–13 μm, broadly elliptical with a germ pore, smooth, dextrinoid.
Occurrence: Solitary or gregarious on the ground in open woods, along paths, woodland borders, pastures, grassy areas, parkland, and waste ground; saprobic; summer-fall; common.
Edibility: The tender caps are edible and highly regarded by many mycophagists.

Comments: Before the cap expands, the Parasol Mushroom looks like a kettledrum stick. At this stage it can be confused with the poisonous Green-gilled Lepiota, *Chlorophyllum molybdites* (p.64), which at maturity has greenish gills and produces greenish spores. The edible *Macrolepiota rhacodes* (not illustrated) is similar to the Parasol Mushroom, but is comparatively squatty in stature, more scaly, and its flesh stains orangish when cut (especially at the base of the stalk). The Parasol Mushroom shown here is common in the Appalachians. It is generally smaller and less scaly than the *Macrolepiota procera* described from Europe, and may represent an altogether different species. Members of the genus *Amanita* also have free gills and white spores, but do not have scales on the cap. *Macrolepiota* means "large scale." *Procera* means "tall."

Latin name: *Psilocybe thrausta* (Schulzer) Bon
Synonym: *Stropharia thrausta* (Schulzer ex Kalchbr.) Sacc.
Common name: Fall Pumpkin Psilocybe
Order: Agaricales
Family: Strophariaceae
Width of cap: 1–2½ in. (2.5–6.5 cm)
Length of the stalk: 2–4 in. (5–10 cm)
Cap: Broadly conical to convex, often with an obtuse umbo, becoming flattened with age; pumpkin-orange to brownish orange; surface viscid, smooth, with flattened whitish scales near the margin; **flesh** whitish, odor and taste not distinctive.
Gills: Adnate, sometimes with a decurrent tooth, moderately close; pale grayish when young, becoming purplish brown with a white edge at maturity.
Stalk: Slender, more or less equal; whitish or colored like the cap, especially toward the base; surface densely covered with whitish scales from the base up to a flaring membranous ring near the apex, powdery to slightly scurfy above the ring; upper surface of the ring striated and white, but soon blackened from the accumulation of discharged spores.
Spore print: Dark purple-brown.
Spores: 12–14 x 6–7 μm, elliptical with an apical germ pore, smooth, brown.
Occurrence: Scattered in small groups on the ground in mixed woods, especially along rivers and streams; appearing terrestrial but may be attached to buried woody debris; saprobic; fall; common.
Edibility: Poisonous.

Comments: When in pristine condition, this is one of the prettiest of the fall mushrooms. It is easily recognized by its pumpkin-orange cap with delicate whitish scales near the margin, and densely scaly stalk. *Psilocybe squamosa* (not illustrated), which is nearly identical except that it has an ochre-brown cap, is also poisonous. *Thrausta* means "brittle," relating to the overall fragile structure of the fruitbody.

Latin name: *Rozites caperatus* (Pers.) P. Karst.
Common name: Gypsy
Order: Agaricales
Family: Cortinariaceae
Width of cap: 2–4 in. (5–10 cm)
Length of the stalk: 2–4½ in. (5–11.5 cm)
Cap: Globose to ovoid in the button stage, becoming convex with a broad umbo to flat; pale ochre to ochraceous brown or orange-brown, covered at first with whitish to pale lilac fibrils that persist on the disc, giving the caps a "frosted" appearance; surface radially wrinkled with age, at least near the margin; **flesh** whitish to yellowish, odor and taste not distinctive.
Gills: Adnate, crowded; yellowish to light brown, becoming rusty brown, covered at first with a white membranous partial veil.
Stalk: Equal to slightly bulbous at the base, solid; dingy white; smooth from the base up to a prominent membranous ring on the upper stalk, more or less scurfy above the ring.
Spore print: Rusty brown to cinnamon-brown.
Spores: 10.5–14 x 7–9 μm, broadly elliptical, slightly roughened with minute warts, pale brown, dextrinoid.
Occurrence: Solitary to scattered in groups on the ground in broad-leaved and conifer woods, especially with oak, pine, and spruce; mycorrhizal; summer-fall; common.
Edibility: Edible.

Comments: The Gypsy is recognized by its "frosted" cap that often becomes wrinkled with age, and well-defined persistent ring on the stalk. *Cortinarius corrugatus* (p.272) also has a brown wrinkled cap, but it lacks a membranous ring on the stalk. Several species of *Agaricus* are similar in form and stature, but they have gills that are free from the stalk, and produce chocolate-brown spores. The genus name *Rozites* honors the French mycologist Ernest Roze (1833–1900). The species epithet *caperatus* means "wrinkled" or "creased."

Latin name: *Squamanita umbonata* (Sum.) Bas
Common name: Knobbed Squamanita
Order: Agaricales
Family: Tricholomataceae
Width of cap: 1½–2½ in. (4–6.5 cm)
Length of the stalk: above ground portion 1–3 in (2.5–7.5 cm)
Cap: Broadly conical to convex, becoming flattened with a low umbo; ochre-brown to orange-brown or bronze; surface dry, radially fibrillose to scaly; **flesh** white, odor and taste not distinctive.
Gills: Adnexed to sinuate, close; white.
Stalk: More or less equal to club-shaped, arising from a thick, swollen, rooting base; white, or white merged with the cap color; surface cottony to scaly, especially near the base; concentric or fragmented bands of coarse scales at ground level; partial veil leaving a cottony brown ring or ring zone on the upper stalk.
Spore print: White.
Spores: 6–9 x 3.5–5.5 μm, elliptical, smooth, hyaline, non-amyloid.
Occurrence: Solitary or in groups of 2 or 3 individuals that originate from a common underground tuberous base; in mixed woods often with oak and pine; saprobic; late summer-fall; rare.
Edibility: Unknown.

Comments: The similar *Squamanita odorata* (not illustrated), which also grows from a swollen underground base, has a purplish brown radially fibrillose cap, and a strong, fruity odor. Some species of *Amanita* that have a swollen, rooting stalk base could be confused with *Squamanita umbonata,* but they have free gills and do not grow in fused clusters. The name *Squamanita* means "scale mushroom." *Umbonata* refers to the umbonate (knobbed) cap.

Latin name: *Stropharia hardii* G.F. Atk.
Common name: Hard's Stropharia
Order: Agaricales
Family: Strophariaceae
Width of the cap: 1½–4 in. (4–10 cm)
Height of the stalk: 1½–3½ in. (4–9 cm)
Cap: Convex at first, becoming flattened or shallowly depressed with age; dingy yellow to ochraceous or yellow-brown; surface with appressed, radiating scales, somewhat viscid to dry; **flesh** whitish, odor and taste not distinctive.
Gills: Adnate to adnexed, close; whitish at first, becoming grayish brown to purplish brown when mature.
Stalk: More or less equal, sometimes with a small bulbous base, usually with prominent white rhizomorphs at the base; whitish to pale ochraceous; partial veil leaving a white membranous ring on the upper stalk; surface scurfy to lightly scaly above and below the ring.
Spore print: Purple-brown.
Spores: 6–9 x 3–5 μm, elliptical with an apical pore, smooth.
Occurrence: Solitary or in small scattered groups on the ground in broad-leaved and mixed woods, frequently occurring in low areas along streams; saprobic; late summer-fall; occasional to locally common.
Edibility: Unknown.

Comments: Hard's Stropharia often has a worn, disheveled appearance even when fresh. Faded specimens of *Stropharia rugoso-annulata* (p.78) can be similarly colored but they have a thick segmented ring on the stalk and often grow gregariously in grassy woodland margins, open areas, along paths, or less often in open woods. *Stropharia hornemannii* (p.43) is more robust and has a densely scaly stalk below the ring. *Agrocybe acericola* (not illustrated) grows on decaying wood, has an ochre to yellow-brown glabrous cap, and brown gills. *Agrocybe praecox* (not illustrated) has a tan to brown glabrous cap, brown gills, and lacks rhizomorphs at the stalk base. It grows on the ground in grassy areas, wood margins, and in open woods. Hard's Stropharia was named in honor of Ohio mycologist Miron Elisha Hard (1849–1914), who authored a popular mushroom guidebook in 1908 that included the first published photograph of this species.

Latin name: *Stropharia rugoso-annulata* Farl. ex Murrill
Common name: Wine-cap; Burgundy-cap
Order: Agaricales
Family: Strophariaceae
Width of cap: 1½–5 in. (4–12.5 cm)
Length of the stalk: 2–5 in. (5–12.5 cm)
Cap: Hemispheric at first, becoming convex to flattened or slightly depressed at the center with age; margin incurved; burgundy to purplish brown when young, but fading to tan, sometimes with olivaceous tones; surface somewhat viscid at first, later dry, often developing cracks or fissures, exposing whitish flesh beneath; **flesh** thick, white, odor and taste not distinctive.
Gills: Adnate or sinuate, crowded; whitish when young, becoming pale gray, then purplish black with age.
Stalk: More or less equal, or with a slightly bulbous base, and white cord-like strands (rhizomorphs) at the base; surface whitish to dingy yellowish, smooth to lightly furrowed; partial veil leaving a thick, white, membranous ring on the upper stalk; ring striate on the upper side, soon blackened from falling spores, coarsely segmented (like a cogwheel) on the underside.
Spore print: Purplish black.
Spores: 10–14 x 7–9 μm, elliptical with an apical germ pore, smooth, pale purplish brown.
Occurrence: Solitary or more often gregarious on soil or wood mulch in grassy areas, gardens, open woods, or woodland borders, especially in moist, sandy soil; saprobic; spring-fall; common.
Edibility: Edible, but best when in the firm, button stage. Older or water-soaked specimens unappealing.

Comments: When the partial veil ruptures it will sometimes adhere to the cap margin as thick membranous fragments instead of forming the usual "cogwheel" ring on the stalk. Age and exposure to sunlight affect the color saturation of the Wine-cap. Specimens with caps that have faded to pale tan might be mistaken for a species of *Agaricus,* however they have free gills and chocolate-brown spores. The similar *Stropharia hardii* (p.77) has an ochre-brown cap and is strictly a woodland species that usually grows singly or in small scattered groups.

The Wine-cap is one of the earliest edible gilled mushrooms to appear in the spring. The spawn (mycelium used for cultivation) is sold commercially under the name King Stropharia. *Rugoso-annulata* means "wrinkled ring."

Latin name: *Tricholoma caligatum* (Viviani) Ricken
Synonym: *Armillaria caligata* (Viviani) Gill.
Common name: Fragrant Armillaria
Order: Agaricales
Family: Tricholomataceae
Width of the cap: 2–6 in. (5–15 cm)
Height of the stalk: 2–4 in. (5–10 cm)
Cap: Hemispherical at first, becoming convex to broadly convex with a low umbo, margin incurved when young; brown to reddish brown, paler near the margin; surface somewhat viscid when wet, cuticle breaking into appressed scales or radiating fibrils, revealing white to cream flesh beneath; margin often ragged with partial veil remnants; **flesh** white, odor spicy or not distinctive, taste usually bitter.
Gills: Adnate, close to crowded; white to cream, developing brownish stains with age; covered at first with a membranous partial veil.
Stalk: Stout, more or less equal or tapering toward the base; sheathed from the base up to a flaring membranous ring; surface below the ring with patches and bands that are colored like the cap, white above the ring; ring white on the upper surface, brownish on the underside.
Spore print: White.
Spores: 6–7.5 x 4.5–5.5 μm, broadly elliptical, smooth, hyaline, non-amyloid
Occurrence: Solitary or gregarious, sometimes in small caespitose clusters on the ground in broad-leaved and mixed woods, especially with oak; mycorrhizal; summer-fall; fairly common in some years.
Edibility: Non-poisonous but most collections are bitter and unpalatable.

Comments: The odor and taste of this mushroom are variable. It is sometimes reported to be a good edible, but most West Virginia collections have intensely bitter tasting flesh. The name *caligatum* comes from the Latin word "caliga," which means "boot" and refers to the sheathing stalk base.

Gilled Mushrooms Growing on the Ground, with Flesh and Gills Exuding a Clear, Milky, or Colored Latex When Cut or Broken

The mushrooms grouped here are members of the genus *Lactarius,* which is characterized by species having white spores, brittle flesh, gills that are attached to the stalk, and—most importantly—tissue that exudes a white, clear, or colored fluid (latex) when the fruitbody is cut or broken. For this reason they are called "milk mushrooms." The taste of the latex is important in the identification of species. It ranges from mild tasting to intensely acrid. A small amount on the tip of the tongue is sufficient to determine this, although it sometimes takes a minute or so for an acrid taste to become noticeable. The quantity of latex that is "leaked" is variable depending on the species. With some milk mushrooms the latex flows in steady drops, while in others it is barely a seep. The freshness of the mushroom also affects the amount of latex exuded. Older and dryer specimens may not "leak" at all, but normally a shallow knife cut at the apex of the stalk or across the gills will produce an observable flow of latex. A few species in the genus *Mycena* will exude colored sap from the stalk when it is broken, but they are small, delicate mushrooms with slender stalks that are less than ¼ inch (6 mm) in diameter. By contrast, most species of *Lactarius* are squatty and robust.

Some milk mushrooms are popular edibles. Only a few are suspected to be poisonous, but many others have flesh that is sharply acrid and unpalatable. Most, if not all, milk mushrooms form mycorrhizal associations with trees and shrubs.

Latin name: *Lactarius aquifluus* Peck
Synonym: *Lactarius helvus* (Fr.) Fr. (according to some authorities)
Common name: Burnt Sugar Lactarius
Order: Russulales
Family: Russulaceae
Width of cap: 1½–5 in. (4–12.5 cm)
Length of the stalk: 1½–3½ in. (4–9 cm)
Cap: Broadly convex to flattened or depressed in the center, with or without a small umbo, margin incurved at first; pale cinnamon to reddish brown, with or without obscure concentric zones; surface smooth to minutely scaly with age; **flesh** aromatic, like maple syrup, burnt sugar, or curry; **latex** watery, transparent, mild tasting.
Gills: Adnate to subdecurrent; crowded; whitish to pale pinkish tan.
Stalk: Equal, becoming hollow; colored like the cap or paler, often with a whitish bloom, and with white mycelium at the base.
Spore print: Creamy white.
Spores: 6–9 x 5.5–7.5 μm, elliptical, surface with a net-like pattern of warts and ridges, hyaline, ornamentation amyloid.
Occurrence: In scattered groups or clusters on the ground in conifer and mixed woods, especially at higher elevations in the mountains; mycorrhizal; summer-fall; common.
Edibility: Not recommended. There are conflicting reports regarding the edibility of this species.

Comments: This mushroom is often found in wet places, especially with sphagnum moss. In older mushroom books it has been called *Lactarius helvus,* a nearly identical and possibly synonymous European species. *Lactarius camphoratus* (p.84) has a similar aromatic odor, but it is a much smaller mushroom with a darker reddish brown cap. Compare also with *Lactarius quietus* var. *incanus* (p.101). *Lactarius mutabilis* (not illustrated) also has a mildly fragrant, sweet odor, but it has a more or less concentrically zoned purplish brown cap, and watery white latex. The name *aquifluus* means "water flow," referring to the clear latex.

Latin name: *Lactarius aspideoides* Burlingham
Common name: None
Order: Russulales
Family: Russulaceae
Width of cap: 1½–3 in. (4–7.5 cm)
Length of the stalk: 1½–2½ in. (4–6.5 cm)
Cap: Convex to flattened with a depressed center, or shallowly funnel-shaped with age; pale buff to pale yellow, often with lilac to violet stains, glabrous, viscid; **flesh** thin, firm, brittle, whitish, odor not distinctive, taste bitter to tardily acrid; **latex** abundant on fresh specimens, white, unchanging but staining tissue violet.
Gills: Decurrent to subdecurrent, close, colored like the cap, staining violet from the latex where injured.
Stalk: Equal, becoming hollow; colored like the cap and gills; surface smooth, glabrous, viscid.
Spore print: Pale yellowish.
Spores: 7–9 x 6.5–8 μm, broadly elliptical, surface with warts and ridges, hyaline, ornamentation amyloid.
Occurrence: Scattered in small groups on the ground in broad-leaved and conifer woods; mycorrhizal; summer-fall; uncommon.
Edibility: Unknown.

Comments: The overall viscid aspect of the fruitbody, its pale color, and the abundant latex that soon stains the tissue violet are distinctive. *Aspideoides* means "shield-like."

Latin name: *Lactarius camphoratus* (Bull.) Fr.
Common names: Fragrant Lactarius; Aromatic Milk Cap
Order: Russulales
Family: Russulaceae
Width of cap: ¾–1¾ in. (2–4.5 cm)
Length of the stalk: 1–2 in. (2.5–5 cm)
Cap: Convex to depressed, usually with an umbo, margin incurved at first; reddish brown, darker at the center; surface smooth, glabrous, moist or dry; **flesh** aromatic, sweet, like curry or burnt sugar; **latex** scant, white to whey-like, mild to bitter tasting.
Gills: Adnate to decurrent; crowded; pale flesh-colored, becoming rusty brown with age.
Stalk: Equal; colored like the cap or paler, sometimes with a purplish tinge; surface smooth.
Spore print: White to pale yellowish.
Spores: 7–8.5 x 6.5–7.5 μm; subglobose; surface with warts and ridges, hyaline, ornamentation amyloid.
Occurrence: Scattered or in small groups in conifer woods or mixed woods; mycorrhizal; summer-fall; common.
Edibility: Non-poisonous. The dried caps are sometimes used as a flavoring.

Comments: The fragrant odor and small size are the best field characters for recognizing this species. The odor is subtle in fresh specimens but becomes increasingly more pronounced upon drying. There are several small, reddish brown to orange-brown species of *Lactarius* found in the central Appalachians that are difficult to distinguish in the field. *Lactarius oculatus* (not illustrated) lacks a characteristic odor and has tardily acrid tasting latex. *Lactarius theiogalus* (p.106) has acrid tasting white latex that slowly turns yellowish on exposure to the air. The brownish red cap of *Lactarius rimosellus* (not illustrated) usually fades and becomes cracked with age. *Lactarius rufus* (p.102) is typically larger than *Lactarius camphoratus,* and has strongly acrid tasting latex. *Camphoratus* means "having a sweet scent."

Latin name: *Lactarius chelidonium* var. *chelidonioides* (A.H. Sm.) Hesler & A.H. Sm.
Synonym: *Lactarius chelidonioides* A.H. Sm.
Common name: None
Order: Russulales
Family: Russulaceae
Width of cap: 1¼–3 in. (3–7.5 cm)
Length of the stalk: ¾–2 in. (2–5 cm)
Cap: Convex, becoming broadly convex with a central depression to nearly funnel-shaped, margin incurved at first; color variable, often a mix of bluish green, grayish green, grayish yellow, dingy orange-brown, or straw yellow, with or without slight concentric zones, especially near the margin; surface glabrous, somewhat viscid when wet; **flesh** bluish with orangish tones above the gills, odor not distinctive, taste slowly and slightly peppery; **latex** scant, yellowish to brownish.
Gills: Short decurrent, close to crowded; dingy yellow to grayish olive, developing green tints or stains with age.
Stalk: More or less equal or swollen toward the base, soon hollow; colored like the cap or paler; flesh tinged orangish.
Spore print: Pale buff.
Spores: 7–9 x 5–6.5 μm, elliptical, surface with warts and some connecting ridges, ornamentation amyloid.
Occurrence: Gregarious on the ground under conifers, especially pine; mycorrhizal; summer-fall; occasional.
Edibility: Edible.

Comments: This mushroom has a short stalk and a particularly squatty stature. The cap often appears to be sitting directly on the ground. Compare with *Lactarius deterrimus* (p.89), which has bright orange latex. The name *chelidonium* comes from the Greek word chelidon, meaning "a swallow." The roundabout connection to this mushroom is that the name *Chelidonium* is also the genus of the flowering plant celandine, which appears about the time swallows arrive in spring. Celandine has a saffron-colored sap, which is a bit like the latex of this species. Whew!

Latin name: *Lactarius corrugis* Peck
Synonym: *Lactarius volemus* var. *subrugosus* Peck
Common name: Corrugated Lactarius
Order: Russulales
Family: Russulaceae
Width of cap: 2–6 in. (5–15 cm)
Length of the stalk: 1½–4 in. (4–10 cm)
Cap: Convex to broadly convex, often with a depressed center, margin incurved at first; orange-brown to dark reddish brown; surface velvety, wrinkled to concentrically corrugated, especially near the margin; **flesh** firm, brittle, white, staining brown, odor mild in young specimens, becoming fishy with age; **latex** copious, sticky, white and unchanging but staining tissue brown, taste mild or astringent.
Gills: Adnate to slightly decurrent, occasionally forked, close to crowded; ochre to cinnamon, staining brown where damaged.
Stalk: Equal; colored like the cap or paler, sometimes with a whitish bloom; smooth, dull.
Spore print: White.
Spores: 9–12 x 8.5–12 μm, subglobose, surface with warts and ridges, hyaline, ornamentation amyloid.
Occurrence: Usually in scattered groups but sometimes solitary on the ground in oak woods or mixed woods; mycorrhizal; summer-fall; common.
Edibility: Edible and very good.

Comments: The dark, earthy colors of this stocky mushroom often blend in with the forest litter. The general aspect is similar to *Lactarius volemus* (p.109), which has a more orange cap with fewer (if any) wrinkles, lighter colored gills, and a strong fishy odor at all stages. These two species are closely related and intermediate color forms are sometimes found. *Corrugis* means "corrugated."

Latin name: *Lactarius croceus* Burlingham
Common name: None
Order: Russulales
Family: Russulaceae
Width of cap: 2–4 in. (5–10 cm)
Length of the stalk: 1–2½ in. (2.5–6.5 cm)
Cap: Broadly convex to depressed, or nearly funnel-shaped, margin incurved at first; bright orange to saffron, becoming duller brownish orange with age; surface smooth, viscid when moist, more or less concentrically zoned or without zonation; **flesh** white, staining orange-yellow when broken, odor fruity; **latex** white, slowly turning orange-yellow on exposure to the air, scant, taste bitter to acrid, sometimes slowly acrid.
Gills: Adnate to decurrent; close to subdistant; cream to pale yellow, slowly staining orange-yellow.
Stalk: Equal, becoming hollow with age, glabrous; pale orange-yellow, or colored like the cap, sometimes spotted with darker shallow pits.
Spore print: Pale yellow.
Spores: 7–9 x 6–7 μm, subglobose to elliptical, surface with lengthwise bands and warts, hyaline, ornamentation weakly amyloid.
Occurrence: Solitary or in groups on the ground in broad-leaved woods, especially with oak; mycorrhizal; summer-fall; common.
Edibility: Unknown. Some other species with acrid white latex that turns yellow on exposure to the air are poisonous.

Comments: *Lactarius psammicola* (p.98) is similar but has a strongly zoned cap and acrid tasting white latex that does not change to orange-yellow on exposure to the air. *Lactarius volemus* (p.109) has a mild tasting latex and a pronounced fishy odor. *Lactarius deliciosus* (not illustrated) has a mild tasting, orange-colored latex and grows in association with conifer trees. Compare also with *Lactarius peckii* (p.97). *Croceus* means "saffron colored."

Latin name: *Lactarius deceptivus* Peck
Common name: Decepitve Lactarius
Order: Russulales
Family: Russulaceae
Width of cap: 2–9 in. (5–23 cm)
Length of the stalk: 2–4 in. (5–10 cm)
Cap: Convex becoming depressed in the center to funnel-shaped; margin strongly inrolled at first with a thick, soft, cottony roll that conceals the gills in young specimens; surface white, smooth to velvety at first, then breaking up into yellowish brown scales from the center outward; **flesh** firm, white, staining brownish, taste acrid, odor mild, becoming pungent with age; **latex** white and unchanging but staining the tissue brown, taste strongly acrid.
Gills: Adnate to decurrent; close to subdistant; white at first, becoming tan with age.
Stalk: Equal or slightly tapering toward the base, short, thick, hard; white, staining brown; surface velvety to slightly scaly with age, especially near the base.
Spore print: White to pale yellowish.
Spores: 9–13 x 7.5–9 µm, broadly elliptical, surface minutely roughened with warts and ridges, hyaline, ornamentation amyloid.
Occurrence: Solitary or in scattered groups on the ground, or sometimes on moss-covered logs and stumps in wet conifer woods, especially with hemlock, also in broad-leaved woods, especially with oak; mycorrhizal; early summer-fall; common.
Edibility: Non-poisonous. According to some reports the acrid taste of the flesh disappears when cooked, but as an edible it is not highly regarded.

Comments: *Deceptivus* means "deceiving," perhaps alluding to the changeable aspect of this large and common mushroom as it ages. Older scaly, brown, and funnel-shaped specimens look very different from the all white, convex, and nearly smooth young fruitbodies. *Lactarius piperatus* (not illustrated) is similar but has densely crowded gills, a firm rather than soft and cottony cap margin, and exceedingly acrid latex. *Lactarius subvellereus* var. *subdistans* (not illustrated) has more widely spaced gills, and a firm, even cap margin. *Russula brevipes* (not illustrated) and *Russula angustispora* (p.217) are medium to large all-white mushrooms of similar size and stature, but they do not exude a latex when cut or broken.

Latin name: *Lactarius deterrimus* Gröger
Synonym: *Lactarius deliciosus* var. *deterrimus* (Gröger) Hesler & A.H. Sm.
Common name: Spruce Saffron Lactarius
Order: Russulales
Family: Russulaceae
Width of cap: 1½–4 in. (4–10 cm)
Length of the stalk: 1–2½ in. (2.5–6.5 cm)
Cap: Convex with incurved margin, becoming depressed in the center; orange-buff, sometimes fused with green, or having green stains, obscurely concentrically zoned or without zones; surface glabrous, smooth, viscid when moist; **flesh** whitish to pale cinnamon, staining vinaceous red then green from the latex, odor and taste not distinctive; **latex** bright orange-red.
Gills: Adnate to decurrent; close; orange with reddish brown or green stains.
Stalk: Equal, becoming hollow; colored like the cap and staining similarly; surface smooth to slightly pitted.
Spore print: Pale buff.
Spores: 7.5–9 x 6–7 μm, broadly elliptical, surface with warts and ridges, hyaline, ornamentation amyloid.
Occurrence: Solitary to scattered in groups on the ground in conifer woods, especially with spruce; mycorrhizal; summer-fall; occasional, or sometimes locally common.
Edibility: Edible.

Comments: *Lactarius deliciosus* (not illustrated) is a similar species that forms mychorrhiza with pine. It has a prominently zoned cap and a pitted stalk. Some mycologists consider *Lactarius deterrimus* to be a variety of *Lactarius deliciosus.* In Europe, especially in northern Spain, *Lactarius deliciosus* is a highly prized edible mushroom. Although edible, the Spruce Saffron Lactarius does not elicit the same level of gastronomic enthusiasm. Compare with *Lactarius chelidonium* (p.85).

Latin name: *Lactarius gerardii* Peck
Common name: Gerard's Lactarius
Order: Russulales
Family: Russulaceae
Width of cap: 1½–4½ in. (4–11.5 cm)
Length of the stalk: 1½–3 in. (4–7.5 cm)
Cap: Convex, often with an umbo, becoming flattened to sunken in the center, margin often scalloped; dark brown to yellow-brown; surface dry, velvety, often radially wrinkled; **flesh** firm, white, odor mild, taste mild or becoming slightly acrid; **latex** white, taste mild to slightly acrid.
Gills: Adnate to decurrent; subdistant to well-spaced; white to cream.
Stalk: Nearly equal to somewhat swollen in the center, becoming hollow; colored like the cap, dull to velvety.
Spore print: White.
Spores: 7–10 x 7.5–9 μm, globose to broadly elliptical, surface reticulated, hyaline, ornamentation amyloid.
Occurrence: Solitary or more often in groups on the ground in broad-leaved and conifer woods; mycorrhizal; summer-fall; common.
Edibility: Edible.

Comments: Other closely related medium-size brownish species include, *Lactarius gerardii* var. *subrubescens* (not illustrated), which is nearly identical, but has latex that stains the broken flesh and gills violaceous pink. *Lactarius subgerardii* (not illustrated), which grows under hemlock, is smaller and has acrid tasting latex. *Lactarius petersenii* (not illustrated) has brown latex that stains damaged tissue brown. *Lactarius fumosus* (not illustrated) has crowded gills, white flesh that stains reddish when broken, tardily acrid tasting latex, and yellow spores. Compare with *Lactarius subplinthogalus* (p.104) and *Lactarius lignyotellus* (p.94). Gerard's Lactarius was named for William Ruggles Gerard (1844–1914), who collected the type specimen on which the scientific description is based.

Latin name: *Lactarius griseus* Peck
Synonym: *Lactarius glabripes* A.H. Sm.
Common name: Gray-brown Lactarius
Order: Russulales
Family: Russulaceae
Width of cap: ½–1½ in. (1.3–4 cm)
Length of the stalk: 1–2 in. (2.5–5 cm)
Cap: Convex to broadly convex with a depressed center, or becoming funnel-shaped with a small umbo; brown to grayish brown or pinkish gray; surface minutely scaly, especially in the center; **flesh** white to yellow, odor not distinctive, taste mild to slowly acrid; **latex** white, drops drying yellowish, taste mild to slowly acrid.
Gills: Adnate to decurrent; close; whitish to creamy yellow.
Stalk: Equal, hollow; colored like the cap or paler, with white hairs at the base.
Spore print: Yellowish.
Spores: 7–9 x 6–7 μm, subglobose to broadly elliptical, surface with warts and ridges, hyaline, ornamentation amyloid.
Occurrence: In small groups or clusters on mossy ground or well-decayed wood, especially moss-covered logs and stumps, also in sphagnum moss bogs; probably mycorrhizal; early summer-fall; common.
Edibility: Not edible.

Comments: The Gray-brown Lactarius is smaller than most other members of the genus. Older specimens, or those found in dry conditions when the presence of latex is no longer apparent, might be mistaken for a species of *Clitocybe*. *Lactarius cinereus* (not illustrated) is somewhat similar but has a larger viscid gray cap and acrid tasting latex. *Griseus* means "gray."

Latin name: *Lactarius hysginus* (Fr.) Fr.
Common name: None
Order: Russulales
Family: Russulaceae
Width of cap: 1½–3 in. (4–7.5 cm)
Length of the stalk: 1¼–2½ in. (3–6.5 cm)
Cap: Convex at first with an incurved margin, then flattened and depressed in the center or nearly funnel-shaped; reddish brown, paler pinkish buff toward the margin, fading with age, sometimes obscurely zoned; surface glabrous, shiny, viscid when wet; **flesh** whitish, odor not distinctive, taste acrid; **latex** white, taste acrid.
Gills: Adnate to subdecurrent; close to crowded; whitish at first, becoming yellowish with age.
Stalk: Equal or tapering downward, soon hollow; colored like the cap or paler, surface viscid when wet, smooth, often spotted with shallow pits.
Spore print: Yellowish.
Spores: 5.5–7 x 6–7.5 µm, subglobose to broadly elliptical, surface with warts and ridges, hyaline, ornamentation amyloid.
Occurrence: Solitary or in small groups on the ground in conifer woods, especially with hemlock; mycorrhizal; summer-fall; occasional.
Edibility: Unknown.

Comments: *Lactarius mucidus* (not illustrated) is similar but has a dark purplish brown to chocolate-brown cap, and whitish to cream gills that become spotted blue-green from the latex. *Hysginus* means "scarlet," which is stretching it a bit for the typical color of this species.

Latin name: *Lactarius indigo* (Schwein.) Fr.
Common name: Indigo Lactarius
Order: Russulales
Family: Russulaceae
Width of cap: 2–5 in. (5–12.5 cm)
Length of the stalk: 1–3 in. (2.5–7.5 cm)
Cap: Broadly convex, usually with a depressed center or funnel-shaped; margin inrolled at first, becoming elevated with age; silvery gray fused with blue, often zoned and dark blue spotted, at least near the margin; surface more or less viscid, glabrous; **flesh** thick, firm, whitish, staining dark blue when cut or broken, odor not distinctive, taste mild to slightly bitter; **latex** deep indigo blue, scant, slowly becoming dark green on exposure to air, and staining injured tissue green, taste mild.
Gills: Adnate to short-decurrent; crowded; indigo blue, fading and sometimes with yellowish tints with age, staining green where damaged.
Stalk: Nearly equal or tapering downward, short, soon hollow; colored like the cap, often spotted with dark blue shallow pits.
Spore print: Cream.
Spores: 7–9 x 5.5–7.5 µm, broadly elliptical to nearly globose, surface with warts and ridges sometimes forming a reticulum, hyaline, ornamentation amyloid.
Occurrence: Solitary or more often gregarious on the ground in woods, especially with oak and pine; summer-fall; occasional to locally common.
Edibility: Edible. The flesh has a coarse, grainy texture.

Comments: Few mushrooms are so unmistakable as the Indigo Lactarius. Its popularity as an edible may be due in part to the fact that it is easy to recognize and there are no poisonous look-alikes. The somewhat similar *Lactarius chelidonium* (p.85) has a grayish green cap that is often tinged with blue and yellow, and yellowish latex that slowly stains the flesh and gills green. *Indigo* means "indigo blue."

Latin name: *Lactarius lignyotellus* A.H. Sm. & Hesler
Common name: Sooty Spruce Lactarius
Order: Russulales
Family: Russulaceae
Width of cap: 1–2½ in. (2.5–6.5 cm)
Length of the stalk: 1½–3 in. (4–7.5 cm)
Cap: Convex with a pointed umbo to nearly conical, becoming flattened or depressed in the center, which usually remains prominently umbonate, margin incurved at first, often pleated; blackish brown, paler with age; surface dry, dull, velvety, usually wrinkled; **flesh** white to cream, unchanging or very slowly staining pinkish rose when cut or broken, odor and taste not distinctive; **latex** white, abundant in fresh specimens, taste mild or slightly bitter.
Gills: Adnate to short-decurrent; close to moderately well-spaced, sometimes forked; white at first, becoming pale ochraceous, edges pale to dark brown.
Stalk: More or less equal; colored like the cap or paler, whitish toward the base; surface dry, dull, velvety.
Spore print: Pale yellowish orange.
Spores: 7–9.5 x 7.5–10.5 μm, globose to subglobose, surface with warts and irregular ridges, hyaline, ornamentation amyloid.
Occurrence: Solitary to gregarious on the ground in high elevation conifer woods, especially with red spruce (*Picea rubens*), often on mossy ground, or infrequently on well-decayed wood; mycorrhizal; summer-fall; common.
Edibility: Reports vary, probably non-poisonous.

Comments: The pinkish rose staining on the cut or broken flesh, if perceivable at all, may not be apparent for several hours. *Lactarius lignyotus* (not illustrated) is larger and does not have brown gill edges (some varieties of *Lactarius lignyotus* do have dark gill edges but these have not been reported from West Virginia). *Lactarius gerardii* (p.90) has a lighter, yellow-brown cap, and widely spaced gills. *Lactarius fumosus* (not illustrated) has crowded gills, a dingy yellow-brown to almost whitish cap without an umbo, flesh and gills that stain reddish on injury, and pinkish buff spores. The name *lignyotellus* is derived from the Latin word lignyot, meaning "sooty."

Latin name: *Lactarius luteolus* Peck
Synonym: *Lactarius foetidus* Peck
Common name: None
Order: Russulales
Family: Russulaceae
Width of cap: 1–2½ in. (2.5–6.5 cm)
Length of the stalk: 1–2½ in. (2.5–6.5 cm)
Cap: Convex to flat, or depressed in the center; white to pale buff, becoming brownish with age; surface dry with a velvety bloom, somewhat wrinkled or uneven; **flesh** white, staining brown, odor mild to fishy, taste mild; **latex** white, copious, sticky, taste mild, staining all parts of the fruitbody brown on contact.
Gills: Adnate to subdecurrent; close; white to cream, staining yellowish to brown.
Stalk: Equal or tapering downward; colored like the cap, staining brown; surface dry with a velvety bloom.
Spore print: White to cream.
Spores: 7–9 x 5.5–7 μm, elliptical, surface with warts and ridges, hyaline, ornamentation amyloid.
Occurrence: Usually in small groups scattered on the ground in broad-leaved and mixed woods, often with oak, especially on sandy soil along streams and rivers; mycorrhizal; late spring-fall; occasional to locally common.
Edibility: Edible.

Comments: The pale color of the fruitbody, abundant sticky latex that stains everything brown that comes in contact with it (including fingers), and fishy odor are distinctive features. The closely related *Lactarius volemus* (p.109) is similar but it has a yellowish orange to orange-brown cap and stalk. *Luteolus* means "yellowish."

Latin name: *Lactarius maculatipes* Burlingham
Common name: None
Order: Russulales
Family: Russulaceae
Width of cap: 1½–4 in. (4–10 cm)
Length of the stalk: 1–3 in. (2.5–7.5 cm)
Cap: Broadly convex to flattened with a depressed center, or funnel-shaped with age, margin incurved when young; white to cream with alternating concentric yellowish zones or spots, becoming ochraceous orange in the center with age; surface glabrous, viscid when wet; **flesh** white, slowly becoming yellowish when exposed, odor not distinctive, taste acrid, sometimes slowly; **latex** white, slowly staining the flesh and gills yellowish; taste acrid, sometimes slowly.
Gills: Decurrent, crowded; whitish to pale pinkish buff, with yellowish to ochraceous stains with age.
Stalk: Equal or tapering toward the base, soon becoming hollow; colored like the cap, developing yellowish stains when bruised or with age; surface spotted with shallow pits, viscid when wet.
Spore print: Buff to yellowish.
Spores: 6.5–8 x 6–7.5 μm, subglobose to elliptical, surface with warts and ridges, hyaline, ornamentation amyloid.
Occurrence: Solitary, or more often in groups, sometimes clustered on the ground in woods and woodland margins, especially under oaks; mycorrhizal; summer-fall; occasional to locally common.
Edibility: Unknown.

Comments: The overall pale color, zonate cap, spotted stalk, and acrid tasting latex are distinctive. *Maculatipes* means "spotted foot."

Latin name: *Lactarius peckii* Burlingham
Common name: Peck's Lactarius
Order: Russulales
Family: Russulaceae
Width of cap: 1½–4 in. (4–10 cm)
Length of the stalk: 1–2½ in. (2.5–6.5 cm)
Cap: Broadly convex, soon depressed in the center, or shallow funnel-shaped; orange-brown to brick red with darker concentric zones (sometimes obscure); surface dry, minutely roughened to glabrous; **flesh** firm, pale vinaceous brown, odor not distinctive, taste strongly acrid; **latex** abundant, white, taste strongly acrid.
Gills: Adnate to decurrent, close to crowded; pale cinnamon to reddish brown or darker on drying, often with rusty brown stains.
Stalk: More or less equal, becoming hollow with age; colored like the cap or paler, sometimes spotted or stained dull orange to reddish brown; surface smooth, with a whitish bloom at first.
Spore print: White.
Spores: 6–7.5 μm, globose, surface more or less reticulated, hyaline, ornamentation amyloid.
Occurrence: Scattered in groups on the ground in broad-leaved and mixed woods, along trails, woodland margins, and parklands where oak is present; mycorrhizal; summer-fall; common.
Edibility: Unknown.

Comments: *Lactarius volemus* (p.109), which sometimes grows in the same habitat, has white-to cream-colored gills, a mild tasting latex that stains the flesh and gills brown, and a pronounced "fishy-smoky" odor. Peck's Lactarius is named in honor of the preeminent nineteenth-century American mycologist Charles Horton Peck (1833–1917).

Latin name: *Lactarius psammicola* f. *glaber* Hesler & A.H. Sm.
Common name: None
Order: Russulales
Family: Russulaceae
Width of cap: 2½–6 in. (6.5–15 cm)
Length of the stalk: ¾–2 in. (2–5 cm)
Cap: Convex with a central depression and arched margin, becoming broadly funnel-shaped with age; dull orange to brownish orange with alternating paler concentric zones; margin at times faintly hairy when young, soon naked; surface glabrous, smooth, viscid; **flesh** thick, whitish, odor not distinctive or somewhat fragrant, taste intensely acrid; **latex** white, slowly staining injured tissue pinkish cinnamon, taste sharply acrid.
Gills: Decurrent, close, infrequently forked; cream to pinkish buff or pale tan, staining pinkish cinnamon where injured.
Stalk: Central or eccentric, equal to slightly tapered downward and narrowed at the base, becoming hollow; whitish; surface dull, usually pitted.
Spore print: Pinkish buff.
Spores: 7.5–9 x 6.5–7.5 μm, broadly elliptical, surface with warts and ridges, hyaline, ornamentation amyloid.
Occurrence: Scattered in small groups on the ground in broad-leaved woods with oak and hickory; mycorrhizal; summer-fall; fairly common.
Edibility: Unknown.

Comments: *Lactarius peckii* (p.97) is similar but smaller and has reddish brown gills. *Lactarius psammicola* f. *psammicola* (not illustrated) has a coarsely fibrillose cap with a prominently hairy margin when young. *Psammicola* means "sand dwelling." *Glaber* means "smooth" or "without hairs."

Latin name: *Lactarius purpureo-echinatus* Hesler & A.H. Sm.
Common name: Purple Scale Lactarius
Order: Russulales
Family: Russulaceae
Width of cap: 1½–3 in. (4–7.5 cm)
Length of the stalk: 1–2½ in. (2.5–6.5 cm)
Cap: Convex, soon flattened and depressed in the center, margin incurved at first; grayish lavender to purple-drab, obscurely concentrically zoned, or without zones; surface dry, matted-fibrillose with erect scales; **flesh** whitish to pale vinaceous, odor not distinctive, taste mild to somewhat acrid; **latex** white to whey-like, taste mild to slightly acrid.
Gills: Adnate to short decurrent, subdistant to well spaced, often with crossveining; whitish to vinaceous buff or dull grayish red.
Stalk: Equal, hollow; colored like the cap or paler, with grayish bloom when young; surface smooth, unpolished.
Spore print: Pale yellow.
Spores: 6.5–8 x 5.5–6.5 μm, broadly elliptical, surface with ridges arranged in bands, hyaline, ornamentation amyloid.
Occurrence: In small groups on the ground in broad-leaved and mixed woods, especially with oak; mycorrhizal; summer-fall; uncommon.
Edibility: Unknown.

Comments: Although not often encountered, the Purple Scale Lactarius is easily recognized by its purplish cap with erect scales. *Lactarius pyrogalus* (p.100) is similarly colored but has a glabrous, distinctly zoned cap and sharply acrid tasting latex. *Purpureo-echinatus* means "purple" and "like a hedgehog."

Latin name: *Lactarius pyrogalus* (Bull.) Fr.
Common name: Fire-milk Lactarius
Order: Russulales
Family: Russulaceae
Width of cap: 1½–4½ in. (4–11.5 cm)
Length of the stalk: 1–2½ in. (2.5–6.5 cm)
Cap: Convex, soon depressed in the center, margin incurved at first; pinkish gray to purplish gray, usually with darker concentric bands, especially near the margin; surface glabrous, slightly viscid; **flesh** white, or grayish beneath the cuticle, slowly becoming pale yellowish when exposed, odor not distinctive or at times somewhat fruity, taste strongly acrid; **latex** abundant, white to cream, often drying yellowish or olivaceous on injured tissue, taste strongly acrid.
Gills: Adnate to short decurrent, well-spaced; pale ochraceous-buff, becoming darker pinkish buff.
Stalk: Equal or tapered downward; colored like the cap or paler, with white mycelium at the base; surface smooth, dry to slightly viscid.
Spore print: Pale ochraceous.
Spores: 6–7.5 x 5–6.5 μm, subglobose to broadly elliptical, surface with ridges arranged in bands, hyaline, ornamentation amyloid.
Occurrence: Scattered in small groups on the ground in broad-leaved and mixed woods, especially near blue beech (*Carpinus*); mycorrhizal; summer-early fall; occasional.
Edibility: Poisonous.

Comments: This mushroom frequently occurs in low, moist woods along rivers and streams under blue beech trees (also known as ironwood). *Lactarius speciosus* (not illustrated) is somewhat similar but has close gills that stain dull lilac in contact with the latex, and a bearded cap margin when young. *Pyrogalus* means "fire milk" in reference to the sharply acrid tasting latex.

Latin name: *Lactarius quietus* var. *incanus* Hesler and A.H. Sm.
Common name: None
Order: Russulales
Family: Russulaceae
Width of cap: 1–4 in. (2.5–10 cm)
Length of the stalk: 1–5 in. (2.5–12.5 cm)
Cap: Broadly convex, becoming flattened with a central depression, margin incurved at first, expanding and sometimes translucent-striate with age; purplish brown to purplish gray, darker reddish brown in the center, sometimes slightly zonate; surface smooth or with small irregular bumps, somewhat hoary at first, becoming glabrous; **flesh** pale pinkish buff, odor faintly sweet, somewhat like maple syrup, taste mild and then slowly and slightly acrid; **latex** whitish or watery, taste mild to bitter or weakly acrid.
Gills: Adnate to slightly notched or subdecurrent; close, sometimes forked; whitish or tinged pale pinkish at first, developing orange-cinnamon stains and becoming cinnamon with age.
Stalk: Equal to enlarged toward the base, becoming hollow with age; colored like the cap or paler; surface dry, often with a whitish bloom when young.
Spore print: Pinkish buff.
Spores: 6.5–9 x 5–7.5 µm; elliptical, surface with warts and ridges sometimes forming a partial reticulum, hyaline, ornamentation amyloid.
Occurrence: Scattered to gregarious on the ground under oaks; mycorrhizal; summer-fall; common.
Edibility: Unknown.

Comments: Similar aromatic species include *Lactarius aquifluus* (p.82), which has a more pronounced maple sugar or curry odor and is usually associated with conifer trees. *Lactarius mutabilis* (not illustrated) is similarly colored but has mild tasting white latex, produces a whitish to yellowish cream spore deposit, and occurs in conifer and mixed woods. The name *quietus* means "quiet" or "calm," perhaps referring to the muted cap color. *Incanus* means "hoary," referring to the whitish bloom that is sometimes evident on young specimens.

Latin name: *Lactarius rufus* (Scop.) Fr.
Common name: Red Hot Lactarius
Order: Russulales
Family: Russulaceae
Width of cap: 1½–3½ in. (4–9 cm)
Length of the stalk: 1½–3½ in. (4–9 cm)
Cap: Convex to broadly convex and depressed in the center, or broadly funnel-shaped, usually with a central umbo, margin inrolled at first; dark reddish brown to brick-red, at times obscurely zoned; surface smooth, hoary at first, drying to a dull finish; **flesh** pale purplish, odor not distinctive, taste acrid; **latex** abundant, white, unchanging, taste strongly acrid (sometimes slowly).
Gills: Adnate to short decurrent; whitish to pinkish tan or darker with age; close to crowded.
Stalk: Equal, becoming hollow; colored like the cap or paler, often with a white bloom at first, base white; surface smooth and dry.
Spore print: Cream to pale yellow.
Spores: 8.5–11 x 6–8 μm, broadly elliptical, surface with warts and ridges forming a partial reticulum, hyaline, ornamentation amyloid.
Occurrence: Often gregarious in conifer woods, especially with spruce at high elevations, or in sphagnum moss; mycorrhizal; late summer-fall; common.
Edibility: Poisonous.

Comments: There are many reddish brown to orange-brown species of *Lactarius.* The strongly acrid taste of the flesh and latex, in combination with the preferred habitat in high elevation spruce forests help to distinguish the Red Hot Lactarius from some of the others. *Lactarius oculatus* (not illustrated) is often found in the same habitat but is smaller and has mild tasting or only slightly peppery latex. Compare also with *Lactarius camphoratus* (p.84) and *Lactarius hysginus* (p.92). *Rufus* means "red-brown."

Latin name: *Lactarius sordidus* Peck
Common name: Dingy Lactarius
Order: Russulales
Family: Russulaceae
Width of cap: 2–5 in. (5–12.5 cm)
Length of the stalk: 1–3 in. (2.5–7.5 cm)
Cap: Broadly convex with a central depression and becoming more or less funnel-shaped with age; yellowish brown merging with dark olive-brown toward the center, sometimes with obscure concentric zones; margin inrolled at first; surface smooth, slightly viscid when wet; **flesh** thick, firm, whitish to yellowish, slowly staining brown where injured, odor not distinctive, taste acrid; **latex** white, staining the gills brown, taste acrid (sometimes slowly).
Gills: Adnate to short-decurrent; crowded; white to yellowish.
Stalk: Short, equal or tapered slightly upward, hollow; colored like the cap, paler toward the apex; surface spotted with shallow brownish pits.
Spore print: White.
Spores: 5.5–7.5 x 5–6 μm, broadly elliptical, surface more or less reticulated, hyaline.
Occurrence: Solitary or gregarious on the ground under spruce and balsam fir, or in other conifer woods; mycorrhizal; late summer-fall; occasional.
Edibility: Unknown.

Comments: KOH applied to the cap cuticle produces a magenta stain. In West Virginia, this species occurs most often in the high elevation spruce forest. *Lactarius atroviridis* (not illustrated), which grows in broad-leaved and conifer woods, is similar but has a dark olive-green to blackish green pitted cap and stalk, and white latex that stains the gills greenish. *Sordidus* means "dirty looking."

Latin name: *Lactarius subplinthogalus* Coker
Common name: None
Order: Russulales
Family: Russulaceae
Width of cap: 1½–3 in. (4–7.5 cm)
Length of the stalk: 1½–3 in. (4–7.5 cm)
Cap: Broadly convex to flattened and depressed in the center, or somewhat funnel-shaped, margin pleated; pale buff to pale tan, more brownish with age; surface dull, smooth or somewhat wrinkled; **flesh** whitish, slowly staining salmon to brownish orange when cut or bruised, odor not distinctive, taste acrid; **latex** white, staining flesh and gills salmon to brownish orange, taste acrid.
Gills: Adnate to short-decurrent, well-spaced; colored like the cap or paler, with salmon-colored stains from contact with the latex.
Stalk: Equal or tapering downward, becoming hollow; colored like the cap, or paler, whitish at the base; surface smooth, dry.
Spore print: Pinkish buff.
Spores: 7.5–9.5 x 7–8 μm; subglobose to broadly elliptical; surface with warts and ridges, hyaline, ornamentation amyloid.
Occurrence: Solitary or in small groups on the ground in broad-leaved and mixed woods, especially under oak; mycorrhizal; summer to early fall; occasional to locally common.
Edibility: Unknown.

Comments: *Lactarius fumosus* (not illustrated) has a darker brown cap, close gills, and flesh that stains reddish when cut or broken; *Lactarius subvernalis* (not illustrated) has a whitish to pale grayish tan cap, close gills, and bright salmon-orange staining on damaged tissue. Compare also with *Lactarius gerardii* (p.90). *Subplinthogalus* means "near brick-colored milk," referring to the staining of the flesh and gills by the latex.

Latin name: *Lactarius subpurpureus* Peck
Common names: Wine-red Lactarius; Variegated Lactarius
Order: Russulales
Family: Russulaceae
Width of cap: 1½–4 in. (4–10 cm)
Length of the stalk: 1–3 in. (2.5–7.5 cm)
Cap: Broadly convex with depressed center, becoming flat to funnel-shaped with age, margin incurved at first; wine red blended with pinkish gray, more or less concentrically zoned, sometimes with greenish spots; surface glabrous, lubricous; **flesh** whitish to pinkish, staining reddish when damaged, odor and taste not distinctive; **latex** scant, dark purple-red, slowly staining tissue reddish then green, taste mild or slightly peppery.
Gills: Adnate to subdecurrent, moderately well-spaced, sometimes forked near the stalk; wine red, sometimes tinged grayish, staining or aging green.
Stalk: Equal, becoming hollow; colored like the cap, often with a whitish bloom, white mycelium at the base; surface more or less spotted with darker shallow pits, viscid when wet.
Spore print: Cream.
Spores: 8–11 x 6.5–8 μm, elliptical, with warts and ridges that form a partial reticulum, hyaline, ornamentation amyloid.
Occurrence: Solitary or gregarious, sometimes clustered on the ground in mixed woods; mycorrhizal with hemlock and other conifers; summer-fall; common.
Edibility: Edible.

Comments: This species is easily recognized by its overall wine red and pinkish gray colors, and the dark purplish red latex that weakly seeps from injured areas. All parts slowly stain green. *Subpurpureus* means "almost purple."

Latin name: *Lactarius theiogalus* (Bull.) Gray
Common name: Sulphur-milk Lactarius
Order: Russulales
Family: Russulaceae
Width of cap: ¾–2 in. (2–5 cm)
Length of the stalk: 1–2 in. (2.5–5 cm)
Cap: Convex, becoming flat to somewhat funnel-shaped, usually with a pointed umbo; cinnamon-orange, orange-brown, or reddish brown, darker in the center, sometimes obscurely zoned, colors fading with age; surface glabrous, smooth; **flesh** firm, brittle, white, becoming yellowish from the latex when cut, odor not distinctive, taste somewhat acrid when fresh, less so with age; **latex** white, often slowly changing to yellow on exposure to air taste variously acrid (sometimes slowly).
Gills: Adnate to short-decurrent, close to crowded; pale pinkish cinnamon, becoming darker with age, often spotted with reddish brown stains.
Stalk: Equal or swollen at the base, becoming hollow; colored like the cap or paler, white mycelium at the base; surface smooth.
Spore print: White to cream.
Spores: 7–9 x 6–7.5 μm, subglobose to broadly elliptical, surface with warts and ridges, hyaline, ornamentation amyloid.
Occurrence: Often gregarious on wet ground, in leaf litter, or in sphagnum moss in broad-leaved and mixed woods, usually associated with birch trees; mycorrhizal; summer-fall; fairly common.
Edibility: Doubtful. Other species having latex that changes from white to yellow are poisonous.

Comments: The color change of the latex from white to yellow is not always apparent, especially with older specimens, but expressed latex will usually stain white paper yellow. Similar species include *Lactarius vinaceorufescens* (p.108), which is usually found in conifer woods. It has a cinnamon-pink to vinaceous brown cap, and white latex that quickly and unambiguously turns yellow on exposure to air. *Lactarius chrysorheus* (not illustrated), which is most often found in broad-leaved woods, has a cream to yellowish tan or pale orange-cinnamon cap, and copious white latex that quickly turns yellow on exposure to the air. *Theiogalus* means "sulphur milk."

Latin name: *Lactarius torminosus* (Schaeff.) Gray
Common names: Wooly Lactarius; Bearded Milk Cap
Order: Russulales
Family: Russulaceae
Width of cap: 2–5 (5–12.5 cm)
Length of the stalk: 1¼–2½ in. (3–6.5 cm)
Cap: Convex to flat with a depressed center, to shallow funnel-shaped, margin strongly inrolled and shaggy at first; pale buff to pinkish beige or salmon-pink, whitish near the margin, more or less concentrically zoned; surface smooth and glabrous in the center, becoming increasingly fleecy with long hairs toward the margin, viscid when wet; **flesh** thick, brittle, white to pinkish tinged, odor not distinctive or slightly fruity, taste acrid; **latex** white, unchanging, scant, strongly acrid tasting.
Gills: Short-decurrent, close, occasionally forking near the stalk; white to cream, often pinkish tinged, becoming pale tan with age.
Stalk: Equal or narrowing at the base, becoming hollow; colored like the cap or paler; surface dry, glabrous or lightly downy.
Spore print: white to cream.
Spores: 7–10 x 6–8 μm, elliptical, surface with scattered warts and ridges, hyaline, ornamentation amyloid.
Occurrence: Solitary or in small groups on the ground in broad-leaved and mixed woods, almost always associated with birch trees; mycorrhizal; summer-fall; occasional.
Edibility: Poisonous.

Comments: There are variations of the species described here that may represent a complex of closely related, macroscopically similar species. *Lactarius pubescens* (not illustrated) is generally paler in color and has smaller spores (6–8.5 x 5–6.5 μm). Although the Wooly Lactarius has a history of use as an edible in parts of Europe, extensive processing is required to render it safe and palatable. *Torminosus* means “tormenting” or “causing colic,” which is a likely consequence from consuming North American collections of this mushroom.

Latin name: *Lactarius vinaceorufescens* A.H. Sm.
Common name: Yellow-staining Milk Cap
Order: Russulales
Family: Russulaceae
Width of cap: 1½–4 in. (4–10 cm)
Length of the stalk: 1½–2½ in. (4–6.5 cm)
Cap: Convex to broadly convex with an incurved margin at first, becoming flattened with age; buff to pale pinkish cinnamon, or vinaceous red to vinaceous brown with age, with or without obscure concentric bands made up of darker spots; surface glabrous, smooth, dry to somewhat viscid; **flesh** white to pinkish, staining yellow when cut or broken, odor not distinctive, taste slightly to distinctly acrid; **latex** white, quickly becoming bright yellow on exposure to the air, taste bitter to acrid.
Gills: Adnate to subdecurrent, close, forked near the stalk; pale whitish to pinkish buff when young, staining and becoming pinkish brown to dark reddish brown with age.
Stalk: Equal or narrowing toward the apex; colored like the cap but usually paler, often with reddish brown stains with age; surface more or less smooth, glabrous.
Spore print: White to yellowish.
Spores: 6.5–9 x 6–7 μm, subglobose to broadly elliptical, surface with warts and ridges, hyaline, ornamentation amyloid.
Occurrence: Gregarious on the ground in conifer woods, especially with pine, also reported to occur under broad-leaved trees; mycorrhizal; summer-fall; common.
Edibility: Possibly poisonous.

Comments: *Lactarius chrysorheus* (not illustrated) has a cream to pale yellowish cinnamon or pale orange-cinnamon cap, and lacks reddish brown stains on the gills. It grows in broad-leaved and mixed woods, especially with oak. *Vinaceorufescens* means "becoming wine reddish."

Latin name: *Lactarius volemus* Fr.
Common names: Bradley; Tawny Lactarius; Apricot Milk Cap; Leatherback
Order: Russulales
Family: Russulaceae
Width of cap: 1½–5 in. (4–12.5 cm)
Length of the stalk: 1½–4 in. (4–10 cm)
Cap: Broadly convex to flattened, becoming depressed in center, margin incurved at first; orange to orange-brown, fading to paler orange-brown or yellowish with age; surface dry, smooth or concentrically wrinkled (especially near the margin), sometimes cracking in dry weather; **flesh** thick, firm, white, staining brown from the latex when cut or broken, odor and taste strong, "fishy" or "smoky"; **latex** abundant, white, quickly staining tissue brown, taste mild.
Gills: Adnate to subdecurrent, close; white to cream, staining brown from contact with the latex.
Stalk: Equal; colored like the cap, or more often paler; surface smooth.
Spore print: White.
Spores: 7.5–10 x 7.5–9 μm, subglobose, surface with warts and ridges that form a reticulum, hyaline, ornamentation amyloid.
Occurrence: Gregarious, or sometimes solitary on the ground in broad-leaved woods, especially with oak; mycorrhizal; summer-early fall; common.
Edibility: Edible and good.

Continued on next page

Comments: The Bradley is one of the more popular edible wild mushrooms in West Virginia. It is easily recognized by its strong odor and copious latex that stains all parts of the fruitbody brown. Despite being somewhat messy to deal with (the latex stains hands, mushroom baskets, and countertops brown), it is considered by many people to be one of the best—and certainly one of the most flavorful—mushrooms in our woods. The origin of the regional common name "Bradley" is probably a corruption of "Bratling," which is the German name for this species, meaning "mushroom for roasting or broiling." The epithet *volemus* is derived from the Latin "volema pira," which is a variety of cultivated pear, presumably one that is orange-colored? *Lactarius hygrophoroides* (not illustrated) is similar but has widely spaced gills, a mild odor, and latex that does not stain damaged tissue brown. It is common in sandy soil along streams and rivers and is also a good edible, though milder in flavor. Compare with *Lactarius corrugis* (p.86), which has a dark reddish brown, prominently wrinkled cap, and *Lactarius peckii* (p.97), which has a zoned cap and exceedingly acrid tasting latex.

Small Mushrooms with Decurrent or Subdecurrent Gills or Gill-like Folds

The mushrooms placed in this section are small to medium-small in size with stalks that are typically no thicker than a standard pencil (¼ inch = 6 mm) at the apex. In some species the gills are not strongly decurrent at first but become more so as the cap expands and the cap margin is elevated. Gills that only slightly extend down the stalk are termed subdecurrent. Also included here are some of the smaller chanterelles that have gill-like folds or blunt ridges on the underside of the cap. Some medium-small sized mushrooms with decurrent gills that have stalks in the ¼-inch thick range are included in Section 1-F. A few of the smaller *Lactarius* species (Section 1-D) are similar in size and stature to the mushrooms grouped here, but they exude a clear, milky, or colored sap from the broken or cut flesh.

Latin name: *Cantharellus appalachiensis* Petersen
Common name: Appalachian Chanterelle
Order: Cantharellales
Family: Cantharellaceae
Width of cap: ½–1½ in. (1.3–4 cm)
Length of the stalk: ¾–2 in. (2–5 cm)
Cap: Convex at first, becoming nearly flat with a central depression to somewhat funnel-shaped, cap margin incurved when young; yellow-brown to ochre, or yellowish orange, darker at the center; **flesh** thin, yellowish brown, odor and taste not distinctive.
Fertile surface: Gill-like, with forked blunt ribs and crossveins; cream to pale yellow or pale tan.
Stalk: More or less equal or flaring near the apex, becoming hollow with age; colored like the cap or paler, base often white; surface smooth.
Spore print: Pale ochraceous salmon.
Spores: 6–10 x 4–6 μm, elliptical, smooth, hyaline, non-amyloid.
Occurrence: Solitary or scattered in groups, sometimes clustered, on the ground in broad-leaved and mixed woods; mycorrhizal; summer; occasional to fairly common locally.
Edibility: Edible.

Comments: *Cantharellus infundibuliformis* (p.115) is similar but has a much darker blackish brown cap with a paler yellowish brown margin. *Hygrocybe pratensis* (p.121) can also be similar in color and size, but it has true gills with sharp edges rather than blunt ribs. The name *appalachiensis* refers to the Appalachian mountain region where this mushroom was first described.

Latin name: *Cantharellus cinnabarinus* Schwein.
Common name: Cinnabar Chanterelle
Order: Cantharellales
Family: Cantharellaceae
Width of cap: ½–1½ in. (1.3–4 cm)
Length of the stalk: ¾–2 in. (2–5 cm)
Cap: Convex, becoming flat to somewhat funnel-shaped with age, cap margin incurved at first; bright red to pinkish orange; surface smooth, dull; **flesh** thin, whitish to colored like the cap surface, odor and taste not distinctive.
Fertile surface: With well-spaced, forked, gill-like ribs having blunt edges and conspicuous crossveins; colored like the cap or paler.
Stalk: Nearly equal or tapering downward; colored like the cap or paler, often white at the base; surface smooth.
Spore print: Pinkish white.
Spores: 6–10.5 x 4–6 μm, elliptical, smooth, hyaline, non-amyloid.
Occurrence: Usually gregarious in broadleaved and mixed woods, especially on sandy soil or among mosses along rivers and streams; mycorrhizal; late spring-fall; common.
Edibility: Edible.

Comments: Several small, reddish species of *Hygrocybe* are similar to the Cinnabar Chanterelle but they have true gills with sharp edges that feel waxy when crushed. *Cinnabarinus* means "cinnabar-red."

Latin name: *Cantharellus minor* Peck
Common name: Small Chanterelle
Order: Cantharellales
Family: Cantharellaceae
Width of cap: ⅜–1 in. (1–2.5 cm)
Length of the stalk: ¾–1½ in. (2–4 cm)
Cap: Broadly convex to funnel-shaped, cap margin incurved at first, later uplifted and wavy; yellow-orange, fading to pale orange; surface smooth, dull; **flesh** thin, pale yellow, odor and taste not distinctive.
Fertile surface: With well-spaced blunt ribs and occasional crossveins, forked near the margin; colored like the cap, or more pinkish.
Stalk: Slender, nearly equal or flaring at the apex, becoming hollow; colored like the cap; surface smooth.
Spore print: Pale yellowish orange.
Spores: 6–11.5 x 4–6.5 μm, elliptical, smooth, hyaline, non-amyloid.
Occurrence: In scattered groups on the ground, often among moss, in broad-leaved woods, grassy woodland margins, and along trails; mycorrhizal; summer-fall; common.
Edibility: Edible.

Comments: Although edible, this small chanterelle has little substance and it is rarely found in quantity. *Cantharellus ignicolor* (not illustrated) has a slightly larger, minutely scurfy cap and orange gill-like folds that develop pinkish gray to pale brownish tones with age. The similar *Hygrocybe vitellina* (not illustrated) has a viscid yellow cap and stalk, and non-forking sharp-edged gills. Other small orange *Hygrocybe* species have waxy gills with sharp-edges. *Minor* means "smaller."

Latin name: *Cantharellus infundibuliformis* (Scop.) Fr.
Common names: Winter Chanterelle; Funnel-shaped Cantharellus
Order: Cantharellales
Family: Cantharellaceae
Width of cap: ½–2 in. (1.3–5 cm)
Length of the stalk: 1–2½ in. (2.5–6.5 cm)
Cap: Convex to flat or funnel-shaped with an uplifted wavy margin with age, often perforated in the center; blackish brown to dark olive-brown, paler on the margin, fading to yellowish brown with age; surface smooth to slightly scaly, sometimes wrinkled or crimped on the margin; **flesh** thin, watery yellow-brown, odor and taste not distinctive.
Fertile surface: With well-spaced, blunt ribs and extensive crossveining, forked near the margin; yellowish gray to pale violaceous gray.
Stalk: Cylindrical or sometimes compressed, more or less equal to tapering downward, hollow; yellow to tan, or olive-brown, sometimes with white mycelium at the base; surface smooth.
Spore print: pale pinkish buff.
Spores: 9–12 x 6.5–8 μm, broadly elliptical to subglobose, smooth, hyaline, non-amyloid.
Occurrence: Scattered on the ground, especially in wet mossy places, or sometimes on well-decayed logs in conifer and mixed woods, also in sphagnum bogs; mycorrhizal; summer-late fall; common.
Edibility: Edible.

Comments: This small Chanterelle can often be found late in the mushroom season at high elevations in red spruce and northern hardwood forests. *Cantharellus tubaeformis* (not illustrated) is nearly identical except it has white spores. Some mycologists consider these two chanterelles to be varieties of the same species. *Infundibuliformis* means "funnel-shaped."

Latin name: *Chroogomphus vinicolor* (Peck) O.K. Miller
Common names: Wine-cap Chroogomphus; Pine Spike
Order: Boletales
Family: Gomphidiaceae
Width of cap: ¾–3 in. (2–7.5 cm)
Length of the stalk: 1–3 in. (2.5–7.5 cm)
Cap: Broadly conical to convex, with or without an umbo; color variable from wine red, reddish brown, orange-brown or yellow-brown; surface glabrous, smooth, often radially streaked, shiny, somewhat sticky when wet; **flesh** orangish to ochraceous, odor and taste not distinctive.
Gills: Subdecurrent to decurrent, moderately well-spaced; ochraceous buff to pale orange when young, becoming smoky brown to nearly black with age, at first covered with a thin, web-like partial veil that soon disappears as the cap expands.
Stalk: Cylindrical, tapering downward; ochraceous to wine red or reddish brown; surface dry, smooth to fibrillose, partial veil sometimes leaving a thin zone of fibrils on the upper stalk.
Spore print: Smoky gray to blackish.
Spores: 17–23 x 4.5–7.5 μm, narrowly elliptical to spindle-shaped, smooth.
Occurrence: Gregarious on the ground under pines, and other conifers; mycorrhizal; fall; common.
Edibility: Edible.

Comments: This small- to medium-sized mushroom is typically found late in the season under pines, often in company with Slippery Jack boletes such as *Suillus luteus* (p.287) and *Suillus brevipes* (not illustrated). *Chroogomphus rutilus* (not illustrated) is nearly identical but generally more brownish overall. However, the color range of these two species overlaps and microscopic examination of the cystidia is necessary in order to distinguish them. For practical purposes there is little difference. Both are edible and rarely insect damaged, but there is not much to recommend them for the table except as "fillers" to include with more flavorful species. *Vinicolor* means "wine red."

Latin name: *Craterellus fallax* A.H. Sm.
Common names: Black Trumpet; Trumpet of Death; Poor Man's Truffle; Horn of Plenty
Order: Cantharellales
Family: Craterellaceae
Fruitbody: 1½–3½ in. (4–9 cm) tall; trumpet-shaped to deep, hollow funnel-shaped with an incurved margin at first that becomes flared, wavy, and often recurved with age; **inner surface** dry, roughened to minutely scaly, dark sooty brown to blackish, paler when dry; **outer fertile surface** smooth, nearly even to wrinkled, ash-gray to blackish brown, developing a salmon-buff tinge as the spores mature; **flesh** thin, brittle, grayish brown, odor like candle wax or fruity, taste mild; **stalk** short, blackish, not well differentiated from the fertile surface.
Spore print: pale salmon-buff to ochraceous buff.
Spores: 11–18 x 7–11 µm, broadly elliptical, smooth, hyaline.
Occurrence: Usually gregarious, often in caespitose clusters on the ground in broad-leaved and mixed woods, especially with oak and beech; mycorrhizal; summer-fall; common.
Edibility: Edible and good, occasionally bitter tasting. Easily preserved by drying.

Comments: Black Trumpets are much appreciated by connoisseurs of wild mushrooms. They are frequently found in large scattered patches but their dark color blends well with the forest floor so they are easily overlooked. The equally edible but seldom encountered *Craterellus cornucopioides* (not illustrated) is nearly identical but has white spores and lacks salmon-buff tones on the mature fertile surface. *Craterellus foetidus* (p.132) is also similar but has more prominent folds or ridges on the fertile surface and a pleasant "maple syrup" odor. The name *fallax* means "false" or "deceptive," perhaps referring to the close similarity to *Craterellus cornucopioides.*

Latin name: *Hygrocybe appalachiensis* (Hesler & A.H. Sm.) comb. nov.
Synonym: *Hygrophorus applachiensis* Hesler & A.H. Sm.
Common name: Appalachian Waxy Cap
Order: Agaricales
Family: Tricholomataceae
Width of cap: ¾–2½ in. (2–6.5 cm)
Length of the stalk: 1½–3 in. (4–7.5 cm)
Cap: Convex to flat with a depressed center, or somewhat funnel-shaped; red to deep purplish red, fading to orangish yellow, often with a whitish margin; surface fibrillose to minutely scaly; **flesh** yellow to orangish, odor and taste not distinctive.
Gills: Decurrent to subdecurrent, well-spaced; colored like the cap or paler with yellowish white gill edges.
Stalk: Cylindrical to compressed, more or less equal, hollow; colored like the cap, often white at the base; surface smooth to slightly scurfy.
Spore print: White.
Spores: 11–18 x 7–10 μm, elliptical, smooth, hyaline, non-amyloid.
Occurrence: Solitary, or more often in small groups and caespitose clusters on the ground in broad-leaved or mixed woods; saprobic; summer-fall; occasional to locally common.
Edibility: Unknown.

Comments: This mushroom is an unusual shade of deep purplish red that distinguishes it from other red Waxy Caps. *Hygrocybe cantharellus* (p.120) is smaller, has a more slender stalk, and is brighter red overall. *Appalachianensis* refers to the Appalachian Mountains, from where this species was first described.

Latin name: *Hygrocybe borealis* Peck
Synonyms: *Camarophyllus borealis* (Peck) Murrill; *Hygrophorus borealis* Peck
Common name: Snow White Waxy Cap.
Order: Agaricales
Family: Tricholomataceae
Width of cap: ½–1¾ in. (1.3–4.5 cm)
Length of the stalk: 1–2½ in. (2.5–6.5 cm)
Cap: Convex to broadly convex, becoming flattened, often depressed in the center; watery white to milk white, sometimes tinged yellowish on the disc; surface glabrous, smooth, lubricous, hygrophanous, margin sometimes faintly striate when moist; **flesh** thin, fragile, white, odor and taste not distinctive.
Gills: Decurrent to subdecurrent, moderately well-spaced; white; thick and waxy, sometimes with crossveins.
Stalk: More or less equal, or tapering downward, sometimes slightly swollen at the base; white; surface smooth to lightly streaked, dry.
Spore print: White.
Spores: 7–9 x 4.5–6.5 µm, elliptical, smooth, hyaline, non-amyloid.
Occurrence: Solitary or in small groups on soil or humus in conifer and mixed woods; saprobic; summer-fall; occasional.
Edibility: Edible, but not worthwhile.

Comments: This clean-looking little white mushroom represents a complex of forms that have been variously interpreted as separate species, including *Hygrophorus niveus* and *Hygrophorus virgineus. Hygrophorus eburneus* (not illustrated) is a larger white waxy cap that has a viscid, glutinous cap and stalk. Similar whitish *Clitocybe* species differ by having thin, non-waxy gills. *Borealis* means "northern."

Latin name: *Hygrocybe cantharellus* (Schwein.) Murrill
Synonym: *Hygrophorus cantharellus* (Schwein.) Fr.
Common name: Chanterelle Waxy Cap
Order: Agaricales
Family: Tricholomataceae
Width of cap: ⅜–1¼ in. (1–3 cm)
Length of the stalk: 1–3½ in. (2.5–9 cm)
Cap: Convex, becoming flattened or depressed in the center with age; scarlet-red, fading to orange or orangish yellow; surface roughened to finely scaly in the center, margin weakly striate or somewhat scalloped; **flesh** thin, yellow to orange, odor and taste not distinctive.
Gills: Well-spaced; pale yellow to orangish; thick but with a sharp edge, waxy.
Stalk: Slender, cylindrical or sometimes compressed, more or less equal or slightly enlarged at the base; colored like the cap or paler, or remaining red after the cap fades, base yellowish to white; surface smooth, glabrous, dull to satiny.
Spore print: White.
Spores: 8.6–10.3 x 5.7–7 μm, elliptical, smooth, hyaline non-amyloid.
Occurrence: Solitary or in small groups on soil, humus, mossy ground, or on well-decayed, often moss-covered logs and stumps; saprobic; summer-fall; common.
Edibility: Non-poisonous.

Comments: This mushroom has various forms and may represent a complex of closely related species. *Hygrocybe turunda* (not illustrated) is very similar but grows in sphagnum bogs and has small, dark brown, erect scales in the center of the cap. *Hygrocybe miniata* (not illustrated) is more compact in stature and has reddish gills that are broadly attached to the stalk. Compare with *Cantharellus cinnabarinus* (p.113), which has blunt gill-like ridges with veining between, rather than sharp-edged true gills beneath the cap. *Cantharellus* means "small vase."

Latin name: *Hygrocybe pratensis* (Pers.) M. Bon
Synonyms: *Camarophyllus pratensis* (Pers.) P. Kumm.; *Hygrophorus pratensis* (Pers.) Fr.
Common names: Meadow Waxy Cap; Salmon Waxy Cap
Order: Agaricales
Family: Tricholomataceae
Width of cap: ¾–2½ in. (2–6.5 cm)
Length of the stalk: 1–2½ in. (2.5–6.5 cm)
Cap: Bell-shaped to convex, becoming flat or depressed with a broad umbo and wavy margin; color highly variable from creamy tan, pinkish buff, peach-colored, yellow-orange or reddish orange; surface dry, dull, smooth to slightly roughened, often cracking on the disc; **flesh** white, thick in the center, thin near the margin, odor and taste not distinctive.
Gills: Subdecurrent at first, appearing more decurrent as the cap expands, thick, well-spaced with crossveins between; white to pale yellowish or pale orange; thick but sharp-edged, waxy.
Stalk: More or less equal to tapering downward; white or like the cap but paler; surface smooth, dry.
Spore print: White.
Spores: 5.5–8 x 3.5–5 μm, elliptical to subglobose, smooth, non-amyloid.
Occurrence: Scattered on the ground, usually in small groups in woods, parklands, grassy woodland margins, and along riverbanks; saprobic; spring-fall; common.
Edibility: Edible.

Comments: This widespread mushroom is variable in color and size. Most West Virginia collections are small to medium-small. Some color forms might be mistaken for small, pale-colored Chanterelles, which have blunt, gill-like ribs and veins on the underside of the cap rather than sharp-edged gills. Compare with *Hygrocybe borealis* (p.119). *Pratensis* means "of the meadow." The Meadow Waxy Cap was first described from Europe, where it primarily occurs in grasslands. Mycologists have differing opinions as to the correct placement and nomenclature for this common mushroom.

Latin name: *Lentinus suavissimus* Fr.
Synonym: *Panus suavissimus* (Fr.) Singer
Common name: Fragrant Lentinus
Order: Polyporales
Family: Polyporaceae
Width of cap: ¾–2 in. (2–5 cm)
Length of the stalk: ⅜–¾ in. (1–2 cm)
Cap: Kidney-shaped to fan-shaped, or nearly circular in outline, depressed in the center with an upturned margin with age; pale yellow to ochraceous; surface smooth to radially fibrillose; **flesh** whitish, tough, odor sweet, anise-like.
Gills: Close to moderately well-spaced; white to yellowish; gill edges serrated.
Stalk: Short to nearly absent, thick, often swollen at the base; eccentric to central; colored like the cap, typically reddish brown at the point of attachment.
Spore print: White to cream.
Spores: 7–8 x 3.4–4 µm, cylindrical, smooth, hyaline, non-amyloid.
Occurrence: Solitary or in small groups on fallen branches and limbs of broad-leaved trees; saprobic; summer-fall; occasional.
Edibility: Not edible.

Comments: The strong anise odor, tough consistency, and serrated gill edges are distinctive. *Lenzites betulina* (not illustrated) is somewhat similar but has an odorless, zoned cap, and an even to slightly scalloped gill-like fertile surface. *Suavissimus* means "sweet-scented."

Latin name: *Leptonia serrulata* (Pers.) Quél.
Synonym: *Entoloma serrulatum* (Fr.) Hesler
Common names: Saw-gilled Leptonia; Blue-toothed Entoloma
Order: Agaricales
Family: Entolomataceae
Width of the cap: ½–1½ in. (1.3–4 cm)
Height of the stalk: 1¼–2½ in. (3–6.5 cm)
Cap: Convex to broadly convex, becoming umbilicate; dark bluish black, fading to grayish, sometimes with brownish tones with age; surface dry, radially fibrillose, sometimes minutely scaly in the center, margin may become somewhat striate with age; **flesh** thin, pale grayish blue, fragile, odor and taste not distinctive.
Gills: Subdecurrent to decurrent, close to moderately well-spaced; pale bluish gray becoming pinkish tinged as the spores mature; gill edges dark bluish black, roughened to finely toothed.
Stalk: Slender, cylindrical to sometimes furrowed or compressed; colored like the cap or paler, with white downy mycelium at the base; cartilaginous; surface smooth, granulose near the apex.
Spore print: Salmon pink.
Spores: 7–11 x 5–7.5 μm, broadly elliptical, angular, smooth, non-amyloid.
Occurrence: Usually in small groups on the ground in woods and semi-open grassy areas, or sometimes on decaying moss-covered logs and stumps; saprobic; early summer-fall; occasional.
Edibility: Not edible.

Comments: The bluish black cap and gill edges are distinctive. *Serrulata* means "finely toothed."

Latin name: *Xeromphalina campanella* (Batsch.) Maire
Synonym: *Omphalia campanella* (Batsch.) Quél.
Common names: Golden Trumpets; Bell Omphalina
Order: Agaricales
Family: Tricholomataceae
Width of cap: ¼–1 in. (0.5–2.5 cm)
Length of the stalk: ¾–2 in. (2–5 cm)
Cap: Bell-shaped to broadly convex with a depressed or umbilicate center, or nearly funnel-shaped with age; orange-yellow, darker tawny-orange in the center; surface smooth, shiny, radially lined from the central depression to the margin; **flesh** thin, yellowish, odor and taste not distinctive.
Gills: Strongly decurrent, moderately well-spaced, with crossveining between; cream-colored to yellowish, aging ochraceous.
Stalk: Slender, more or less equal with slightly flaring apex and somewhat bulbous base, tough; reddish brown to date-brown, paler toward the apex; surface smooth to roughened, or scurfy on the lower portion, with a tuft of orange-brown mycelium at the base.
Spore print: White.
Spores: 5–7 x 3–4 μm, elliptical, smooth, hyaline, amyloid.
Occurrence: Densely clustered on decaying stumps, logs, and fallen limbs of conifers; saprobic; spring-fall; common.
Edibility: Non-poisonous.

Comments: This common mushroom often grows in massive clusters, at times almost entirely covering old tree stumps. It seems to have a special affinity for moss-covered logs and stumps. *Xeromphalina kauffmanii* (not illustrated) is nearly identical but grows on decaying wood of broad-leaved trees. In mixed woods these two species can be difficult to distinguish since the woody substrate type is not always recognizable. The genus name *Xeromphalina* means "dry navel." *Campanella* means "bell-shaped."

Medium to Large Mushrooms with Decurrent Gills or Blunt Gill-like Folds

This section includes true gilled mushrooms, larger chanterelles, and one unusual bolete that has a gill-like hymenium. All have central or eccentric, ringless stalks that are typically more than ¼ inch (6 mm) thick at the apex. The gills or folds on the underside of the cap may be either slightly or strongly decurrent. With some chanterelles, the fertile undersurface is composed of shallow wrinkles, or in some instances it can be nearly smooth. Similar mushrooms that exude sap from cut or broken flesh (milk mushrooms) are placed in Section 1-D. Smaller mushrooms that have decurrent gills are placed in Section 1-E.

Latin name: *Cantharellus cibarius* Fr.
Common names: Chanterelle; Golden Chanterelle; Pfifferling
Order: Cantharellales
Family: Cantharellaceae
Width of cap: 1–4½ in. (2.5–11.5 cm)
Length of the stalk: ¾–2½ in. (2–6.5 cm)
Cap: Convex to flat with a depressed center, becoming funnel-shaped with an uplifted, lobed, wavy margin; chrome-yellow to egg yolk yellow-orange, paler with age or when the caps are more exposed to sunlight; surface smooth, sometimes with a bloom when young; **flesh** thick, white, odor fruity (like apricots) or mild, taste somewhat peppery when raw.
Fertile surface: Composed of decurrent, thick ribs or folds with blunt edges, usually with extensive forking and crossveins; colored like the cap or paler (occasionally brighter than faded caps).
Stalk: Cylindrical or compressed, tapering downward, central or eccentric, solid; more or less colored like the fertile surface, staining ocraceous where handled; surface smooth to slightly roughened.
Spore print: Pale yellow to ocraceous.
Spores: 7.5–10.5 x 5–6.5 μm, elliptical, smooth, hyaline, non-amyloid.
Occurrence: Usually gregarious on the ground in broad-leaved and conifer woods, especially with oak; mycorrhizal; summer-fall; common.
Edibility: Edible and choice.

Comments: This chanterelle is well-known and much appreciated in many parts of the world for its esculent qualities. The equally delicious *Cantharellus lateritius* (not illustrated) is similar but has thinner flesh and a wrinkled to nearly smooth fertile surface. *Canthatellus persicinus* (p.128) is also similar but salmon pink to peach-colored. The False Chanterelle, *Hygrophoropsis aurantiaca* (p.138), is orange to brownish orange overall, has sharp-edged, forked, true gills, and grows on well-decayed wood or lignin-rich humus. Compare with the poisonous Jack O' Lantern Mushroom, *Omphalotus illudens* (p.146), which is orange to brownish orange, has non-forking true gills, and typically grows in large clusters on decaying wood (sometimes buried). Compare also with *Gomphus floccosus* (p.136) and *Gomphus kauffmanii* (p.137), both of which have distinctly scaly caps. *Cibarius* means "pertaining to food."

Latin name: *Cantharellus persicinus* R.H. Petersen
Common name: Peach Chanterelle
Order: Cantharellales
Family: Cantharellaceae
Width of cap: ½–2 in. (1.3–5 cm)
Length of the stalk: ¾–2½ in. (2–6.5 cm)
Cap: Hemispherical at first, becoming convex to flattened; orange-buff to salmon or peach-colored, fading with the loss of moisture; surface smooth, dull; **flesh** off-white, odor fruity, taste mild.
Fertile surface: Composed of decurrent, thick ribs or folds with blunt edges, developing crossveining at maturity; pale salmon to buff at first, becoming ochraceous salmon with age.
Stalk: More or less cylindrical, often tapering downward, solid; colored like the cap, base whitish.
Spore print: White to pale pinkish salmon.
Spores: 10.4–11.5 x 5.8–7.2 µm, narrow elliptical, smooth, hyaline, non-amyloid.
Occurrence: Solitary or in small groups on the ground in mixed woods, especially with oak and hemlock; mycorrhizal; summer-fall; uncommon.
Edibility: Edible.

Comments: The Peach Chanterelle closely resembles *Cantharellus cibarius* (p.126) in size and stature. However it differs in color and has larger spores. Compare also with *Hygrocybe pratensis* (p.121). *Persicinus* means "peach-colored."

Latin name: *Clitocybe clavipes* (Pers.) Fr.
Common name: Club-footed Clitocybe
Order: Agaricales
Family: Tricholomataceae
Width of cap: 1¼–3 in. (3–7.5 cm)
Length of the stalk: 1¼–2½ in. (3–6.5 cm)
Cap: Broadly convex to flat or depressed in the center, at times shallow funnel-shaped, usually with a slight obtuse umbo; pale grayish brown to medium brown, sometimes with olive tones, margin paler; surface smooth, glabrous; **flesh** soft, thick at the center, white, odor of cherry bark or bitter almond, taste not distinctive.
Gills: Decurrent, close to moderately well-spaced; white to cream.
Stalk: Usually swollen or club-shaped at the base; colored like the cap or paler, with white mycelium at the base; stuffed, spongy, surface smooth, somewhat silky fibrillose streaked.
Spore print: White.
Spores: 6–10 x 3.5–5 µm, broadly elliptical, smooth, hyaline, non-amyloid.
Occurrence: Usually gregarious on the ground in conifer woods, especially with pine and spruce, also in broad-leaved woods; saprobic; summer-fall; common.
Edibility: Edible with caution. Some individuals experience skin rash and headaches after consuming this mushroom at the same meal with alcohol.

Comments: The Club-footed Clitocybe will sometimes fruit prolifically in conifer plantations. *Clitocybe gibba* (p.130) also has a cherry bark odor, but it is a more slender mushroom with a thin, pinkish tan cap. There are several brownish species of *Lactarius* that are similar in size and stature, but they exude latex when cut or broken. *Clavipes* means "club foot."

Latin name: *Clitocybe gibba* (Pers.) P. Kumm.
Synonym: *Clitocybe infundibuliformis* (Schaeff.) Fr.
Common names: Funnel Cap; Funnel Clitocybe
Order: Agaricales
Family: Tricholomataceae
Width of cap: 1¼–3 in. (3–7.5 cm)
Length of the stalk: 1½–2½ in. (4–6.5 cm)
Cap: Flat to shallowly depressed, becoming funnel-shaped, with or without a small central umbo; ochre to pinkish tan, fading paler with age; surface smooth; **flesh** thin, white, odor of bitter almond, taste mild.
Gills: Long-decurrent, close; white to pale buff.
Stalk: Slender, more or less equal to slightly enlarged at the base; colored like the cap or paler, often with white mycelium at the base; surface smooth to lightly streaked, dry.
Spore print: White.
Spores: 5–10 x 3.5–6 μm, elliptical, smooth, hyaline, non-amyloid.
Occurrence: Solitary or in small groups, sometimes growing in fairy rings on soil in broad-leaved and conifer woods, grassy woodland margins, and roadsides; saprobic; summer-fall; common.
Edibility: Edible with caution. It can easily be confused with a variety of other mushrooms of unknown edibility.

Comments: The Funnel Clitocybe is quite common in some years. The pronounced odor of bitter almond and thin, pinkish tan cap are the best field characters. Compare with *Clitocybe clavipes* (p.129), which also has an odor of bitter almonds. The name *gibba* means "swollen" or "humped."

Latin name: *Clitopilus prunulus* (Scop.) Fr.
Common name: Sweetbread Mushroom
Order: Agaricales
Family: Entolomataceae
Width of cap: 1–3 in. (2.5–7.5 cm)
Length of the stalk: 1–2½ in. (2.5–6.5 cm)
Cap: Convex with an incurved margin at first, becoming nearly flat or somewhat depressed with an irregular, wavy margin in age; dull white to grayish with a bloom; surface dry, felt-like, or slightly tacky; **flesh** white, fragile, odor of yeast or bread dough, taste mild.
Gills: Decurrent or subdecurrent, close; whitish at first, becoming dull pink; easily separated from the cap.
Stalk: Equal or tapering in either direction, often eccentric, solid; white; surface smooth to slightly downy.
Spore print: Salmon-pink.
Spores: 9–12 x 5–7 μm. elliptical, longitudinally furrowed, non-amyloid.
Occurrence: Solitary or in small groups, sometimes clustered on the ground in broad-leaved and mixed woods, grassy woodland margins, and parklands; saprobic; summer-fall; common.
Edibility: Edible.

Comments: Use caution when collecting the Sweetbread Mushroom for eating. There are several whitish mushrooms of similar size and stature with which it could be confused, including the poisonous *Clitocybe dealbata* (not illustrated), a grassland species with a white cap and stalk, white to buff-colored gills, and white spores. The best field characters for recognizing the Sweetbread Mushroom are its distinctive odor of bread dough (or like cucumber to some people), fragile flesh, and decurrent gills that become pinkish at maturity. Compare with *Rhodocybe mundula* (p.151), which has a concentrically cracked cap surface and bitter tasting flesh. Also compare with *Entoloma abortivum* (p.133), a more robust mushroom that typically grows in clusters on or near decaying stumps. *Prunulus* means "small plum" in reference to the whitish bloom that is more or less evident on the caps of fresh specimens.

Latin name: *Craterellus foetidus* A.H. Sm.
Common name: Fragrant Black Trumpet
Order: Cantharellales
Family: Craterellaceae
Width of cap: 1½–3½ in. (4–9 cm)
Length of the stalk: 1–3 in. (2.5–7.5 cm)
Cap: Asymmetrically funnel-shaped with a lobed, wavy margin; gray with blackish brown radial fibrils or scales; **flesh** thin, fibrous, odor sweet, like maple sugar, taste mild.
Fertile surface: Consisting of decurrent shallow veins and blunt ridges, often forking and running together with crossveins to form a net-like pattern, at times becoming almost poroid, especially near the margin; pale ash-gray to bluish gray, sometimes with dull orangish tints or stains.
Stalk: More or less cylindrical, tapering upward from the base and merging with the fertile surface; colored like the fertile surface or paler; surface uneven.
Spore print: White.
Spores: 8.5–12 x 5–7 μm, elliptical, smooth, hyaline, non-amyloid.
Occurrence: Scattered, often in small caespitose clusters on the ground in broad-leaved woods, especially with oak; mycorrhizal; summer-fall; occasional.
Edibility: Edible.

Comments: The similar and more common Black Trumpet, *Craterellus fallax* (p.117), has a comparatively smooth to wrinkled fertile surface, a pleasant fruity odor, and ochraceous buff-colored spores. Although the odor of the Fragrant Black Trumpet can at times be cloying, the epithet *foetidus,* which means "bad smelling," seems misapplied.

Latin name: *Entoloma abortivum* (Berk. & M.A. Curtis) Donk
Synonym: *Clitopilus abortivus* Berk. & M.A. Curtis
Common name: Aborted Entoloma
Order: Agaricales
Family: Entolomataceae
Width of cap: normal form 2–5 in. (5–12.5 cm)
Length of the stalk: 1–3½ in. (2.5–9 cm)
Cap: Normal forms are convex with an inrolled margin at first, becoming flattened with age; gray to grayish brown; surface dry, silky to minutely scaly; **flesh** thick, white, odor of bread dough or cucumber, taste not distinctive.
Gills: Short or long decurrent, close; pale gray becoming dingy-salmon at maturity.
Stalk: More or less equal or enlarged at the base, central or eccentric; colored like the cap or paler, usually with cottony white mycelium at the base; surface dry, slightly scurfy.
Spore print: Salmon-pink.
Spores: 7.5–10 x 5–6.5 μm, elliptical, angular, smooth, hyaline, non-amyloid.
Occurrence: Sometimes solitary, but more typically in groups or clusters, usually consisting of normal and aborted forms growing together, but either form may occur independently, on or near decaying wood of broad-leaved trees, especially stumps and well-rotted logs, or from buried wood; saprobic? (see comments below); late summer-fall; common.
Edibility: Edible with caution.

Continued on next page.

Comments: The Aborted Entoloma is easily recognized when it occurs in mixed clusters that include aborted fruitbodies. The traditional view of the relationship is that aborted Entoloma fruitbodies develop when its mycelium is infected by mycelium of the Honey Mushroom (*Armillaria* spp.), causing normal *Entoloma* fruitbodies to become distorted and non-functional. However, recent studies by mycologist T.J. Volk and others have shown that the aborted fruitbodies are actually aborted Honey Mushrooms, and that the *Entoloma* rather than the *Armillaria* is the parasite. The whitish to pinkish white aborted fruitbodies are 1–3 in. (2.5–7.5 cm) in diameter, have pinkish-marbled watery flesh, and are also edible. Care must be taken not to confuse *Entoloma abortivum* with others in the genus, many of which are known to be poisonous. Compare with *Clitopilus prunulus* (p.131) and *Rhodocybe mundula* (p.151). *Abortivum* means "with malformed parts."

Latin name: *Gomphus clavatus* (Pers.) Gray
Synonym: *Cantharellus clavatus* Fr.
Common name: Pig's Ears
Order: Phallales
Family: Gomphaceae
Width of cap: individual caps 1¼–4 in. (3–10 cm)
Length of the stalk: ¾–2 in. (2–5 cm)
Fruitbody: Usually a compound cluster consisting of individual branches that are somewhat club-shaped at first, then expand to funnel-shaped with a lobed, wavy margin; upper surface purplish gray, becoming dingy yellow to yellow-brown, smooth to slightly scaly; **flesh** thick, white, odor and taste not distinctive; **fertile surface** consisting of shallow, often forked blunt ridges and crossveins, at times nearly poroid; pale violet to grayish violet, becoming ochraceous with age; **stalk** short, thick, solid, tapering at the base, often branching, merging with, and more or less colored like the fertile surface.
Spore print: Pale ochraceous.
Spores: 10–13 x 4–6 µm, narrow-elliptical, obscurely warted, hyaline, non-amyloid.
Occurrence: Solitary or in groups on the ground under conifers, especially hemlock, sometimes growing in rings or arcs; mycorrhizal; late summer-fall; uncommon.
Edibility: Edible.

Comments: In the eastern U.S., this mushroom is usually associated with the northern boreal forest, but its range also extends southward at the higher elevations of the Appalachians. Compare with *Gomphus kauffmanii* (p.137), which is similar but has a coarsely scaly cap. See also *Gomphus floccosus* (p.136), which has a scaly orange cap. *Clavatus* means "club-like" referring to the shape of young specimens.

Latin name: *Gomphus floccosus* (Schwein.) Singer
Synonym: *Cantharellus floccosus* Schwein.
Common name: Scaly Chanterelle
Order: Phallales
Family: Gomphaceae
Width of cap: 2–5 in. (5–12.5 cm)
Height of the fruitbody: 2½–6 in. (6.5–15 cm)
Fruitbody: Cylindrical with a sunken center at first, soon expanding and becoming funnel-shaped or vase-shaped, margin often wavy in age; inner surface yellowish orange to reddish orange, more or less scaly, especially toward the center; **flesh** thick, white, fibrous, odor and taste not distinctive; **fertile surface** consisting of long-decurrent blunt ridges and wrinkles, often forking, and at times almost elongated poroid; yellowish to cream, becoming pale ochraceous with age; **stalk** short, somewhat rooting, continuous with and often indistinguishable from the fertile portion.
Spore print: Ochre-yellow.
Spores: 11.5–14.5 x 7–8 μm, elliptical, minutely warted, hyaline, non-amyloid.
Occurrence: Scattered and often gregarious on the ground, or in moss in conifer woods and mixed woods, especially under hemlock; mycorrhizal; spring-fall; common.
Edibility: Not recommended.

Comments: The very young fruitbodies look like small yellow cylinders, but they soon expand at the top and become vase-shaped. *Gomphus kauffmanii* (p.137) is similar but is tan to tawny-brown on the inner surface, and the scales are much coarser. Apparently some people can consume the Scaly Chanterelle with impunity, but many others experience gastrointestinal disorders of varying intensity from eating it. *Floccosus* means "woolly."

Latin name: *Gomphus kauffmanii* (A.H. Sm.) Corner
Synonym: *Cantharellus kauffmanii* A.H. Sm.
Common name: Kauffman's Scaly Chanterelle
Order: Phallales
Family: Gomphaceae
Width of cap: 2–7 in. (5–18 cm)
Height of the fruitbody: 3–7 in. (7.5–18 cm)
Fruitbody: Funnel-shaped to vase-shaped, margin expanding and becoming wavy with age; inner surface with dense creamy tan to tawny ochre coarse scales; **flesh** fibrous, white, odor and taste not distinctive; **fertile surface** long-decurrent, blunt ridges and wrinkles, often forking; cream to pinkish buff, sometimes with purplish stains; **stalk** short, somewhat rooting, continuous with and often indistinguishable from the fertile surface.
Spore print: Ochre-yellow.
Spores: 12.5–18.5 x 6–7.5 μm, elliptical to spindle-shaped, minutely warted, hyaline, non-amyloid.
Occurrence: Solitary to scattered in groups, sometimes growing in arcs or fairy rings, on the ground in mixed woods and conifer woods, especially under hemlock; mycorrhizal; summer-fall; occasional.
Edibility: Not recommended.

Comments: As with the closely related *Gomphus floccosus* (p.136), this species is edible for some individuals but causes gastrointestinal distress in others. The specific epithet honors the renowned American agaricologist Calvin H. Kauffman (1869–1931).

Latin name: *Hygrophoropsis aurantiaca* (Wulfen) Maire
Synonym: *Clitocybe aurantiaca* (Wulfen) Fr.
Common name: False Chanterelle
Order: Boletales
Family: Hygrophoropsidaceae
Width of cap: 1½–3½ in. (4–9 cm)
Length of the stalk: 1½–3½ in. (4–9 cm)
Cap: Convex at first, becoming flat to depressed in the center, or shallow funnel-shaped with an incurved margin; orange to orange-brown, or sometimes yellow-brown; surface smooth or downy, dry; **flesh** thin, soft, brownish orange, odor and taste not distinctive.
Gills: Long-decurrent, crowded, thin, repeatedly forked; pale yellow-orange to bright or dark orange.
Stalk: More or less equal, often curved, central or eccentric; colored like the cap or paler; tough; surface dry, velvety.
Spore print: White.
Spores: 5–8 x 2.5–4.5 μm, broadly elliptical, smooth, hyaline, dextrinoid.
Occurrence: Solitary or in groups on humus-rich ground or well-decayed woody debris, often near old stumps, in broad-leaved and conifer woods; saprobic; summer-fall; common.
Edibility: Not recommended. Apparently edible for some, but gastrointestinal disturbances from consuming this species have been reported.

Comments: The False Chanterelle is sometimes mistaken for the edible Chanterelle, *Cantharellus cibarius* (p.126), which is more yellowish and has relatively blunt, gill-like ribs rather than true gills beneath the cap. The poisonous *Omphalotus illudens* (p.146) has unforked gills, and typically grows in large caespitose clusters on decaying wood. *Aurantiaca* means "orange-colored."

Latin name: *Hygrophorus chrysodon* (Batsch) Fr.
Common name: Golden Speckled Waxy Cap
Order: Agaricales
Family: Tricholomataceae
Width of cap: 1–3 in. (2.5–7.5 cm)
Length of the stalk: 1¼–2½ in. (3–6.5 cm)
Cap: Convex, often with a low umbo, becoming flattened with age, margin inrolled when young; white with scattered golden yellow flecks or granules, especially on the disc and margin; surface smooth, viscid when moist; **flesh** thick, white, odor and taste not distinctive.
Gills: Decurrent to subdecurrent, moderately well-spaced; white; thick and waxy.
Stalk: Equal; white with golden yellow scurfy flecks or granules concentrated near the apex, sometimes forming a distinct ring zone; surface viscid when moist.
Spore print: White.
Spores: 7–10 x 3.5–6 μm, elliptical, smooth, hyaline, non-amyloid.
Occurrence: Solitary or more often scattered in small groups on the ground beneath conifers in mixed woods, grassy woodland margins; open woods, parklands, campgrounds and picnic areas, also reported to occur under oak and other broad-leaved trees; mycorrhizal; summer-fall; occasional.
Edibility: Edible but not worthwhile.

Comments: This mushroom is easily recognized by the yellow flecks on the cap and stalk, although this feature may be less obvious in rainy weather. *Chrysodon* means "golden tooth," in reference to the scurfy granules that sometimes form a jagged yellow margin on the cap.

Latin name: *Hygrophorus flavodiscus* Frost
Common name: Butterscotch Waxy Cap
Order: Agaricales
Family: Tricholomataceae
Width of cap: 1–2½ in. (2.5–6.5 cm)
Length of the stalk: 1½–3 in. (4–7.5 cm)
Cap: Hemispheric with an inrolled margin when young, becoming convex to flattened with age; ochraceous yellow to ochraceous orange at first, becoming whitish as the cap expands, except in the center, which remains butterscotch-yellow; surface glutinous when fresh, smooth beneath the gluten, more or less radially streaked on the disc; **flesh** thick, white, odor and taste not distinctive.
Gills: Adnate to short-decurrent, close to moderately well-spaced; white or tinged very pale pinkish; thick but tapering to a sharp edge, waxy, clean-looking, at first enclosed by a glutinous veil.
Stalk: More or less equal, solid; white to somewhat yellowish stained; surface sheathed with gluten from the base upward and leaving an inconspicuous ring zone on the upper stalk where the glutinous sheath ends.
Spore print: White.
Spores: 6–8.5 x 3.5–5 μm, elliptical, smooth, hyaline, non-amyloid.
Occurrence: Scattered to clustered in small groups on the ground in pine woods; mycorrhizal; late fall-early winter; common.
Edibility: Edible.

Comments: This "cold weather" species fruits late in the mushroom season, including after the first hard frost in late fall. Although a bit tedious and messy to deal with because of its glutinous quality, it is a fair edible that can be gathered when few other edible wild mushrooms are available. It is preferable, but not essential, to remove the glutinous coating before cooking. *Russula earlei* (p.223) is similar in color and has a waxy-looking cap, but its gills are broadly attached to the stalk. The name *flavodiscus* means "yellow disc."

Latin name: *Hygrophorus fuligineus* Frost
Common name: Sooty Hygrophorus
Order: Agaricales
Family: Tricholomataceae
Width of cap: 1¼–3½ in. (3–9 cm)
Length of the stalk: 1½–3½ in. (4–9 cm)
Cap: Subglobose to hemispheric with an inrolled margin at first, becoming convex to flattened with age; nearly black when young, becoming blackish brown, sometimes with olive tones, darker in the center, paler on the margin; surface smooth, covered with a thick, transparent gluten when fresh and moist, shiny when dry; **flesh** thick, white, odor and taste not distinctive.
Gills: Decurrent to subdecurrent, or adnate when young, moderately well-spaced; white, or with a slight pinkish tinge, clean looking; covered at first with a transparent glutinous veil.
Stalk: More or less equal, sometimes slightly swollen in the middle, solid; white to pale brownish; surface sheathed with gluten from the base upward and leaving an inconspicuous ring zone on the upper stalk where the gluten ends.
Spore print: White.
Spores: 7–9 x 4.5–5.5 µm, short-elliptical, smooth, hyaline, non-amyloid.
Occurrence: Solitary to scattered in groups on the ground under conifers, especially pine; mycorrhizal; late fall-early winter; occasional to locally common.
Edibility: Edible.

Comments: When young and fresh the entire fruitbody is covered in a thick, sometimes dripping, transparent gluten. This distinctive mushroom is found late in the season, often in company with *Hygrophorus flavodiscus* (p.140). *Hygrophorus hypothejus* (p.142) is similar but has a brown, glutinous cap and yellowish gills. *Fuligineus* means "sooty."

Latin name: *Hygrophorus hypothejus* Fr.
Common names: Herald of Winter; Late Fall Waxy Cap
Order: Agaricales
Family: Tricholomataceae
Width of cap: 1–2½ in. (2.5–6.5 cm)
Length of the stalk: 1–3 in. (2.5–7.5 cm)
Cap: Convex with an incurved margin at first, becoming flattened to somewhat funnel-shaped with age, with or without a low umbo; olive-brown to yellow-brown, becoming yellowish orange to tawny toward the margin with age, darker in the center; surface viscid to glutinous, smooth, sometimes obscurely streaked, cuticle separable; **flesh** pale yellow, odor and taste not distinctive.
Gills: Subdecurrent to decurrent, or adnate in very young specimens, moderately well-spaced; white to pale yellow, becoming yellow-orange with age or after freezing, thick but tapering to a sharp edge, clean looking.
Stalk: Equal or tapering downward; yellow to yellowish orange; surface sheathed with gluten from the base upward and leaving an inconspicuous ring zone on the upper stalk where the gluten ends.
Spore print: White.
Spores: 7.5–10 x 4–5 µm, elliptical, smooth, hyaline, non-amyloid.
Occurrence: Often gregarious on the ground under pines in woods, woodland margins, and parkland; mycorrhizal; late fall-early winter; common.
Edibility: Edible.

Comments: The appearance of this species in late fall is an indication that the end of the Appalachian mushroom season is near. For a fleshy mushroom, it is amazingly tolerant of freezing weather, often not fruiting until after the first frost. The color of the gills and stalk intensify after being subjected to frost. Compare with *Hygrophorus fuligineus* (p.141), which has a darker blackish brown cap, white gills, and a white stalk. The name *hypothejus* is derived from the Greek word hypotheios, meaning "sulphur-yellow beneath," which likely pertains to the color of the stalk and gills.

Latin name: *Hygrophorus purpurascens* (Alb. & Schwein.) Fr.
Common name: Veiled Purple Hygrophorus
Order: Agaricales
Family: Tricholomataceae
Width of cap: 2–4½ in. (5–11.5 cm)
Length of the stalk: 1½–3½cm)
Cap: Convex with an incurved margin at first, becoming flattened with an upturned margin with age; reddish purple, becoming whitish toward the margin; surface glabrous or slightly scaly, often streaked, somewhat viscid when young; **flesh** thick at the center, white, odor and taste not distinctive.
Gills: Subdecurrent to short decurrent, close to moderately well-spaced; white to ivory, often developing vinaceous stains; covered at first with a cottony-fibrillose veil.
Stalk: More or less equal or tapered in either direction; white or streaked with reddish purple; surface glabrous or slightly scurfy toward the apex; partial veil leaving a fleeting, inconspicuous whitish fibrillose ring near the apex.
Spore print: White.
Spores: 5.5–8 x 3–4.5 μm, elliptical, smooth, hyaline, non-amyloid.
Occurrence: Solitary or in loosely scattered groups on the ground in conifer woods; mycorrhizal; fall; occasional.
Edibility: Edible but poor.

Comments: *Hygrophorus russula* (not illustrated) is nearly identical but lacks a partial veil and grows in association with oak. *Hygrophorus erubescens* (not illustrated), which grows in conifer woods, is also similar but lacks a partial veil. *Purpurascens* means "becoming purple."

Latin name: *Lentinus strigosus* (Schwein.) Fr.
Synonym: *Panus rudis* Fr.
Common names: Hairy Panus; Ruddy Panus
Order: Polyporales
Family: Polyporaceae
Width of cap: 1–3 in. (2.5–7.5 cm)
Length of the stalk: Up to 1 in. (2.5 cm)
Cap: Kidney-shaped to fan-shaped in outline, at times more or less funnel-shaped, margin inrolled and often lobed; purplish at first, becoming reddish brown to pinkish tan; surface densely hairy; **flesh** thin, tough, white, odor not distinctive, taste slightly bitter.
Gills: Decurrent, close to crowded; violaceous to whitish or colored like the cap; gill edges even.
Stalk: Short, eccentric to laterally attached, or central when growing on top of a horizontal surface, sometimes absent; colored like the cap; densely hairy; tough.
Spore print: White to cream.
Spores: 4.5–7 x 2.5–3 µm, elliptical, smooth, hyaline, non-amyloid.
Occurrence: Solitary or more often in clusters on decaying wood of broad-leaved trees; saprobic; spring-early winter; occasional.
Edibility: Non-poisonous. Sometimes described as edible, but the hairy caps and tough, bitter flesh discourage most mycophagists from experimenting.

Comments: This common and widespread species is often one of the first gilled mushrooms to colonize freshly cut stumps or fallen logs. *Lentinus torulosus* (p.145) is similar but larger and has a smooth to scaly cap. *Strigosus* means "with rigid hairs or bristles."

Latin name: *Lentinus torulosus* (Pers.:Fr.) C.G. Lloyd
Synonym: *Panus conchatus* (Bull.) Fr.
Common name: None
Order: Polyporales
Family: Polyporaceae
Width of cap: 2–4 in. (5–10 cm)
Length of the stalk: ½–1¼ in. (1.3–3 cm)
Cap: Broadly convex at first, becoming flattened to depressed in the center or funnel-shaped, margin inrolled at first, often lobed; violet when young, becoming pinkish cinnamon with violaceus tones to vinaceous brown, paler with age; surface smooth to more or less scaly as the cap expands; **flesh** thin, tough, whitish, odor and taste not distinctive.
Gills: Decurrent, moderately close, often forking near the stalk, gill edges even; whitish with a pale violet tinge to yellowish.
Stalk: Short, solid, tapering downward, eccentric to lateral, or sometimes central when attached to horizontal surfaces; violet to pale tan; surface velvety at least at the base.
Spore print: White.
Spores: 6–7 x 3–3.5 μm, elliptical, smooth, hyaline, non-amyloid.
Occurrence: Solitary or in small groups or clusters on decaying logs and stumps of broad-leaved trees; saprobic; spring-fall; occasional.
Edibility: Edible but tough.

Comments: *Pleurotus ostreatus* (p.25) is similar but white to grayish brown overall, and typically grows in overlapping shelf-like clusters. *Lentinus strigosus* (p.144) is smaller and has a densely hairy cap and stalk. *Torulosus* means "thickened at intervals."

Latin name: *Omphalotus illudens* (Schwein.) Sacc.
Synonym: *Clitocybe illudens* (Schwein.) Sacc.
Common name: Jack O' Lantern Mushroom
Order: Agaricales
Family: Marasmiaceae
Width of cap: 2–7 in. (5–18 cm)
Length of the stalk: 2–8 in. (5–20 cm)
Cap: Convex at first, becoming flattened or shallowly depressed with age, with or without an umbo, margin incurved when young, later expanding and becoming wavy; bright orange to yellow-orange, becoming brownish orange with age; surface smooth; **flesh** white to yellowish, odor not distinctive.
Gills: Decurrent, close to moderately well-spaced; colored like the cap; luminescent.
Stalk: Variable from nearly equal to spindle-shaped, usually narrowed at the base, sometimes twisted, solid, central to eccentric; colored like the cap or paler; surface smooth.
Spore print: Cream to pale yellowish.
Spores: 4–5 x 3.5–4.5 μm, more or less globose, smooth, hyaline, non-amyloid.
Occurrence: Typically growing in large clusters, often with multiple stalk bases originating from one point, on or around well-decayed stumps and logs of broad-leaved trees (especially oak), at the base of standing dead or dying trees, or from decaying underground roots; saprobic; summer-fall; common.
Edibility: Poisonous.

Comments: The Jack O' Lantern is surely one of the most spectacular of all mushrooms. It sometimes occurs in enormous clusters that can be seen from a distance, or even spotted from a moving vehicle. The clusters sometimes grow in lawns and grassy places from the remaining root systems of trees that are no longer present, leading to confusion with edible Chanterelles, *Cantharellus cibarius* (p.126), which are not associated with decaying wood. The Jack O' Lantern can further be distinguished from Chanterelles by having sharp-edged true gills rather than blunt gill-like folds or ridges beneath the cap. Also, Chanterelles rarely occurs in clusters of more than 2 or 3 fruitbodies, and they are more "egg-yolk" yellow. *Armillaria tabescens* (not illustrated) is similar to the Jack O' Lantern in form and growth habit but is brown rather than orange. Compare also with *Lactarius croceus* (p.87).

The gills of fresh specimens of the Jack O' Lantern will glow in the dark, hence the common name. To observe this phenomenon take the mushrooms into a completely dark room. After ones eyes have adjusted to the dark, an eerie glow can be seen emanating from the gills. The intensity of the luminescence depends on the condition of the specimens. The name *illudens* means "deceiving."

Latin name: *Paxillus atrotomentosus* (Batsch) Fr.
Synonym: *Tapinella atrotomentosa* (Batsch) Sutara
Common names: Velvet Paxillus; Velvet-footed Paxillus
Order: Boletales
Family: Paxillaceae
Width of cap: 2–6 in. (5–15 cm)
Length of the stalk: 1½–5 in. (4–12.5 cm)
Cap: Convex with an inrolled margin at first, becoming flattened to depressed in the center with age; yellow-brown to reddish brown; surface dry, dull to velvety; **flesh** cream to ocraceous, odor not distinctive, taste more or less bitter or acidic.
Gills: Short or long decurrent, close, often forked near the stalk; cream to yellow-ochre; easily separable from the cap.
Stalk: Thick, stout, more or less equal or bulbous at the base, sometimes rooting, central or eccentric; surface densely covered with dark brown to nearly black velvety or woolly hairs.
Spore print: Yellowish to brownish.
Spores: 5–6 x 3–5 μm, broadly elliptical, smooth, pale brown.
Occurrence: Solitary or in small groups on or around well-decayed stumps, logs, and fallen limbs, or on the ground from buried wood of conifer and broad-leaved trees, especially pine and hemlock; saprobic; summer-fall; common.
Edibility: Not edible.

Comments: This handsome species is easily recognized by its dull brownish cap, dark velvety stalk, and association with decaying wood. *Paxillus involutus* (p.149) is less robust, has gills that bruise brown, and a smooth stalk. *Atrotomentosus* means "black" and "densely hairy," in reference to the stalk.

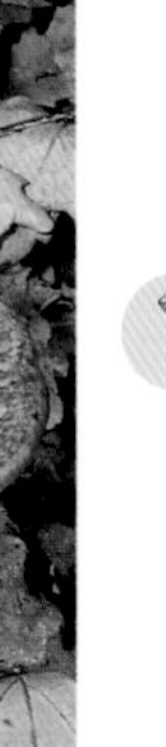

Latin name: *Paxillus involutus* (Batsch) Fr.
Common names: Poison Paxillus, Brown Roll-rim
Order: Boletales
Family: Paxillaceae
Width of cap: 1½–4½ in. (4–11.5 cm)
Length of the stalk: 1–2½ in. (2.5–6.5 cm)
Cap: Convex at first, becoming flattened to depressed in the center, margin inrolled; olive-brown to yellow-brown or reddish brown; surface smooth to slightly fibrillose-matted, somewhat tacky when wet, margin hairy when young, often obscurely grooved; **flesh** pale ochraceous, staining reddish brown when cut, odor not distinctive or somewhat sour.
Gills: Decurrent, crowded, often forked near the stalk, easily separated from the cap; pale ochre to yellowish brown, becoming cinnamon-brown to reddish brown, developing reddish brown stains with age or when rubbed.
Stalk: More or less equal or tapering downward, solid; colored like the cap, often darker at the base, staining rusty brown when handled; surface smooth.
Spore print: Brown.
Spores: 8–10 x 5–7 μm, elliptical, smooth, pale brown.
Occurrence: Solitary or in groups on the ground, especially near decaying wood, or on decaying logs and stumps in conifer and mixed woods; saprobic, but also capable of forming mycorrhiza; summer-fall; common.
Edibility: Dangerously poisonous.

Comments: Although this species has a history of use as an edible mushroom in parts of Europe, it contains a cumulative toxin that is potentially dangerous, and fatalities have occurred from eating it. *Paxillus atrotomentosus* (p.148) is more robust and has a dark brown velvety stalk. *Involutus* means "rolled inward" referring to the cap margin.

Latin name: *Phylloporus rhodoxanthus* (Schwein.) Bres.
Synonym: *Paxillus rhodoxanthus* (Schwein.) Ricken
Common name: Gilled Bolete
Order: Boletales
Family: Boletaceae
Width of cap: 1½–4 in. (4–10 cm)
Length of the stalk: 1½–3 in. (4–7.5 cm)
Cap: Convex with an incurved margin at first, becoming flattened to shallowly depressed with age; dull red to reddish brown, reddish yellow, or olive-brown; surface dry, somewhat velvety, often cracked, exposing yellowish flesh in the cracks; **flesh** pale yellow, or just below the cuticle, colored like the cap, thick in the center, odor and taste not distinctive.
Gills: Decurrent to subdecurrent, moderately well-spaced, separable from the cap; deep yellow to olivaceous yellow; often wrinkled, and sometimes with crossveining that results in a nearly poroid appearance.
Stalk: Cylindrical, usually tapered toward the base, solid, firm; yellow to more or less colored like the cap, with yellow mycelium at the base; surface often dotted, sometimes with lines extending down from the gills.
Spore print: Olivaceous yellow-brown.
Spores: 9–14 x 3.5–5 μm, elliptical to spindle-shaped, smooth, pale yellow.
Occurrence: Usually solitary but sometimes in small groups on the ground in broad-leaved woods, especially with oak, and conifer woods; mycorrhizal; summer-fall; common.
Edibility: Edible.

Comments: This mushroom has the general aspect of a bolete from above, and inexperienced collectors are often surprised to discover "gills" on the underside of the cap. However, the microscopic structure and biochemistry clearly align it with the boletes even though it has a gill-like fertile surface. Some collections show strong, almost poroid crossveining between the gills, thus appearing to be intermediate between a gilled mushroom and a bolete. *Phylloporus leucomycelinus* (not illustrated) is nearly identical except that it has white mycelium at the base of the stalk. The generic name *Phylloporus* means "leaf-pore," in reference to the gill-like fertile surface. *Rhodoxanthus* means "rose or red and yellow."

Latin name: *Rhodocybe mundula* (Lasch) Singer
Synonym: *Clitopilus noveboracensis* Peck
Common name: Cracked-cap Rhodocybe
Order: Agaricales
Family: Entolomataceae
Width of cap: 1¼–3 in. (3–7.5 cm)
Length of the stalk: 1–2½ in. (2.5–6.5 cm)
Cap: Convex to broadly convex, becoming nearly flat or depressed in the center, margin inrolled at first, expanded and wavy with age; dingy white to grayish brown; surface concentrically cracked, at least with age, sometimes slightly zonate near the margin; **flesh** thin, white, odor of flour, taste bitter.
Gills: Decurrent to subdecurrent, close to crowded, usually with some forking; whitish at first, becoming dull pink to brownish gray as the spores mature.
Stalk: Slender, more or less equal or tapering downward; colored like the cap or paler; surface somewhat downy; base with white mycelial threads.
Spore print: Dull pink to pale salmon-pink.
Spores: 4–6 x 4–4.5 μm, globose to subglobose, smooth, somewhat lumpy, angular at the ends, hyaline, non-amyloid.
Occurrence: Usually in groups, scattered or clustered on the ground in woods and open areas; saprobic; summer-fall; occasional.
Edibility: Unknown.

Comments: Typical specimens of this rather drab mushroom are recognized by the concentric cracking pattern on the cap. The name *mundula* means "little world." Its application here is uncertain, perhaps referring to the orbital cracking pattern on the cap.

Small Mushrooms Growing on Wood, Gills Not Decurrent

The mushrooms in this section have central, ringless stalks that are ¼ inch (6 mm) or less thick at the apex. The gills are free from the stalk or variously attached but not decurrent. They are wood-rotting saprobes that often grow in groups or clusters and are frequently found on and around decaying logs and stumps, but also on wood chips and wood mulch in landscaped places. The woody substrate is usually obvious, but bear in mind that small twigs and branches as well as roots may be partly or wholly buried and mushrooms that are attached to them can appear to be growing on soil. When groups of mushrooms are concentrated around decaying tree stumps, it is likely that they are associated with the decaying root system.

Latin name: *Baeospora myriadophylla* (Peck) Singer
Synonym: *Collybia myriadophylla* Peck
Common name: Lavender Baeospora
Order: Agaricales
Family: Tricholomataceae
Width of cap: ½–1½ in. (1.3–4 cm)
Length of the stalk: 1–2 in. (2.5–5 cm)
Cap: Broadly convex to flat or depressed in the center; lavender or brownish lavender, becoming dull brown to ochraceous buff with age or in drying; surface smooth, hygrophanous; **flesh** thin, grayish, odor and taste not distinctive.
Gills: Adnate to adnexed, or nearly free from the stalk, crowded; more or less colored like the cap, fading with age.
Stalk: Slender, cylindrical or sometimes compressed, equal, tough, becoming hollow; colored like the cap; surface smooth, often with a whitish bloom when young.
Spore print: White.
Spores: 3.5–4.5 x 2–3 μm, broadly elliptical, smooth, hyaline, amyloid.
Occurrence: In small groups on decaying wood of broad-leaved and conifer trees, especially hemlock; saprobic; spring-fall; occasional.
Edibility: Not edible.

Comments: This small mushroom is fairly distinctive when fresh and moist, but once the lavender color fades it becomes a more generic looking "LBM" (little brown mushroom). *Myriadophylla* means "many leaves," referring to the crowded gills.

Latin name: *Callistosporium purpureomarginatum* R. Fatto & A.E. Bessette
Common name: None
Order: Agaricales
Family: Tricholomataceae
Width of cap: ⅝–1½ in. (1.5–4 cm)
Length of the stalk: ¾–1½ in. (2–4 cm)
Cap: Convex to nearly flat, sometimes with a shallow central depression; purplish red to brownish violet, drying to pinkish tan from the center outward, but retaining a brownish violet band on the margin; surface smooth, hygrophanous; **flesh** thin, yellowish, odor not distinctive, taste slowly bitter with a metallic aftertaste.
Gills: Adnexed to notched at the stalk, close to crowded; yellowish pink, tinged purplish; gill edges reddish purple, somewhat uneven or eroded.
Stalk: Cylindrical, equal, usually curved, hollow; at first colored like the cap or paler, fading to yellowish or brownish with age; surface smooth to slightly fibrillose-streaked.
Spore print: White.
Spores: 4–6 x 3–4 μm, elliptical, smooth, hyaline, non-amyloid.
Occurrence: Solitary or in groups on decaying oak logs; saprobic; summer; rarely reported.
Edibility: Unknown.

Comments: The reddish purple gill edges and hygrophanous cap—which on drying leaves a brownish purple band at the margin—are distinguishing features for this little-known mushroom. It was first described from New Jersey in 1996. The distribution and frequency of occurrence for this species is yet to be determined. The name *purpureomarginatum* refers to the purple gill edges.

Latin name: *Coprinus disseminatus* (Pers.) Gray
Synonym: *Pseudocoprinus disseminatus* (Pers.) Kühner
Common names: Non-inky Coprinus; Little Helmets
Order: Agaricales
Family: Coprinaceae
Width of cap: ¼–¾ in. (0.5–2 cm)
Length of the stalk: ½–1½ in. (1.3–4 cm)
Cap: Hemispheric to bell-shaped or convex, prominently furrowed from the margin nearly to the center; pale yellowish at first, becoming brownish gray with yellow-brown on the disc; surface with fine hairs and minute glistening particles when young, nearly glabrous with age; **flesh** thin, fragile, whitish, odor and taste not distinctive.
Gills: Adnate to barely attached, well-spaced; white at first, becoming grayish to blackish with age, not liquefying, or only slightly so.
Stalk: Slender, equal, hollow, fragile; white to gray; surface smooth to lightly scurfy; downy or with white bristly hairs at the base.
Spore print: Blackish.
Spores: 7–10 x 4–5 μm, elliptical with a truncate apex, smooth.
Occurrence: Usually in large troops, densely clustered on decaying wood of broad-leaved trees, especially on and around stumps, or at the base of standing dead trees. Also in grassy places where there is buried remains of old stumps and root systems; saprobic; spring-fall; common.
Edibility: Non-poisonous.

Comments: This delicate mushroom is typically found in large colonies that sometimes extend several feet from a decaying stump or tree trunk. *Coprinus plicatilis* (not illustrated), which is larger and has a deeply grooved cap with an orange-brown disc, grows singly or loosely scattered on the ground in grassy places. Compare with *Coprinus micaceus* (p.156). *Disseminatus* means "dispersed."

Latin name: *Coprinus micaceus* (Bull.) Fr.
Common names: Glistening Coprinus; Mica Cap
Order: Agaricales
Family: Coprinaceae
Width of cap: ¾–1¾ in. (2–4.5 cm)
Length of the stalk: 1–3 in. (2.5–7.5 cm)
Cap: Oval to bell-shaped, expanding to broadly convex, margin upturned with age; color variable from tawny brown, reddish brown, to ochre-brown, fading with age and becoming grayish near the margin; surface covered at first with glistening particles that easily wash or wear off; margin prominently lined nearly to the center; **flesh** thin, fragile, pale brown, odor and taste not distinctive.
Gills: Barely attached to the stalk, crowded; whitish to grayish brown, ultimately becoming black and partially dissolving.
Stalk: Slender, equal, hollow, fragile, easily splitting; whitish; surface smooth to silky.
Spore print: Black.
Spores: 7–10 x 4.5–6 μm, elliptical with an apical germ pore, smooth, blackish.
Occurrence: Typically clustered in large troops on and around decaying tree stumps or at the base of standing dead or dying trees, in woods, urban areas, parklands, sometimes originating from buried wood in open areas; saprobic; spring-fall; common.
Edibility: Edible.

Comments: This common mushroom is one of the first edible gilled mushrooms to appear in spring, and may reappear at the same location at various times throughout the growing season. Although small and thin fleshed, it has a mild, pleasant flavor. Related look-alike species such as *Coprinus domesticus* (not illustrated) and *Coprinus radians* (not illustrated) grow singly or in small groups on wood from a dense woolly mat of orange mycelium. Compare with *Coprinus disseminatus* (p.155), which also grows in large troops on wood but has a smaller, grayish ochre cap. *Coprinus atramentarius* (not illustrated), which is a larger, gray species that grows in dense clusters on stumps or on the ground from buried wood, lacks glistening particles on the cap, and its cap and gills readily dissolve into black inky fluid at maturity. *Micaceus* means "glistening."

Latin name: *Hypholoma fasciculare* (Huds.) Quél.
Synonym: *Naematoloma fasciculare* (Huds.) P. Karst.
Common name: Sulphur Tuft
Order: Agaricales
Family: Strophariaceae
Width of cap: ¾–2½ in. (2–6.5 cm)
Length of the stalk: 1½–4 in. (4–10 cm)
Cap: Convex to bell-shaped at first, becoming broadly convex to flattened with age, margin incurved when young; color variable, bright yellow to greenish yellow, orange-yellow to orange-brown, especially on the disc; surface dry, glabrous, sometimes with remnants of a thin web-like partial veil on the margin; **flesh** yellow to greenish yellow, odor not distinctive, taste bitter in most collections.
Gills: Adnate, crowded; greenish yellow to greenish gray, eventually becoming purplish brown as the spores mature.
Stalk: Slender, more or less equal, becoming hollow with age; pale-yellow to yellow or tan, brownish near the base, often tinged with purple-brown spore deposits; surface smooth, fibrillose; partial veil web-like, sometimes leaving a thin fibrillose annular zone near the apex.
Spore print: Purple-brown.
Spores: 6–8 x 4–4.5 μm, elliptical, with a germ pore, smooth, pale purple-brown, non-amyloid.
Occurrence: In large or small clusters on and around decaying stumps and logs of conifers and broad-leaved trees, also on wood chips, and on the ground from buried wood; saprobic; early spring-late fall; common.
Edibility: Poisonous.

Comments: This widespread recycler of deadwood is usually found in clusters in a variety of habitats from woodlands to city streets and parks. *Hypholoma capnoides* (not illustrated), which nearly always grows on conifer wood, is similar but lacks olivaceous tones on the cap, and has bluish gray to smoky brown gills, and mild tasting flesh. *Hypholoma sublateritium* (p.208) has a brick-red cap and only occurs on decaying wood of broad-leaved trees. *Pholiota spumosa* (p.160) has a viscid, sticky cap, and brown spores. *Fasciculare* means "tufted" or "clustered in bundles."

Latin name: *Marasmius rotula* (Scop.) Fr.
Common names: Pinwheel Marasmius; Little Wheel
Order: Agaricales
Family: Marasmiaceae
Width of cap: ¼–¾ in. (0.5–2 cm)
Length of the stalk: 1–2 in. (2.5–5 cm)
Cap: Hemispheric to broadly convex with a shallow central depression, often with a very slight umbo, margin scalloped; white to pale yellowish, brown at the center; surface radially furrowed, glabrous, dry; **flesh** thin, whitish, odor not distinctive, taste not distinctive or slightly bitter.
Gills: Broadly attached to a collar that is completely free from the stalk (at times the collar is collapsed against the stalk and less obvious); white to pale yellowish; widely spaced.
Stalk: Thin, wiry, hollow; whitish above, graduating downward to tawny, then dark blackish brown from the middle to the base; surface shiny.
Spore print: White.
Spores: 6.5–9 x 3–5 μm, elliptical, smooth, hyaline, non-amyloid.
Occurrence: In groups or small clusters on decaying wood of broad-leaved trees (often on moss-covered logs and stumps; saprobic; spring-fall; common.
Edibility: Not edible.

Comments: This delicately beautiful mushroom occurs in a variety of woodland habitats. Like many others in the genus *Marasmius,* the caps shrivel up and becomes inconspicuous during dry weather, then revive and reappear rather quickly after rainfall. *Marasmius capillaris* (not illustrated) is somewhat similar but has a pale tan cap with a whitish center, and grows on decaying oak leaves and twigs. *Mycena corticola* (not illustrated) is much smaller and duller than the Pinwheel Marasmius. It has a pale pinkish brown cap and grows singly or gregariously on bark near the base of living trees. The epithet *rotula* is derived from the Latin word "rota," meaning "wheel."

Latin name: *Mycena leaiana* Berk.
Common name: Orange Mycena
Order: Agaricales
Family: Tricholomataceae
Width of cap: ¾–1¾ in. (2–4.5 cm)
Length of the stalk: 1–2½ in. (2.5–6.5 cm)
Cap: Broadly bell-shaped to convex, often shallowly depressed in the center, sometimes with a slight umbo; brilliant orange fading to yellow-orange, or paler as the pigments deteriorate or are washed out; surface viscid and shiny; margin striate; **flesh** watery, whitish to orangish, odor and taste not distinctive.
Gills: Adnate to sinuate, moderately close to crowded; pinkish orange to pinkish yellow, bruising yellow-orange; gill edges deep reddish orange.
Stalk: More or less equal, hollow, tough; orange to yellow; surface smooth, viscid and tacky, with dense orange hairs at the base.
Spore print: White.
Spores: 7–10 x 5–6 µm, elliptical, smooth, hyaline, amyloid.
Occurrence: Usually in caespitose clusters on decaying wood of broad-leaved trees, especially beech; saprobic; spring-fall; common.
Edibility: Not edible.

Comments: The Orange Mycena is a striking mushroom that can be found throughout the growing season but is especially noticeable in the springtime when few other colorful species are present. When handled, the juicy flesh will stain fingers orange. Some bright orange "waxcaps" in the genus *Hygrocybe* bear a superficial resemblance to the Orange Mycena, but they grow on soil. The specific epithet honors Ohio mycologist Thomas Gibson Lea (1785–1844).

Latin name: *Pholiota spumosa* (Fr.) Singer
Common name: Cornsilk Pholiota
Order: Agaricales
Family: Strophariaceae
Width of cap: ¾–2½ in. (2–6.5 cm)
Length of the stalk: 1¼–2½ in. (3–6.5 cm)
Cap: Convex with a broad umbo, becoming nearly flat; olive-brown to tawny in the center, merging with yellow to greenish yellow toward the margin; surface glabrous, smooth, viscid when wet, shiny, obscurely radially streaked on the disc, margin at first with web-like remnants of a partial veil, soon disappearing; **flesh** thin, yellowish, odor of cornsilk, or not distinctive, taste mild.
Gills: Adnate to notched, close; pale yellow to greenish yellow, becoming brownish with age; covered at first with a thin, web-like partial veil.
Stalk: More or less equal, becoming hollow with age; yellow to greenish yellow above, brown on lower portion; surface fibrillose, at times with a faint zone of fibrils near the apex (partial veil remnants).
Spore print: Brown.
Spores: 6–9 x 4–4.5 μm, elliptical with an apical germ pore, smooth, pale brown.
Occurrence: Scattered to clustered on decaying stumps and logs of conifer trees, also on wood chips, sawdust piles, and on the ground from buried wood; saprobic; spring-fall; occasional.
Edibility: Unknown.

Comments: The cornsilk odor of this mushroom is a good field character, although it isn't apparent in all collections. Compare with *Hypholoma fasciculare* (p.157), which has a dry cap and purple-brown spores. *Spumosa* means "frothy" or "foamy," but its meaning here is obscure, perhaps relating to the viscid, almost glutinous surface of the cap, especially when wet.

Latin name: *Pluteus flavofuligineus* G.F. Atk.
Common name: Smoky Yellow Pluteus
Order: Agaricales
Family: Pluteaceae
Width of cap: 1–3 in. (2.5–7.5 cm)
Length of the stalk: 1½–3 in. (4–7.5 cm)
Cap: Bell-shaped to convex, becoming flattened with age; at first golden yellow with a radiating sooty brown center, sometimes with olivaceous tones, becoming dull yellow as the cap expands except on the disc, which remains brownish; surface velvety, more or less wrinkled at the center, margin faintly striate; **flesh** whitish to pale yellowish, odor and taste not distinctive.
Gills: Free from the stalk; whitish at first, becoming pink as the spores mature.
Stalk: Equal or tapering toward the apex; white to pale pink or yellowish with age; surface smooth, often with lengthwise twisted streaks.
Spore print: Pink.
Spores: 6–8 x 5.5–6.5 μm, subglobose, smooth, non-amyloid.
Occurrence: Solitary or in small groups on or around decaying logs, stumps, and fallen branches of broad-leaved trees; saprobic; spring-fall; common.
Edibility: Unknown.

Comments: This handsome mushroom first appears in late April or early May. *Pluteus admirabilis* (not illustrated), which is similar and can also be found in springtime, is smaller and has a much brighter, yellow to orange-yellow cap that becomes yellow-brown with age. *Flavofuligineus* means "sooty yellow."

Latin name: *Pluteus longistriatus* Peck
Common name: Pleated Pluteus
Order: Agaricales
Family: Pluteaceae
Width of cap: ¾–1¾ in. (2–4.5 cm)
Length of the stalk: 1–2½ in. (2.5–6.5 cm)
Cap: Convex to nearly flat, sometimes shallowly depressed in the center, with or without a low umbo; grayish brown to pale reddish brown; surface slightly granular to minutely scaly on the disc, margin with long striations reaching nearly to the center of the cap; **flesh** soft, whitish, odor and taste not distinctive.
Gills: Free from the stalk, close; whitish at first, then pink.
Stalk: Slender, equal, solid; whitish; surface smooth to scurfy, usually with light lengthwise lines.
Spore print: Salmon pink.
Spores: 6–7.5 x 5–5.5 μm, subglobose, smooth, non-amyloid.
Occurrence: Solitary or in small groups on decaying stumps, logs, and woody debris of broad-leaved trees; saprobic; summer-early fall; occasional.
Edibility: Edible, but of little substance.

Comments: The Pleated Pluteus is recognized by its delicate fruitbody, long striations on the cap, free gills, and pink spore deposit. This mushroom will sometimes grow indoors on damp wood. *Longistriatus* means "having long lines."

Latin name: *Pluteus lutescens* (Fr.) Bres.
Synonym: *Pluteus romellii* (Britzelm.) Lapl.
Common name: Yellow-stalk Pluteus
Order: Agaricales
Family: Pluteaceae
Width of cap: 1–2 in. (2.5–5 cm)
Length of the stalk: ¾–1½ in. (2–4 cm)
Cap: Convex to flat, with or without an umbo; brown to olive-brown; surface wrinkled at least at the center, margin obscurely or distinctly furrowed; **flesh** thin, yellowish, odor and taste weakly radish-like or not distinctive.
Gills: Free from the stalk, moderately spaced to close; pale yellowish becoming pink at maturity.
Stalk: More or less equal; yellow at the base, becoming paler yellow toward the apex; smooth, shiny.
Spore print: Pink to dull salmon-pink.
Spores: 5.4–7.6 x 5–6.7 μm, subglobose, smooth, non-amyloid.
Occurrence: Solitary or in small groups on decaying wood or woody debris of broad-leaved trees; saprobic; spring-fall; uncommon.
Edibility: Unknown. Probably edible, but too small and infrequently encountered to be of interest.

Comments: *Pluteus chrysophaeus* (not illustrated) is similar but has a whitish stalk with longitudinal streaks. Macroscopic features of closely related species overlap, making confident field identification difficult. *Lutescens* means "turning yellow."

Latin name: *Pluteus thomsonii* (Berk. & Broome) Dennis
Common name: Ribbed-cap Pluteus
Order: Agaricales
Family: Pluteaceae
Width of cap: ½–1½ in. (1.3–4 cm)
Length of the stalk: 1–2½ in. (2.5–6.5 cm)
Cap: Broadly bell-shaped to convex, becoming nearly flat with a slight umbo; blackish brown to dark brown, paler toward the margin; surface prominently wrinkled to reticulate-veined, especially on the disc; margin striate; **flesh** thin, white, odor and taster not distinctive.
Gills: Free from the stalk, close; white to grayish, becoming pinkish at maturity.
Stalk: More or less equal or somewhat swollen at the base, hollow; grayish to grayish brown; surface pruinose, often streaked.
Spore print: Dull pink.
Spores: 6–8 x 5.5–6 μm, subglobose, smooth, hyaline.
Occurrence: Solitary or in small groups on decaying wood, woody debris, wood mulch, or on the ground from buried wood of broad-leaved trees; saprobic; summer-fall; uncommon.
Edibility: Unknown.

Comments: The strongly reticulate-veined cap is distinctive. *Pluteus granularis* (not illustrated) is somewhat similar, having a dark brown velvety cap that lacks prominent veining, and smaller spores (5–6.5 x 4–5 μm). According to Dr. Scott Redhead, this mushroom was named in honor of a Dr. Thomson, who was an associate of the great nineteenth-century British mycologist Miles J. Berkeley.

Latin name: *Xeromphalina tenuipes* (Schwein.) A.H. Sm.
Synonym: *Collybia tenuipes* Schwein.
Common name: None
Order: Agaricales
Family: Tricholomataceae
Width of cap: 1–2½ in. (2.5–6.5 cm)
Length of the stalk: 1½–3 in. (4–7.5 cm)
Cap: Convex to bell-shaped, becoming flat or shallowly depressed in the center, often with a broad umbo; orange-brown to yellow-brown, sometimes with olivaceous tones; surface dry, velvety, becoming wrinkled, margin striate (sometimes obscurely), often wavy with age; **flesh** watery brown, odor not distinctive, taste bitter.
Gills: Adnate to sinuate; white at first, soon pale yellow.
Stalk: More or less equal, cylindrical or sometimes compressed, often flaring at the apex and swollen at the base, tough, hollow; colored like the cap; surface velvety to densely hairy.
Spore print: White.
Spores: 7–9 x 4.5–5 μm, sausage-shaped, smooth, amyloid.
Occurrence: Solitary or in small groups or clusters on or around decaying logs, stumps, and fallen branches of broad-leaved trees; saprobic; spring-summer; fairly common in some years.
Edibility: Unknown.

Comments: This handsome mushroom first appears early in the spring. Small groups often grow in lines alongside fallen branches on the ground. *Flammulina velutipes* (not illustrated) has similar colors and a velvety brown stalk, however it has a smooth, viscid cap, and it normally fruits in cool weather from fall through spring. *Tenuipes* means "slender foot."

Small Mushrooms That Grow on the Ground or on Other Non-woody Substrates, Having White to Pinkish Buff Spores, and Gills That Are Not Decurrent

The mushrooms in this section have central, ringless stalks that are usually no thicker at the apex than a standard pencil (¼ inch = 6 mm). The gills are free to variously attached to the stalk, but typically not decurrent.

In order to reduce the possibilities when searching for a correct match, Section 1-H is divided into two parts based on the color of the spore deposit. Similar mushrooms that grow on non-woody substrates but have dark-colored spores are placed in Section 1-H(b). If the stalk thickness is borderline in the ¼-inch range, also review Section 1-K(a), which includes somewhat larger mushrooms that produce light-colored spores.

Latin name: *Calocybe persicolor* (Fr.) Singer
Common name: Clustered Pink Calocybe
Order: Agaricales
Family: Tricholomataceae
Width of cap: ¾–1¾ in. (2–4.5 cm)
Length of the stalk: ¾–1¾ in. (2–4.5 cm)
Cap: Convex with an incurved margin at first, becoming flattened to depressed in the center, sometimes with a low umbo; pink to vinaceous pink or pale pinkish tan; surface glabrous, more or less smooth, dry; **flesh** thin, white, odor and taste not distinctive.
Gills: Adnate to notched with a decurrent tooth, close to crowded, ragged with age; white.
Stalk: More or less equal, cylindrical or sometimes compressed; yellowish brown to concolorous with the cap, paler at the apex, with whitish hairs at the base; surface smooth.
Spore print: White.
Spores: 4–6 x 2–3 μm, elliptical, smooth, hyaline, non-amyloid.
Occurrence: Usually in groups or caespitose clusters on the ground in grassy areas, pastures, woods, and woodland margins; saprobic; summer-fall; uncommon.
Edibility: Unknown.

Comments: This pretty little mushroom is not often recorded, perhaps because it is short and squatty, and often hidden in grass. *Persicolor* means "peach-colored."

Latin name: *Clitocybe tarda* Peck
Synonyms: *Tricholoma sordidum* Fr.; *Lepista sordida* (Fr.) Singer
Common name: None
Order: Agaricales
Family: Tricholomataceae
Width of cap: ¾–2½ in. (2–6.5 cm)
Length of the stalk: 1–2½ in. (2.5–6.5 cm)
Cap: Convex at first, later becoming flattened or depressed in the center, sometimes with an obtuse umbo, margin incurved when young, uplifted and wavy with age; color variable from pinkish brown to pale grayish violet to violaceous brown, fading to pale cinnamon or nearly white; surface smooth, at times with faint striations on the margin, hygrophanous; **flesh** thick in the center, thin toward the margin, watery, whitish to pinkish brown, odor and taste not distinctive.
Gills: Adnate to notched, moderately close; pinkish to pale violet.
Stalk: Slender, more or less equal; colored like the cap or paler, base with white mycelium; surface often with brownish streaks.
Spore print: Pinkish buff.
Spores: 6–8 x 3.5–5 μm, elliptical, minutely warted, hyaline, non-amyloid.
Occurrence: In groups or clusters, sometimes forming fairy rings or arcs in open grassy areas, mulched ground, or on manure-enriched ground; saprobic; summer-fall; occasional.
Edibility: Reported to be edible.

Comments: This variable mushroom looks like a slender, pale version of *Clitocybe nuda* (p.244). *Tarda* means "late" or "tardy."

Latin name: *Gymnopus confluens* (Pers.) Antonin, Halling & Noordel.
Synonym: *Collybia confluens* Pers.
Common name: Tufted Collybia
Order: Agaricales
Family: Tricholomataceae
Width of cap: ¾–2 in. (2–5 cm)
Length of the stalk: 1½–4 in. (4–10 cm)
Cap: Convex with an inrolled margin at first, becoming broadly convex to nearly flat, sometimes with a low or sunken umbo; pale brown to reddish brown, fading to pale pinkish buff or whitish; surface glabrous, hygrophanous, slightly striate when moist; **flesh** thin, flexible, whitish, odor and taste pleasant but otherwise not distinctive.
Gills: Adnate to adnexed or nearly free from the stalk, thin, narrow, crowded; cream colored to pinkish buff or pinkish cinnamon.
Stalk: Slender, cylindrical to compressed, more or less equal, tough, elastic, hollow; surface dry, covered with grayish hairs, reddish brown beneath.
Spore print: White to cream.
Spores: 7–9.2 x 3.5–4.2 μm, elliptical, smooth, hyaline, non-amyloid.
Occurrence: Usually gregarious, often clustered, sometimes in fairy rings or arcs on the ground in conifer and broad-leaved woods; saprobic; late spring-fall; common.
Edibility: Edible, but discard the tough stalks.

Comments: The Tufted Collybia has the ability to revive after drying when wet conditions recur. *Gymnopus subnudus* (not illustrated) is similar but generally smaller, with thinner flesh, a darker reddish brown cap, and more widely spaced gills. *Gymnopus polyphyllus* (p.172) is more compact in stature and has a distinct "spoiled cabbage" odor. *Confluens* means "flowing together."

Latin name: *Gymnopus dryophilus* (Bull.) Murrill
Synonym: *Collybia dryophila* (Bull.) Quél.
Common names: Common Collybia; Oak-loving Collybia
Order: Agaricales
Family: Tricholomataceae
Width of cap: 1–2½ in. (2.5–6.5 cm)
Length of the stalk: 1–3 in. (2.5–7.5 cm)
Cap: Convex at first, becoming broadly convex to flat; reddish brown to orange-brown, fading to yellowish tan or paler when dry, margin often with a narrow white band; surface glabrous, more or less smooth, at times faintly striate on the margin, moist; **flesh** thin, whitish, odor and taste not distinctive.
Gills: Adnexed to nearly free from the stalk, crowded; white to pale pinkish buff.
Stalk: Slender, more or less equal, or swollen toward the base, soon becoming hollow, tough, cartilaginous; colored like the cap but usually paler near the apex, often with cottony white mycelium and rhizomorphs at the base; surface smooth, sometimes with faint vertical lines.
Spore print: White to creamy white.
Spores: 5.5–6.5 x 3–3.5 μm, elliptical, smooth, hyaline, non-amyloid.
Occurrence: Scattered in groups or clusters, sometimes in fairy rings on soil or well-decayed woody debris in broad-leaved and conifer woods, and grassy woodland margins; saprobic; spring-fall; common.
Edibility: The caps are generally regarded as edible for most people, but reports vary (see comments below).

Comments: The mixed reviews regarding the edibility of this species may be due to environmental factors. It is reported to accumulate heavy metals such as mercury. Care should be taken not to gather these mushrooms (or any others) for eating from areas that may be polluted. There are several very similar and closely related species including *Gymnopus subsulphurea* (not illustrated), which has a more yellowish cap and stalk, pale yellow gills, and pinkish rhizomorphs at the stalk base. *Gymnopus earleae* (not illustrated) grows more sparsely in early spring and has a pale tan cap. *Rhodocollybia butyracea* (not illustrated) has a reddish brown to chestnut-colored cap with a lubricous, "buttery" feel, and a club-shaped stalk with vertical striations. *Gymnopus acervatus* (not illustrated) grows in clusters on decaying wood. It has smaller reddish brown caps that fade to pale grayish tan on drying. The Oak-loving Collybia is sometimes infected with *Syzygospora mycetophila*, a parasite that causes a convoluted, jelly-like growth on the cap and stalk. *Dryophilus* means "oak-loving," but this mushroom occurs in a variety of woodlands with or without oaks being present.

Latin name: *Gymnopus polyphyllus* (Peck) Halling
Synonym: *Collybia polyphylla* (Peck) Singer ex Halling
Common name: Bad Cabbage Collybia
Order: Agaricales
Family: Tricholomataceae
Width of cap: 1–2½ in. (2.5–6.5 cm)
Length of the stalk: 1½–2½ in. (4–6.5 cm)
Cap: Convex with an incurved margin at first, becoming broadly convex to flattened or shallowly depressed with age; cinnamon to pale reddish brown, soon becoming pinkish buff to white, except on the disc, which remains brownish; surface glabrous, with a whitish bloom when young; **flesh** thin, white, odor and taste variously described as like garlic, onions, or spoiled cabbage.
Gills: Adnate to notched, or nearly free from the stalk, crowded; white.
Stalk: Equal, hollow; more or less colored like the cap; surface smooth to cottony near the base, at first covered with a whitish bloom.
Spore print: White.
Spores: 5.5–7.5 x 3–4 μm, elliptical to tear-drop shaped, smooth, hyaline, non-amyloid.
Occurrence: In groups or clusters on humus and leaf litter, or around well-decayed stumps in broad-leaved and mixed woods; saprobic; spring-fall; occasional.
Edibility: Unknown. The unpleasant odor of the flesh discourages experimentation.

Comments: The pronounced odor of rotting cabbage is the best field character to distinguish this species. *Gymnopus dryophilus* (p.170) lacks a distinctive odor and has a dark reddish brown to chestnut-brown cap that becomes paler with age but is never white. *Gymnopus confluens* (p.169) is taller, more slender, and has a densely hairy stalk. *Rhodocollybia maculata* (p.253) is white overall when young, but soon develops reddish brown spots. *Polyphyllus* means "many leaves" in reference to the crowded gills.

Latin name: *Hygrocybe conica* (Scop.) P. Kumm.
Synonym: *Hygrophorus conicus* (Scop.) Fr.
Common name: Witches' Hat
Order: Agaricales
Family: Tricholomataceae
Width of cap: ¾–2½ in. (2–6.5 cm)
Length of the stalk: 1–3½ in. (2.5–9 cm)
Cap: Bluntly to acutely conical at first, expanding and becoming nearly flat with an umbo; color variable from bright red, blood-red, orange-red, to orange, sometimes with olive tints, turning blackish with age; surface smooth to fibrous or slightly scurfy, dry to slightly viscid; **flesh** thin, more or less colored like the cap cuticle, odor and taste not distinctive.
Gills: Adnexed and narrowly attached or free from the stalk, close to moderately well-spaced; whitish to yellow or orangish; thick, waxy; gill edges eroded.
Stalk: Equal, hollow, easily splitting; yellow to greenish yellow or greenish orange, white at the base, blackening when handled or with age; surface smooth to fibrous, sometimes twisted-striate.
Spore print: White.
Spores: 8–14 x 5–7 μm, elliptical, smooth, hyaline, non-amyloid.
Occurrence: Solitary or in small in groups on the ground in woods, grassy open places, woodland margins, parklands, burned areas, and along railroad tracks; saprobic; spring-fall; fairly common.
Edibility: Not recommended. Sometimes reported to be poisonous.

Comments: This variably colored mushroom may represent a complex of closely related species, but the eventual blackening of all parts of the fruitbody is consistent. Compare with *Hygrocybe cuspidata* (p.174), which has a bright red, viscid cap with a striate margin, and does not blacken from handling or with age. *Conica* means "cone-shaped."

Latin name: *Hygrocybe cuspidata* (Peck) comb. nov.
Synonym: *Hygrophorus cuspidatus* Peck
Common name: Candy-apple Waxy Cap
Order: Agaricales
Family: Tricholomataceae
Width of cap: ½–2 in. (1.3–5 cm)
Length of the stalk: 1–2½ in. (2.5–6.5 cm)
Cap: Conical to broadly conical at first, expanding and becoming flattened to depressed with an umbo and an upturned margin; deep scarlet, fading to orange-red; surface smooth, glabrous, viscid, and shiny; margin translucent-striate; **flesh** thin, white to yellowish, odor and taste not distinctive.
Gills: Free from the stalk, or nearly free, close to crowded; yellow to orange.
Stalk: Equal; colored like the cap or paler, white at the base; surface smooth and dry, vertically streaked or twisted-streaked.
Spore print: White.
Spores: 8–12 x 4–6.5 μm, elliptical, smooth, hyaline, non-amyloid.
Occurrence: Usually in groups on the ground in mixed woods, disturbed ground, roadsides, and burnsites; saprobic; early summer-fall; common.
Edibility: Non-poisonous.

Comments: This striking red mushroom frequently grows on hard-packed soil and on charred ground such as campfire sites. In these situations, considerable debris adheres to the sticky caps. *Hygrocybe conica* (p.173) is similar but does not have a viscid cap and all parts turn black with age or when bruised. *Hygrocybe punicea* (p.247) is more robust with a bright red, broadly conical to bell-shaped cap. *Cuspidata* means "abruptly pointed."

Latin name: *Hygrocybe flavescens* (Kauffm.) Singer
Synonym: *Hygrophorus flavescens* (Kauffm.) A.H. Sm. & Hesler
Common name: Golden Waxy Cap
Order: Agaricales
Family: Tricholomataceae
Width of cap: ¾–2½ in. (2–6.5 cm)
Length of the stalk: 1¼–2½ in. (3–6.5 cm)
Cap: Convex to broadly convex, becoming flat or shallowly depressed with an upturned margin, sometimes with a broad umbo; yellow-orange, paler on the margin; surface smooth, viscid and shiny, margin translucent-striate; **flesh** thin, yellowish, odor and taste not distinctive.
Gills: Adnexed to sinuate, sometimes separating from the stalk with age, close to moderately well-spaced; pale yellow to pale orange; waxy.
Stalk: More or less equal, cylindrical to compressed, soon becoming hollow, fragile and easily splitting; colored like the cap or paler, sometimes white at the base; surface smooth, glabrous, moist to lubricous.
Spore print: White.
Spores: 7–9.5 x 4.5–6 μm, elliptical, smooth, hyaline, non-amyloid.
Occurrence: Solitary to gregarious on the ground in woods, grassy open woods, along woodland trails, or on mossy ground; saprobic; spring-fall; common.
Edibility: Edible.

Comments: *Hygrocybe chlorophana* (not illustrated) has a lemon-yellow cap and a viscid stalk. Because there are intermediate forms between *Hygrocybe flavescens* and *Hygrocybe chlorophana,* some mycologists consider them to be variations of the same species. *Flavescens* means "turning yellow."

Latin name: *Hygrocybe laeta* (Pers.) P. Kumm.
Synonym: *Hygrophorus laetus* (Pers.) Fr.
Common name: Slippery Skunk Waxy Cap
Order: Agaricales
Family: Tricholomataceae
Width of cap: ½–1½ in. (1–4 cm)
Length of the stalk: 1–2½ in. (2.5–6.5 cm)
Cap: Hemispheric at first, becoming convex to flattened, often with a depressed center; color highly variable but usually with pinkish, vinaceous, or brownish orange tones, sometimes tinged olivaceous, fading to pale pinkish cinnamon or pinkish orange; surface smooth, glabrous, viscid and slippery when fresh, margin translucent-striate; **flesh** colored like the cap, odor skunk-like to fishy or sweet, sometimes odorless.
Gills: Adnate to decurrent, moderately well-spaced; pinkish to pale purple or colored like the cap; waxy.
Stalk: More or less equal; colored like the cap or paler, sometimes with an olivaceous tinge at the apex; surface smooth, viscid, slippery.
Spore print: White.
Spores: 6–8 x 4–5 μm, elliptical, smooth, hyaline, non-amyloid.
Occurrence: Usually in small groups on the ground in woods, grassy places, or in moss; saprobic; early summer-fall; common.
Edibility: Unknown.

Comments: Although variable in color this mushroom can be recognized by its exceedingly slippery cap and stalk that make it nearly impossible to grasp, and the skunk-like odor of most collections. Compare with *Hygrocybe perplexa* (p.178), which has a brownish orange cap and lacks a distinctive odor. *Laeta* means "pleasing."

Latin name: *Hygrocybe marginata* var. *marginata* (Peck) Murrill
Synonym: *Hygrophorus marginatus* Peck
Common name: Orange-gilled Waxy Cap
Order: Agaricales
Family: Tricholomataceae
Width of cap: ½–1½ in. (1.3–4 cm)
Length of the stalk: 1–2½ in. (2.5–6.5 cm)
Cap: Bluntly conical to convex, becoming flattened with a low umbo, margin often uplifted with age; orange to reddish orange, fading to pale orange or yellow; surface glabrous, smooth, hygrophanous, sometimes obscurely striate on the margin; **flesh** thin, whitish to pale yellow or colored like the cap surface, odor and taste not distinctive.
Gills: Ascending-adnate to notched with a decurrent tooth, moderately well-spaced; bright orange; thick but with a sharp edge.
Stalk: Cylindrical or sometimes compressed, more or less equal, hollow, fragile, easily splitting; colored like the cap or paler, especially toward the base, fading to pale orange-yellow; surface smooth.
Spore print: White.
Spores: 7–11.5 x 4–6 µm, broadly elliptical, smooth, hyaline, non-amyloid.
Occurrence: Usually in groups on the ground in broad-leaved and mixed woods, especially in low, wet areas; saprobic; late spring-fall; common.
Edibility: Edible but flavorless.

Comments: The bright orange cap of this species soon fades, but the gills remain brilliant orange for some time. Similar species include the nearly identical *Hygrocybe marginata* var. *concolor* (not illustrated), which has chrome-yellow gills. *Hygrocybe marginata* var. *olivascens* (not illustrated) has an orange cap with a brownish olive center. *Entoloma salmoneum* (not illustrated) is salmon colored throughout and has salmon-pink spores. *Marginata* means "having a distinct margin," referring to the gill edge that often remains more deeply colored than the faded gill face of older specimens.

Latin name: *Hygrocybe perplexa* (A.H. Sm. & Hesler) Arnolds
Synonym: *Hygrophorus perplexus* A.H. Sm. & Hesler
Common name: None
Order: Agaricales
Family: Tricholomataceae
Width of cap: ½–1¼ in. (1.3–3 cm)
Length of the stalk: 1¼–2½ in. (3–6.5 cm)
Cap: Broadly conic to bell-shaped, becoming flattened with an umbo; color variable in the brownish orange to brick-red range, sometimes with olivaceous tones, becoming paler when dry; surface smooth, glabrous, viscid and slippery, translucent-striate; **flesh** thin, colored like the cap, odor and taste not distinctive.
Gills: Adnexed, moderately well-spaced; pallid to pale pinkish at first, becoming yellow to orange-yellow with age.
Stalk: Equal or slightly tapering upward, hollow, easily splitting; pale golden yellow to a paler shade of the cap color, sometimes pinkish or olivaceous at the apex; surface smooth, viscid and slippery.
Spore print: White.
Spores: 6–8 x 4–5 μm, elliptical, smooth, hyaline, non-amyloid.
Occurrence: Solitary or in small groups on the ground in broad-leaved woods; saprobic; summer-fall; uncommon.
Edibility: Unknown.

Comments: *Hygrocybe psittacina* (not illustrated) is similar but has a green cap at first, which soon changes to orange, and the stalk is often green at the apex. Some authorities consider *Hygrocybe perplexa* to be a variety of *Hygrocybe psittacina.* Both are very slippery and nearly impossible to grasp. *Perplexa* means "perplexing," which suggests the confusion with faded specimens of *Hygrocybe psittacina.*

Latin name: *Hygrocybe subovina* (Hesler & A.H. Sm.) comb. nov.
Synonym: *Hygrophorus subovina* Hesler & A.H. Sm.
Common name: Brown Sugar Waxy Cap
Order: Agaricales
Family: Tricholomataceae
Width of cap: ¾–2 in. (2–5 cm)
Length of the stalk: 1¼–2½ in. (3–6.5 cm)
Cap: Convex to broadly convex, becoming flattened with age; dark grayish brown to nearly black; surface dry, dull, smooth to slightly scaly; **flesh** whitish to brownish, brittle, thin at the margin, thicker on the disc, odor of brown sugar, taste somewhat alkaline.
Gills: Adnexed to notched, well-spaced; whitish at first, soon becoming dark grayish brown, often with dull reddish stains; thick, waxy.
Stalk: Cylindrical or sometimes compressed, equal, hollow; colored like the cap; surface smooth.
Spore print: White.
Spores: 5–6 x 5–5.7 μm, globose to subglobose, smooth, hyaline, non-amyloid.
Occurrence: Scattered in small groups on the ground in broad-leaved and mixed woods; saprobic; early summer-fall; occasional to common in some years.
Edibility: Unknown.

Comments: This dark, somber species is easy to overlook, or it might be dismissed as an old, decaying mushroom unless closely examined. The sweet, pleasant odor and darker color distinguish *Hygrophorus subovina* from *Hygrocybe nitrata* (not illustrated), which has a pungent ammonia odor and a snuff-brown cap and stalk. *Ovina* means "of sheep." The specific epithet *subovina* means "close to *Hygrocybe ovina*" (not illustrated), a similar species having an unpleasant ammonia or "sheep-like" odor.

Latin name: *Hygrocybe unguinosa* (Fr.) P. Karst.
Synonyms: *Hygrophorus unguinosus* (Fr.) Fr.; *Hygrocybe irrigata* (Pers.) Bon
Common name: None
Order: Agaricales
Family: Tricholomataceae
Width of cap: ¾–1½ in. (2–4 cm)
Length of the stalk: 1–2½ in. (2.5–6.5 cm)
Cap: Hemispheric to bell-shaped or broadly conical at first, becoming convex to flattened, with or without an umbo; blackish brown, becoming grayish brown with age; surface glabrous, viscid to glutinous, shiny, often wrinkled when dry, margin translucent-striate; **flesh** watery, thin, colored like the cap, odor and taste not distinctive.
Gills: Adnate, sometimes with a decurrent tooth, moderately well-spaced with veining between the gills; whitish to grayish; thick, waxy.
Stalk: Cylindrical or sometimes compressed, nearly equal to tapering at the base, becoming hollow; colored like the cap or paler; surface viscid to glutinous, slippery.
Spore print: White.
Spores: 6–9.5 x 4–6 μm, broadly elliptical, smooth, hyaline, non-amyloid.
Occurrence: Solitary, or in small groups on soil or in moss in broad-leaved and conifer woods; saprobic; early summer-fall; occasional.
Edibility: Non-poisonous.

Comments: It is easy to overlook this small, dark mushroom that blends with the forest litter. The exceedingly slippery fruitbodies are nearly impossible to grasp. Compare with *Hygrophorus fuligineus* (p.141), which is similiarly colored and covered with glutin but much more robust. *Hygrocybe subovina* (p.179) is also similarly colored but has a dry, dull cap and stalk, and a sweet odor. *Unguinosa* means "anointed" or "oily," referring to the glutinous quality of the fruitbodies.

Latin name: *Laccaria bicolor* (Maire) P.D. Orton
Common name: Lilac-foot Laccaria
Order: Agaricales
Family: Hydnangiaceae
Width of cap: ½–1½ in. (1.3–4 cm)
Length of the stalk: 1½–3½ in. (4–9 cm)
Cap: Hemispherical to convex at first, becoming more or less flattened with age, with or without a central depression; cinnamon-brown to pinkish brown, fading to ochraceous; surface dry, smooth to scurfy-scaly; **flesh** thin, watery, ochraceous to pale vinaceous, odor and taste not distinctive.
Gills: Adnate to subdecurrent; moderately well-spaced; pale lilac at first, becoming more whitish as the spores mature.
Stalk: More or less equal to slightly bulbous, becoming hollow; colored like the cap with lilac cottony tomentum at the base; surface fibrous, tough, sometimes twisted-striate.
Spore print: White.
Spores: 7–9 x 6–8 μm, broadly elliptical, spiny, non-amyloid.
Occurrence: In scattered groups or clusters on the ground under conifers in woods, woodland margins and parklands; mycorrhizal; summer-fall; occasional.
Edibility: Edible but uninteresting.

Comments: Although it will fade with age, the copious lilac-colored tomentum at the stalk base of fresh specimens is a good field character. *Laccaria laccata* (not illustrated) is similar but does not have lilac-tinged gills, and the mycelium at the stalk base is white. *Laccaria amethystea* (not illustrated) is entirely purplish or brownish purple. The thick, waxy-looking gills of *Laccaria* species may lead to confusion with wax caps (*Hygrophorus* and *Hygrocybe*), but they have smooth rather than spiny spores. This mushroom is toxic to springtail insects (*Collembola* spp.) that feed on it. The toxins kill the insects, which are then digested by the fungus as a source of nitrogen. *Bicolor* means "two colored."

Latin name: *Marasmius oreades* (Bolton) Fr.
Common names: Scotch Bonnet; Fairy Ring Mushroom
Order: Agaricales
Family: Marasmiaceae
Width of cap: ¾–2½ in. (2–6.5 cm)
Length of the stalk: 1–3 in. (2.5–7.5 cm)
Cap: Bell-shaped to convex, sometimes with a scalloped margin, becoming expanded, usually with an umbo, center sunken and the margin uplifted with age; tan to pale cinnamon or buff, fading to pale tan or nearly white; surface smooth or becoming wrinkled with age; **flesh** thin, whitish, reviving when moistened; odor of bitter almond, taste not distinctive.
Gills: Adnexed to nearly free from the stalk, moderate to well-spaced; cream to pale yellowish tan.
Stalk: Slender, equal, tough; whitish to more or less colored like the cap; surface smooth to minutely downy.
Spore print: Creamy white.
Spores: 7–10 x 3.5–6 μm, elliptical, smooth, hyaline, non-amyloid.
Occurrence: Gregarious, often in fairy rings or arcs in grassy places, lawns, pastures, roadsides, and parks; saprobic; spring-fall; common.
Edibility: Edible and good, discard the tough stalks.

Comments: A characteristic feature of this mushroom, as well as with others in the genus *Marasmius,* is the capacity of the fruitbodies to withstand repeated drying cycles yet remain viable when revived by moisture. Although small, the Fairy Ring Mushroom is highly prized as an edible in many parts of the world. The dried caps are a standard in the mushroom markets of Europe. Some have suggested that it should be eaten in moderation due to the presence of hydrocyanic acid, which gives it a distinctive bitter almond odor. Care should be taken not to confuse the Fairy Ring Mushroom with the poisonous *Clitocybe dealbata* (not illustrated), a small white to grayish white mushroom that grows in the same habitat but lacks a distinctive odor, and has close, broadly attached or subdecurrent gills. *Conocybe lactea* (p.191) is another small white grassland species that is poisonous. It has a cone-shaped cap, brown gills, and a slender, fragile stalk.

The name *oreades* comes from the Greek word *oreias,* which means "mountain nymph." The Fairy Ring Mushroom derives its common name from the habit of growing in rings, or incomplete circles (arcs). Before the biology of this phenomenon was understood, the peculiar growth pattern was attributed to the supernatural. It was believed that fairies and elves frequented the places where the rings occurred, and danced within the circles on moonlit nights. This is a quaint notion, but in reality the rings are formed by the radially expanding mycelium that originates from a point of origin and grows outward in all directions unless impeded by physical obstacles, or limited by the exhaustion of available nutrients. The Fairy Ring Mushroom is a saprobe, feeding on dead grasses and forbs. As this resource is depleted, the center of the initial "disc" of mycelium dies, thus creating a ring that continually expands in search of nutrients at the outer edge. Each year a fairy ring becomes a few centimeters larger in diameter than the year before. By measuring its diameter it is possible to determine the approximate age of a fairy ring. Some are calculated to be several hundred years old. Fairy rings are most conspicuous when the underground mycelium sends forth a crop of mushrooms at its edge. However, even when mushrooms are absent certain fairy rings can sometimes be seen in pastures and lawns as a circle of more vigorous, deeper green grass. The darker green zone is a result of localized fertilization brought about as the mycelium releases nutrients in the process of decomposition.

Latin name: *Marasmius pyrrhocephalus* Berk.
Synonym: *Marasmius elongatipes* Peck
Common name: Fire Cap Marasmius
Order: Agaricales
Family: Marasmiaceae
Width of cap: ⅜–1 in. (1–2.5 cm)
Length of the stalk: 1½–3½ in. (4–9 cm)
Cap: Hemispherical to convex, often flat or slightly depressed in the center; orange-brown to reddish brown, paler with age, darker at the center; surface glabrous, margin striate; **flesh** thin, whitish, odor and taste not distinctive.
Gills: Adnate to abruptly adnexed, moderately well-spaced; white at first, pale yellowish with age.
Stalk: Slender, equal, tough, rooting; blackish brown, paler brown to whitish near the apex; surface scurfy with tawny fibrils.
Spore print: White.
Spores: 6–10 x 3–5 µm, elliptical to spindle-shaped, smooth, hyaline, non-amyloid.
Occurrence: Solitary to scattered on humus and leaf litter in broad-leaved and mixed woods; saprobic; spring-fall; common.
Edibility: Unknown. Too small to be of interest.

Comments: In dry conditions this small mushroom shrivels up and goes unnoticed until it revives and expands soon after a rainfall. *Marasmius sullivantii* (not illustrated) is larger, more brightly orange-colored, has close gills and a smooth stalk. *Pyrrhocephalus* means "fire head" in reference to the cap color.

Latin name: *Mycena acicula* (Schaeff.) Fr.
Synonym: *Marasmiellus acicula* (Schaeff.) Singer
Common name: Coral Spring Mycena
Order: Agaricales
Family: Tricholomataceae
Width of cap: ⅛–½ in. (3–12 mm)
Length of the stalk: 1½–2½ in. (4–6.5 cm)
Cap: Bell-shaped to convex, or nearly flat with age; reddish orange fading to yellow-orange with a reddish center; surface with a powdery bloom at first, then smooth, margin striate to grooved; **flesh** thin, whitish to yellowish, odor and taste not distinctive.
Gills: Narrowly adnate to adnexed, moderately close; white to pale yellow.
Stalk: Slender, equal; translucent yellow to pale orange with white hairs at the base; surface smooth to slightly powdery-scurfy.
Spore print: White.
Spores: 8–12 x 2.5–4 μm, cylindric to spindle-shaped, smooth, non-amyloid.
Occurrence: Solitary or in small groups on decaying leaves, small twigs and forest litter, in wet woods, seeps, along streams and river banks; saprobic; most often fruiting in spring, but can be found throughout the growing season; common.
Edibility: Unknown. Too small to be of interest.

Comments: The dainty Coral Spring Mycena first appears in early May. Although the caps are brightly colored they are easy to miss because of their small size. *Rickenella fibula* (not illustrated), which typically grows among mosses, is similar but has an orange to brownish orange cap with a depressed center, and decurrent gills. *Acicula* means "sharply pointed" or "small needle." The relevance of the epithet as it applies to this mushroom is a bit obscure.

Latin name: *Mycena pura* (Pers.) Sacc.
Common name: Purple Radish Mycena
Order: Agaricales
Family: Tricholomataceae
Width of cap: ¾–1½ in. (2–4 cm)
Length of the stalk: 1¼–3 in. (3–7.5 cm)
Cap: Bell-shaped to convex, becoming flattened, with or without an obtuse umbo, margin often upturned with age; color variable from pale lilac to rose-pink, pinkish gray, purple, or purplish brown; surface hygrophanous, often concentrically zoned, smooth to slightly wrinkled, margin distinctly striate; **flesh** grayish to lilac, odor and taste of radishes.
Gills: Adnate to adnexed, sometimes with a decurrent tooth, at times separating from the stalk with age; whitish to grayish or pale lilac, edges whitish; close to moderately well-spaced with crossveining between.
Stalk: Cylindrical to compressed, equal or tapering upward, hollow; whitish to more or less colored like the cap, with white downy mycelium at the base; surface smooth or with scattered downy flecks, sometimes with straight or twisted vertical lines.
Spore print: White.
Spores: 6–10 x 3–3.5 μm, elliptical, smooth, hyaline, amyloid.
Occurrence: Solitary or in loosely scattered groups, sometimes in clusters on the ground in broad-leaved and conifer woods; saprobic; early summer-fall; common.
Edibility: Poisonous.

Comments: This common, widely distributed mushroom is fairly easy to recognize by its overall purplish to lilac tones and distinct radish-like odor. *Mycena pelianthina* (not illustrated) is similar and also has a radish-like odor, but is duller purplish gray to sordid yellowish overall, and has dark gill edges. *Pura* means "pure" or "unblemished."

Latin name: *Mycena sanguinolenta* (Alb. & Schwein.) Quél.
Common name: Smaller Bleeding Mycena
Order: Agaricales
Family: Tricholomataceae
Width of cap: ¼–⅝ in. (0.5–1.5 cm)
Length of the stalk: 1½–3 in. (4–7.5 cm)
Cap: Conic to bell-shaped, often with an umbo; reddish brown to reddish purple, paler toward the margin; surface smooth, dull, or slightly velvety on the disc, margin prominently grooved; **flesh** thin, more or less colored like the cap, "bleeding" when injured, odor and taste not distinctive.
Gills: Adnate to sinuate, well-spaced; pallid to pinkish, with a dark reddish purple edge.
Stalk: Slender, equal, hollow, fragile; colored like the cap or paler; surface smooth; exuding a dark reddish purple juice when broken.
Spore print: White.
Spores: 8–11 x 4–6 μm, elliptical, smooth, amyloid.
Occurrence: Solitary or in small groups on the ground in conifer and mixed woods, especially low wet woods, less frequently on moss-covered stumps; saprobic; early summer-fall; fairly common.
Edibility: Unknown. Too small to be of interest.

Comments: When fresh the broken cap and stalk of this delicate mushroom will exude a dark reddish purple sap. The similar *Mycena haematopus* (not illustrated) is somewhat larger, grows on decaying wood, usually in caespitose clusters, and lacks dark gill edges. *Sanguinolenta* means "bleeding slowly."

Latin name: *Xerula megalospora* (Clem.) Redhead, Ginns, & Shoemaker
Synonym: *Collybia radicata* var. *pusilla* Peck
Common name: Lesser Rooting Xerula
Order: Agaricales
Family: Marasmiaceae
Width of cap: ¾–3 in. (2–7.5 cm)
Length of the stalk: Above ground portion 3–5 in. (8–12.5 cm), plus up to 5 in. (12.5 cm) additional length below ground.
Cap: Broadly conical to convex, becoming flattened to depressed in the center with an upturned margin, umbonate; smoky brown to brownish gray, or buff to cream, darker in the center; surface viscid when moist, usually wrinkled, margin striate; **flesh** white, odor and taste not distinctive, or sometimes interpreted as reminiscent of geraniums or carrots.
Gills: Adnate, sometimes with a decurrent tooth, widely spaced; white.
Stalk: Slender, swollen at ground level; colored like the cap or paler, white near the apex; surface smooth to striate or twisted-striate, dry.
Spore print: White.
Spores: 18–23 x 11–14 μm, broadly elliptical to lemon-shaped with a prominent apiculus, surface finely roughened, non-amyloid.
Occurrence: Solitary or in small groups on the ground near well-decayed stumps of broad-leaved trees, in woods, grassy areas, and parkland; saprobic; spring-fall; common.
Edibility: Edible.

Comments: This mushroom usually grows on the ground but is always associated with decaying wood, often occurring near decaying stumps, or attached to buried roots. The long rooting stalk, which is easily broken off when attempting to extract it, is a distinctive feature. *Xerula furfuracea* (p.265) is larger, has a darker brown cap, a scurfy brown stalk, and smaller spores. *Megalospora* means "having large spores."

Small Mushrooms That Grow on Soil or Other Non-woody Substrates, Having Pink, Brown, or Blackish Spores, and Gills That Are Not Decurrent

The species in this section have central, ringless stalks that are usually no thicker at the apex than a standard pencil (¼ inch = 6 mm). The gills are nearly free or variously attached to the stalk, but not decurrent.

In order to reduce the number of possibilities when searching for a correct match, Section 1-H is divided into two parts based on the color of the spore deposit. Similar small mushrooms with light-colored spores are placed in Section 1-H(a). If the stalk thickness is marginally in the ¼-inch range, review also the species in Section 1-K(b), which are similar but generally larger mushrooms that also produce dark-colored spores.

Latin name: *Agrocybe semiorbicularis* (Bull.) Fayod
Common name: Hemispheric Agrocybe
Order: Agaricales
Family: Bolbitiaceae
Width of cap: ½–1¼ in. (1.3–3 cm)
Length of the stalk: ¾–2 in. (2–5 cm)
Cap: Hemispheric to convex; straw-yellow to tawny, or orange-brown, darker at the center; surface smooth, tacky when moist, sometimes cracking; **flesh** thin, whitish to pale yellow, odor farinaceous, taste mealy, sometimes bitter.
Gills: Adnate, at times separating from the stalk with age, moderately close; cream at first, becoming grayish brown, gill edge whitish.
Stalk: Equal to slightly enlarged at the base, becoming hollow; colored like the cap, often with white rhizomorphs at the base; surface fibrillose.
Spore print: Dark brown.
Spores: 11–15 x 7–9.5 μm, elliptical with an apical germ pore, smooth.
Occurrence: Gregarious, sometimes in caespitose clusters on the ground in lawns, grasslands, pastures, roadsides, manured ground, gardens, and cultivated areas; saprobic; spring-fall; common.
Edibility: Not recommended. Some closely related species are reported to be edible, but confusion with similar species of unknown edibility is possible or even likely.

Comments: *Stropharia semiglobata* (not illustrated) is somewhat similar but typically larger overall. It grows on dung and dung-enriched soil, has a thin ring zone on the stalk, and purplish black spores. *Psilocybe coprophila* (p.204) also grows on dung and dung-enriched substrates. It has a dark brown, viscid cap and purple brown spores. *Marasmius oreades* (p.182) has a bitter almond odor and white spores. Some authors place *Agrocybe pediades* (not illustrated) in synonymy with *Agrocybe semiorbicularis.* The differences between these two taxa are slight and may simply reflect the range within one variable species. *Semiorbicularis* means "half circular" in reference to the cap shape.

Latin name: *Conocybe lactea* (J.E. Lange) Métrod
Common name: White Dunce Cap
Order: Agaricales
Family: Bolbitiaceae
Width of cap: ½–1¼ in. (1.3–3 cm)
Length of the stalk: 1¼–3 in. (3–7.5 cm)
Cap: Conic to bell-shaped, margin sometimes flaring with age; creamy white with a cream-yellow to ochraceous center; **flesh** thin, fragile, whitish, odor and taste not distinctive.
Gills: Barely attached, close; pale cinnamon, becoming darker as the spores mature.
Stalk: Slender, equal with a small bulb at the base, hollow, fragile; whitish; surface smooth to slightly scurfy.
Spore print: Cinnamon-brown.
Spores: 13–16 x 7–9 μm, elliptical with an apical germ pore, smooth, pale brown.
Occurrence: Usually gregarious but loosely scattered on the ground in lawns, open grasslands, and parks; saprobic; spring-fall; common.
Edibility: Not edible.

Comments: This delicate "urban" mushroom is common in grassy places after rain, or when there has been a heavy dew the night before. The fruitbodies are delicate and short-lived, sometimes lasting only a few hours before wilting into obscurity. *Lactea* means "milk white." The genus name *Conocybe* means "cone head."

Latin name: *Cortinarius bolaris* (Pers.) Fr.
Common name: Saffron-foot Cortinarius
Order: Agaricales
Family: Cortinariaceae
Width of cap: 1–2 in. (2.5–5 cm)
Length of the stalk: 1½–2½ in. (4–6.5 cm)
Cap: Hemispheric to convex with an incurved margin at first, with or without a low umbo, becoming flattened or slightly depressed with age; surface dry with cinnabar-red to brownish red scales over a whitish to ochraceous ground color; margin at first with web-like remnants of a partial veil that soon disappear; **flesh** whitish to ochraceous, odor and taste not distinctive.
Gills: Adnate, close; pale ochraceous when young, becoming cinnamon-brown; covered at first with a whitish web-like partial veil (cortina).
Stalk: Nearly equal to tapering upward from a swollen base; colored like the cap; surface fibrillose-scaly with the scales often arranged in irregular bands, partial veil leaving an indistinct fugacious, cottony, web-like ring zone on the upper stalk; flesh at the base bruising saffron yellow-orange.
Spore print: Rusty brown.
Spores: 6–7 x 5–6 μm, oval to subglobose, surface slightly roughened, pale brown, non-amyloid.
Occurrence: In small groups on the ground or well-decayed wood in broad-leaved and mixed woods, especially with oak and beech; mycorrhizal; summer-fall; occasional.
Edibility: Poisonous.

Comments: The stalk base will instantly stain deep, rich yellow-orange when cut or scratched. *Cortinarius armillatus* (p.270) is a much more robust mushroom that does not stain yellow-orange at the stalk base. *Bolaris* means "a small clod."

Latin name: *Cortinarius iodes* Berk. & M.A. Curtis
Common names: Spotted Cortinarius; Viscid Violet Cortinarius
Order: Agaricales
Family: Cortinariaceae
Width of cap: 1–2 in. (2.5–5 cm)
Length of the stalk: 1½–3 in. (4–7.5 cm)
Cap: Bell-shaped to convex, becoming nearly flat with age; dark violet with scattered yellowish to cream colored spots, or yellowish to cream on the disc, fading with age; surface viscid when wet, sticky; **flesh** thin, pale violet to whitish, odor and taste not distinctive.
Gills: Adnate, close to crowded; violet when young, becoming grayish cinnamon as the spores mature.
Stalk: Equal or tapering upward from a swollen base, solid; colored like the cap, but usually somewhat paler or sometimes almost white; surface viscid, smooth, partial veil (cortina) sometimes leaving an evanescent rusty brown web-like ring zone near the apex.
Spore print: Rusty brown.
Spores: 8–12 x 5–6.5 μm, elliptical, minutely roughened, pale brown, non-amyloid.
Occurrence: Solitary or in small groups on the ground in broad-leaved and mixed woods, especially with oak; mycorrhizal; summer-fall; common.
Edibility: Not recommended. Sometimes reported to be edible, but related species contain toxins.

Comments: When in fresh condition before the colors fade, the deep purple, viscid cap with creamy spots makes this mushroom easy to recognize. *Cortinarius iodeoides* (not illustrated) is nearly identical except that it has a bitter tasting cap pellicle. The poisonous *Inocybe lilacina* (not illustrated) is similarly colored but has a dry, silky cap with a distinct central knob. *Iodes* means “violet-like.”

Latin name: *Cortinarius semisanguineus* Gillet
Common name: Red-gilled Cortinarius
Order: Agaricales
Family: Cortinariaceae
Width of cap: 1–2 in. (2.5–5 cm)
Length of the stalk: 1–2½ in. (2.5–6.5 cm)
Cap: Bell-shaped to convex or nearly flat, usually with an umbo; ochraceous yellow to cinnamon-brown, paler toward the margin; surface smooth, fibrillose-silky, rarely breaking up into fine scales; **flesh** yellowish white to ochraceous, odor not distinctive, taste mild or bitter.
Gills: Adnate to notched, crowded; rusty red to blood red.
Stalk: More or less equal; pale yellow to tawny; surface fibrillose, partial veil (cortina) sometimes leaving an evanescent, web-like ring zone on the upper stalk.
Spore print: Rusty brown.
Spores: 6–8 x 3–4.5 μm, elliptical, surface minutely roughened, pale brown, non-amyloid.
Occurrence: Solitary or in scattered groups or caespitose clusters on soil or mossy ground in conifer and mixed woods; mycorrhizal; summer-fall; common.
Edibility: Not edible. Possibly poisonous.

Comments: This small- to medium-sized mushroom is prized as a source of pigment for dying natural fibers. *Cortinarius cinnamomeus* (not illustrated) is similar but has yellowish cinnamon gills. *Cortinarius sanguineus* (not illustrated) is entirely blood red. Some authors place all three of these pigment-rich *Cortinarius* species into the segregate genus *Dermocybe. Semisanguineus* means "half blood red," referring to the red gills.

Latin name: *Entoloma luteum* Peck
Common name: None
Order: Agaricales
Family: Entolomataceae
Width of cap: ½–1 in. (1.3–2.5 cm)
Length of the stalk: 2–3½ in. (5–9 cm)
Cap: Acutely or broadly conical; olivaceous yellow-brown to ochre-yellow; surface dry, smooth to slightly roughened in the center; **flesh** thin, odor and taste not distinctive.
Gills: Ascending adnate, narrowly attached to the stalk, close; yellowish at first, becoming dull pinkish as the spores mature.
Stalk: Slender, more or less equal, fragile; colored like the cap or paler, with white cottony mycelium at the base; surface fibrous, often twisted-striate.
Spore print: Pink.
Spores: 9–13 x 8–12 µm; angular, hyaline, non-amyloid.
Occurrence: Solitary or in small clusters on the ground in mixed woods; saprobic; summer-fall; uncommon.
Edibility: Unknown, possibly poisonous.

Comments: This graceful mushroom is not difficult to recognize once the gills turn pinkish. *Entoloma murrayi* (p.196) is similar but has a yellow to orange-yellow cap with a pointed umbo. Numerous brownish *Mycena* species are similar in size and stature but they have white spores. *Luteum* means "yellowish."

Latin name: *Entoloma murrayi* (Berk. & M.A. Curtis) Sacc.
Synonym: *Nolanea murrayi* (Berk. & M.A. Curtis) Dennis
Common name: Yellow Unicorn Entoloma
Order: Agaricales
Family: Entolomataceae
Width of cap: ½–1¼ in. (1.3–3 cm)
Length of the stalk: 1½–3 in. (4–7.5 cm)
Cap: Bell-shaped to sharply conical with a pointed central umbo; bright yellow to orange-yellow, fading with age; surface glabrous, smooth, silky; **flesh** thin, pale yellow, odor and taste not distinctive.
Gills: Ascending adnate, narrowly attached to the stalk, moderately well-spaced; yellow, becoming yellowish pink as the spores mature; gill edges uneven.
Stalk: Slender, equal, hollow; pale yellow, surface fibrous, often twisted-striate.
Spore print: Salmon pink.
Spores: 9–12 x 8–10 μm, angular, smooth, hyaline, non-amyloid.
Occurrence: Usually in groups on the ground or in moss, in conifer and broad-leaved woods; saprobic; summer-fall; common.
Edibility: Not edible, possibly poisonous.

Comments: *Entoloma salmoneum* (not illustrated) is very similar but salmon-orange overall. Both species are graceful and attractive, particularly when growing in moss. Some of the wax caps, such as *Hygrocybe marginata* var. *concolor* (not illustrated), are similar but have waxy gills and white spores. *Entoloma murrayi* was named after its collector Dennis Murray of Massachusetts.

Latin name: *Inocybe calamistrata* (Fr.) Gillet
Common name: Green-foot Inocybe
Order: Agaricales
Family: Cortinariaceae
Width of cap: ½–1½ in. (1.3–4 cm)
Length of the stalk: 1½–3½ in. (4–9 cm)
Cap: Bell-shaped to convex; dark brown to reddish brown; surface dry, densely covered with recurved, pointed scales; **flesh** white to reddish tinged; odor spermatic.
Gills: Adnate to adnexed, sometimes separating from the stalk with age, close to crowded; colored like the cap, gill margin whitish.
Stalk: Slender, equal to slightly tapering upward, solid, firm; colored like the cap, with a bluish green base; surface scaly.
Spore print: Dull brown.
Spores: 9.5–12 x 4.5–6.5 μm, elliptical, smooth, brownish, non-amyloid.
Occurrence: Solitary or in small groups on the ground in broad-leaved and conifer woods, especially in low, wet places, and along streams; mycorrhizal; summer-fall; occasional.
Edibility: Poisonous.

Comments: Other scaly Inocybes lack the bluish green stalk base. *Inocybe tahquamenonensis* (not illustrated) grows on the ground in broad-leaved woods, is more squat in stature, has a dark purplish brown to nearly blackish scaly cap and stalk, and has moderately well-spaced gills. *Pouzarella nodospora* (not illustrated) has a dark brown scaly cap, a slender grayish brown hairy stalk with stiff brown hairs at the base, and pinkish angular spores. *Calamistrata* means "a curling iron." The name probably relates to the curled-back scales on the cap and stalk.

Latin name: *Inocybe sindonia* (Fr.) P. Karst.
Common name: None
Order: Agaricales
Family: Cortinariaceae
Width of cap: 1–2 in. (2.5–5 cm)
Length of the stalk: 1¼–2½ in. (3–6.5 cm)
Cap: Broadly conical to bell-shaped, expanding to broadly bell-shaped with a distinct umbo, margin somewhat hairy at first with veil remnants that soon disappear; nearly white or pale cream to yellowish straw; surface smooth, silky, somewhat radially streaked, especially near the margin; **flesh** white, thick at the center, odor more or less spermatic.
Gills: Adnexed to notched and narrowly attached, close; pale grayish white, becoming pale grayish brown with whitish edges; covered at first with a whitish web-like veil.
Stalk: More or less equal or slightly enlarged at the base; colored like the cap or paler; surface smooth to fibrillose-silky or lightly striate, often pruinose on the upper stalk.
Spore print: Pale brown.
Spores: 8–11 x 4.5–5 μm, broadly elliptical, smooth, brownish.
Occurrence: In small groups on the ground in conifer woods and woodland margins, also reported to occur in broad-leaved woods; mycorrhizal; late summer-fall; occasional.
Edibility: Poisonous.

Comments: The similar *Inocybe geophylla* (not illustrated) grows in woods or grassy places and has a dull white to grayish white acutely umbonate cap. *Psathyrella candolleana* (p.201) is somewhat similar but has a thin, fragile, pale yellowish cap, crowded gills that are grayish brown at maturity, and purplish brown spores.

Latin name: *Leptonia incana* (Fr.) Gillet
Synonym: *Entoloma incanum* (Fr.) Hesler
Common name: Green Leptonia
Order: Agaricales
Family: Entolomataceae
Width of cap: ½–1½ in. (1.3–4 cm)
Length of the stalk: 1–2 in. (2.5–5 cm)
Cap: Convex to broadly convex with a central depression; bluish green to bright olive-yellow, or with age becoming yellowish brown, usually darker near the center; surface smooth or slightly scaly at the center, margin more or less striate; **flesh** thin, fragile, pale greenish, bruising bluish green, with a rancid or "mousy" odor.
Gills: Adnate to subdecurrent, moderately well-spaced; whitish to pale greenish, becoming pinkish as the spores mature.
Stalk: More or less equal, cylindrical or compressed, hollow; greenish yellow often tinged with green, white mycelium at the base, bruising brilliant bluish green; surface smooth.
Spore print: Salmon-pink.
Spores: 11–14 x 8–9 μm, elliptical, angular, smooth, hyaline, non-amyloid.
Occurrence: Usually in groups or clusters on disturbed soil, roadsides, grassy places, along unpaved forest roads and woodland trails; saprobic; late spring-fall; locally common.
Edibility: Unknown, possibly poisonous.

Comments: This small greenish mushroom is easy to overlook when growing among herbaceous green plants. However, the nearly flourescent bluish green staining of the bruised tissue, especially on the stalk, and the pronounced odor of mouse dwellings are distinctive features. The specific epithet literally means "hoary gray" (like a mouse), but better applies here to the odor of the fruitbodies.

Latin name: *Panaeolus foenisecii* (Pers.) J. Schröt.
Synonyms: *Psathyrella foenisecii* (Pers.) A.H. Sm.; *Panaeolina foenisecii* (Pers.) Maire
Common names: Haymaker's Mushroom; Lawn Mower's Mushroom
Order: Agaricales
Family: Bolbitiaceae
Width of cap: ½–1 in. (1.3–2.5 cm)
Length of the stalk: 1½–3 in. (4–7.5 cm)
Cap: Hemispheric to bell-shaped or convex; dull smoky brown to reddish brown, hygrophanous, progressively fading as moisture is lost, often producing concentric zones of tan or grayish brown; surface smooth or sometimes cracked, margin more or less striate when moist; **flesh** thin, fragile, watery brown, odor and taste not distinctive.
Gills: Adnate, often separating from the stalk with age, close to moderately well-spaced; pallid at first but soon becoming dark purple-brown mottled as the spores mature, then finally blackish, gill edges whitish.
Stalk: Slender, equal, hollow, fragile; watery brown with a whitish bloom; surface smooth, satiny, at times with light vertical lines.
Spore print: Dark purple-brown to blackish.
Spores: 12–17 x 7–9 μm, oval to elliptical with an apical germ pore, surface warted, purplish brown, non-amyloid.
Occurrence: Usually gregarious in lawns, parks, ball fields, and other short grass areas, especially soon after mowing; saprobic; spring-fall; common.
Edibility: Possibly poisonous. Reports vary, but some collections may contain low levels of hallucinogenic compounds.

Comments: This widely distributed mushroom is common and sometimes abundant (especially in spring and early summer) in grassy places after rain, or in the morning following a heavy dew. The fruitbodies have little substance and will often wilt and disappear in less than a day. *Panaeolus papilionaceus* (not illustrated) is somewhat similar but has a toothed cap margin (hanging veil remnants), and grows on dung, or on dung-enriched ground. *Foenisecii* is derived from the Latin word meaning "mower" or "harvester."

Latin name: *Psathyrella candolleana* (Fr.) Maire
Common name: Common Psathyrella
Order: Agaricales
Family: Coprinaceae
Width of cap: 1–3 in. (2.5–7.5 cm)
Length of the stalk: 1½–3½ in. (4–9 cm)
Cap: Broadly conical to bell-shaped at first, becoming convex or nearly flat, with or without a low umbo; color variable from ivory to yellow-ochre or tan, at times with pale brownish lilac tones toward the margin, fading with age but usually retaining original color on the disc; surface smooth to somewhat radially wrinkled; margin at first fringed with hanging fragments of a thin partial veil that disappear with age, faintly striate when moist; **flesh** thin, yellowish to white, fragile, odor and taste not distinctive.
Gills: Adnate to adnexed, often separating from the stalk with age, close to crowded; whitish at first, then grayish purple to grayish brown.
Stalk: Slender, more or less equal, hollow, fragile; white; surface smooth, shiny, or lightly pruinose near the apex, partial veil rarely leaving a delicate, ephemeral ring on the upper stalk.
Spore print: Purple-brown.
Spores: 7–10 x 4–5 μm, elliptical with an apical germ pore, smooth, pale brown.
Occurrence: Usually in groups on or around well-decayed stumps of broad-leaved trees, in wood mulch beds, on the ground in grassy places, woods, woodland margins, and roadsides; saprobic; spring-early fall; common.
Edibility: Edible.

Continued on next page

Comments: The Common Psathyrella is probably a "species complex," comprised of several closely related species or varieties, all of which approximately fit the description given here. Some authors regard *Psathyrella incerta* (not illustrated) to be a distinct species because it has slightly smaller spores (6–7.5 x 3.5–4 μm). Some members of the genus *Bolbitius* have a similar appearance, but they have free gills, rusty brown or ochre-brown spores, and mostly grow on dung, or manure-enriched substrates. Although the flesh of the Common Psathyrella is thin and fragile, it has good flavor. The specific epithet honors the nineteenth-century French botanist Augustin Pyramus de Candolle (1778–1841), who in 1813 laid the foundation for the rules of botanical nomenclature.

Latin name: *Psathyrella velutina* (Pers.) Singer
Synonyms: *Lacrymaria velutina* (Pers.) Konrad & Maubl.; *Lacrymaria lacrymabunda* (Bull.) Pat.
Common names: Weeping Widow; Velvety Psathyrella
Order: Agaricales
Family: Coprinaceae
Width of cap: 1¼–3 in. (3–7.5 cm)
Length of the stalk: 1½–3½ in. (4–9 cm)
Cap: Broadly bell-shaped at first, becoming convex to flattened with age, often with a low umbo, margin usually fringed with veil remnants; ochre-brown to tawny, or dark reddish brown; surface radially fibrillose-hairy to finely scaly; **flesh** thin, dingy yellowish to watery brown, odor and taste not distinctive.
Gills: Adnate to notched, close to crowded; yellowish brown at first, soon mottled dark brown to nearly black with whitish gill edges; covered at first with a white to buff-colored web-like veil; gill edges sometimes beaded with moisture droplets that become blackened from the spores, and remain as black dots when dried.
Stalk: Slender, equal or slightly enlarged toward the base, hollow; more or less colored like the cap, especially near the base, whitish at the apex; surface fibrillose to somewhat scaly, partial veil leaving a fugacious zone of spore-blackened fibrils on the upper stalk.
Spore print: Blackish brown.
Spores: 9–12 x 6–7 µm, elliptical with an apical projection and germ pore, minutely roughened, pale brown.
Occurrence: Usually gregarious, often in clusters on the ground in open grassy woods, along trails, woodland margins, parks, lawns, roadsides, and disturbed ground; saprobic; spring-fall; common.
Edibility: Edible.

Comments: The hairy-fibrillose cap of this mushroom is reminiscent of species of *Inocybe,* which have whitish to light brown gills and pale brown spores. *Velutina* means "velvety."

Latin name: *Psilocybe coprophila* (Bull.) P. Kumm.
Common name: Dung-loving Psilocybe
Order: Agaricales
Family: Strophariaceae
Width of cap: ⅜–1 in. (1–2.5 cm)
Length of the stalk: ¾–2 in. (2–5 cm)
Cap: Hemispheric to convex or broadly convex or nearly flat with age, sometimes with an obscure low umbo; dark orange-brown to reddish brown, hygrophanous, fading to ochre-brown or tan; surface smooth, sometimes with white cottony patches at the margin, viscid when moist, shiny, margin translucent-striate; **flesh** thin, colored like the cap, odor not distinctive.
Gills: Adnate to subdecurrent, well-spaced; grayish brown becoming purple-brown to nearly black, gill edges whitish.
Stalk: Nearly equal; reddish brown, paler toward the apex; surface at first covered with scurfy whitish fibrils, becoming smoother with age.
Spore print: Purple-brown.
Spores: 11–15 x 6.5–9 μm, elliptical with an apical germ pore, smooth, pale brown.
Occurrence: In groups or clusters on horse and cow dung, or heavily manured soil and other dung-enriched substrates; saprobic; spring-fall; occasional.
Edibility: Not edible. Some collections may contain low levels of hallucinogenic compounds.

Comments: *Psilocybe merdaria* (not illustrated) is nearly identical except that it has a slight ring or ring zone on the stalk. *Stropharia semiglobata* (not illustrated) has a sticky yellowish cap on a long, slender stalk, and a blackish ring zone on the upper portion of the stalk. It sometimes occurs together with *Psilocybe coprophila* on the same "deposit." *Coprophila* means "dung-loving."

Medium to Large Mushrooms Growing on Wood or Associated with Decaying Wood, Gills Not Decurrent, Stalk Ring Lacking

The mushrooms in this section are wood-rotting saprobes that often grow in groups or clusters. They are commonly found on and around decaying logs and stumps, but also on wood chips and wood mulch.

Latin name: *Coprinus variegatus* Peck
Synonym: *Coprinus quadrifidus* Peck
Common name: Scaly Ink Cap
Order: Agaricales
Family: Coprinaceae
Width of cap: 1–3 in. (2.5–7.5 cm)
Length of the stalk: 1–4 in. (2.5–10 cm)
Cap: Oval at first, becoming bell-shaped, then expanding at the base with the margin recurving as the spores mature; yellowish gray to grayish brown, or charcoal gray near the margin; surface at first covered with white to yellowish brown cottony patches that easily wash off; **flesh** thin, odor not distinctive, taste more or less bitter.
Gills: Free from the stalk, crowded to densely packed; white at first, becoming purplish, then black, dissolving into black fluid at maturity.
Stalk: More or less equal, hollow; white; surface somewhat scurfy, often with a cottony rim near the base, and brownish rhizomorphs at the base.
Spore print: Black.
Spores: 7.5–10 x 4–5 μm, elliptical with an apical germ pore, smooth, grayish.
Occurrence: In large clusters on or near well-decayed stumps and logs of broad-leaved trees, especially elm and ash, or sometimes in lawns and grassy places from buried wood; saprobic; spring-summer; common.
Edibility: Not recommended (see comments below).

Comments: This mushroom first appears in early June. It often grows in large clusters of several dozen individuals that might tempt the mushroom hunter in search of edibles. Although it is safely eaten by some, there are reports of gastrointestinal upsets when consuming alcohol at the same meal, or even several hours before or afterward. As with many species of *Coprinus,* the cap and gills of the Scaly Ink Cap dissolve into a black, inky fluid as the spores mature. *Coprinus atramentarius* (not illustrated), which also grows in clusters on decaying wood or on the ground from buried wood, is similar but has a smooth or slightly scaly gray cap. Compare with *Coprinus comatus* (p.268), which is typically larger than the Scaly Ink Cap and grows on the ground in grassy places and along roadsides. Compare also with *Coprinus micaceus* (p.156). *Variegatus* means "irregularly colored, variegated."

Latin name: *Hypholoma sublateritium* (Cooke) Sacc.
Synonym: *Naematoloma sublateritium* (Fr.) P. Karst.
Common name: Brick Cap
Order: Agaricales
Family: Strophariaceae
Width of cap: 1½–3½ in. (4–9 cm)
Length of the stalk: 1½–3½ in. (4–9 cm)
Cap: Convex to broadly convex, becoming nearly flat, margin incurved at first; brick red, becoming pale pinkish buff to pale yellow toward the margin; surface smooth in the center, with scattered silky fibrils near the margin that may wash or wear off with age; **flesh** thick, cream to pale yellowish brown, odor not distinctive, taste mild to bitter.
Gills: Adnate to notched, close; whitish to yellowish, becoming purplish gray, sometimes tinted olivaceous, finally becoming purple-brown as the spores ripen, at first covered with a thin, white, web-like veil.
Stalk: Equal, becoming hollow with age; whitish above, reddish brown toward the base; surface fibrillose-silky, with a thin, hairy ring zone on the upper stalk that becomes conspicuous after accumulating a deposit of purple-brown spores.
Spore print: Purple-brown.
Spores: 6–7.5 x 4–4.5 μm, elliptical with an apical germ pore, smooth, pale brown.
Occurrence: In groups or clusters on decaying stumps, logs, or standing dead trunks of broad-leaved trees; saprobic; fall-early winter; common.
Edibility: Edible but sometimes bitter.

Comments: This late-season mushroom is sometimes found in large clusters or rows along the length of fallen tree trunks. Although edible, the flesh is often insect damaged. Compare with the poisonous *Hypholoma fasciculare* (p.157), which has a yellow cap and greenish yellow gills. *Hypholoma capnoides* (not illustrated) has a yellow to brownish orange cap and grows on decaying conifer wood. *Sublateritium* means "nearly brick-colored."

Latin name: *Megacollybia platyphylla* (Pers.) Kotlaba & Pouzar
Synonyms: *Tricholomopsis platyphylla* (Pers.) Singer; *Collybia platyphylla* Fr.
Common name: Broad Gill
Order: Agaricales
Family: Tricholomataceae
Width of cap: 2–6 in. (5–15 cm)
Length of the stalk: 1½–6 in. (4–15 cm)
Cap: Convex to broadly convex, becoming flattened or shallowly depressed in the center, with or without a low umbo, margin incurved when young; dark ash-gray to smoky brown or gray-brown; surface smooth, radially fibrous-streaked, splitting in dry weather; **flesh** thin, watery white, odor and taste not distinctive.
Gills: Adnexed to sinuate, moderately well-spaced, broad, gill edges uneven to ragged with age; white to pale grayish.
Stalk: More or less equal or tapering upward, tough, fibrous; white, or sometimes with grayish brown fibrils; typically having thick white strands (rhizomorphs) at the base; surface smooth to striate.
Spore print: White.
Spores: 7–10 x 5–7 μm, broadly elliptical to oval, smooth, hyaline, non-amyloid.
Occurrence: Solitary or in small groups on decaying wood or on the ground from buried wood, especially in the vicinity of well-decayed logs and stumps of broad-leaved trees; saprobic; spring-fall; common.
Edibility: Edible with caution when fresh and well-cooked (see comments below).

Continued on next page

Comments: The Broad Gill is one of the first large gilled mushrooms to appear in the spring and can be quite common at that time. Later in the season it occurs less frequently. *Pluteus atricapillus* (p.211) is also an early season large brown mushroom that grows in association with decaying wood, but it has free gills that turn from white to pinkish as its spores mature. *Tricholoma portentosum* (p.259) has a similar dark gray fibrillose cap, but it grows on the ground under pines in the fall. *Tricholoma virgatum* (not illustrated) has a dark gray, broadly conical cap that becomes nearly flat with a pointed umbo with age, and has bitter to acrid tasting flesh. *Entoloma* species have attached gills that turn pinkish as the spores mature, and with few exceptions they do not grow on woody substrates. Although the Broad Gill has been reported to cause gastrointestinal type poisonings, it is eaten without ill effects by many persons. The flavor is a bit like an Oyster Mushroom, *Pleurotus ostreatus* (p.25). *Platyphylla* means "broad leaf," referring to the broad gills.

Latin name: *Pluteus atricapillus* (Batsch) Fr.
Synonym: *Pluteus cervinus* (Schaeff.) P. Kumm.
Common names: Deer Mushroom; Fawn Pluteus
Order: Agaricales
Family: Pluteaceae
Width of cap: 1½–4½ in. (4–11.5 cm)
Length of the stalk: 2–4½ in. (5–11.5 cm)
Cap: Bell-shaped to convex at first, becoming flattened with a broad umbo; color variable from dark umber-brown when young to grayish brown or dingy tan, darker in the center; surface smooth, satiny, somewhat lubricous, with or without radial streaks, sometimes minutely scaly on the disc; **flesh** white, soft, odor and taste of radishes.
Gills: Free from the stalk, close to crowded; white at first, becoming salmon-pink; soft.
Stalk: Equal or tapering upward, solid; white to grayish brown; surface smooth to somewhat scurfy, often streaked; fibrous and solid, but somewhat fragile.
Spore print: Salmon-pink.
Spores: 5.5–7 x 4–5 μm, broadly elliptical, smooth, hyaline, non-amyloid.
Occurrence: Solitary or more often in small groups or clusters on or around decaying stumps and logs of broad-leaved trees, also on wood chips, and sawdust piles; saprobic; spring-fall; common.
Edibility: Edible, but mediocre.

Continued on next page

Comments: This common wood-rotting mushroom is recognized by its brown cap, radish-like odor, and dull pinkish gills that are remotely free from the stalk. It has a long fruiting period but first appears fairly early in the spring. *Pluteus atromarginatus* (not illustrated) is nearly identical but has dark brown gill edges and grows on conifer wood. *Pluteus petasatus* (not illustrated) is also similar but has a white cap with minute brownish scales in the center. *Volvariella bombycina* (not illustrated) has a white silky-scaly cap and a large membranous volva at the stalk base. *Entoloma* species typically grow on the ground and have gills that are attached to the stalk. The name *atricapillus* is a combination of two Latin words meaning "dark" and "hair," referring to the minute scales—or radiating fibrils—that are often but not always present on the cap. The common name, Deer Mushroom, alludes to the brown cap that resembles the color of deer hair.

Latin name: *Psathyrella delineata* (Peck) A.H. Sm.
Common name: None
Order: Agaricales
Family: Coprinaceae
Width of cap: 1¼–4 in. (3–10 cm)
Length of the stalk: 2–4 in. (5–10 cm)
Cap: Broadly conical to bell-shaped, becoming convex to nearly flat with age, usually with a broad umbo; dark reddish brown to rusty brown, fading to orange-brown or tan; surface radially wrinkled, often strongly so, at first covered with a thin layer of silky white fibrils that soon disappear; margin for a time retaining a thin remnant flap of a partial veil; **flesh** thick, fragile, watery brown, odor and taste not distinctive.
Gills: Adnate, close to crowded; grayish brown becoming darker brown, usually with a whitish gill edge; covered at first with a thin, cottony partial veil.
Stalk: More or less equal or tapering upward from an enlarged base, hollow, fragile; whitish to pale brown; surface fibrillose, scurfy.
Spore print: Purple-brown.
Spores: 6.5–9 x 4.5–5.5 μm, oval to elliptical with an apical germ pore, smooth, pale brown.
Occurrence: Solitary or in small groups on well-decayed logs, stumps, and larger fallen limbs of broad-leaved trees; saprobic; spring-fall; common.
Edibility: Presumed to be edible (see comments below).

Comments: In some years this mushroom is quite common in the spring, often first appearing in early June. The edible *Psathyrella rugocephala* (not illustrated) has larger spores (9–11 x 6–8 μm) but is otherwise nearly identical. Since the differences are microscopic, both species have undoubtably been collected and eaten indiscriminately. *Delineata* means "with visible lines," referring to the radially wrinkled cap.

Latin name: *Tricholomopsis rutilans* (Schaeff.) Singer
Common names: Variegated Tricholomposis; Plums and Custard
Order: Agricales
Family: Tricholomataceae
Width of cap: 2–5 in. (5–12.5 cm)
Length of the stalk: 1½–4 in. (4–10 cm)
Cap: Convex first, becoming broadly convex to flattened with age, with or without a low umbo, margin incurved when young, upturned in older specimens; wine red to purplish red when young, cuticle soon separating into reddish brown to yellowish brown scales or tufts, exposing a pale yellowish ground color; surface dry, felt-like when young, soon becoming more or less scaly; **flesh** thick, firm, yellowish, odor and taste unpleasant, like rotting wood.
Gills: Adnate to notched, close to crowded; yellow, gill edges often darker yellow.
Stalk: Equal or enlarged toward the base, often curved, becoming hollow with age; colored like the cap or more yellowish; surface scurfy to somewhat scaly.
Spore print: White.
Spores: 5–7 x 3–4.5 μm, elliptical, smooth, hyaline, non-amyloid.
Occurrence: Solitary or more often in clusters on or around decaying conifer stumps, logs, trunks, and wood chips; saprobic; summer-fall, occasionally in spring; fairly common.
Edibility: Non-poisonous.

Comments: Older specimens (as shown in the photo) may lose most—if not all—of the reddish purple tones that are characteristic of fresh specimens. *Tricholomopsis decora* (not illustrated) is similar but has a golden yellow cap covered with minute blackish scales. *Rutilans* means "reddish."

Medium to Large Terrestrial Mushrooms Having Brittle Flesh That Does Not Exude Latex When Broken

The mushrooms included here are in the genus *Russula,* a large and difficult group with several dozen species occurring in the Central Appalachians. Their stature is typically short and stocky, with the width of the cap usually exceeding the length of the stalk. Although it is difficult to identify many species based solely on field features, it is relatively easy to recognize members of the genus by their squatty stature and the brittle nature of the fruitbody. Their stalks will break cleanly (like a piece of chalk), and with a few exceptions the gills will crumble or flake when rubbed. The brittleness of the flesh and gills is due to the presence of globose cells called sphaerocysts. These globose cells are integrated with the binding filamentous cells that provide the structure of most other mushrooms. Milk mushrooms in the genus *Lactarius* (Section 1-D) are members of the Russulaceae family, and they too have characteristically brittle flesh.

Russulas are a favorite food for a variety of wildlife, especially squirrels and other rodents. It is not uncommon to find partially eaten russulas or tell-tale bits and pieces on the ground where these animals have been feeding. Almost all members of the Russulaceae form mycorrhizal associations with trees.

Latin name: *Russula amoenolens* Romagnesi
Common name: None
Order: Russulales
Family: Russulaceae
Width of cap: 1–3½ in. (2.5–9 cm)
Length of the stalk: 1–2½ in. (2.5–6.5 cm)
Cap: Convex, becoming flattened to depressed in the center; dingy grayish yellow to yellow-brown, often with dark reddish brown spots; surface viscid when moist, soon dry, margin prominently grooved; **flesh** white to pale yellowish, odor spermatic or rancid (sometimes likened to Jerusalem artichokes), taste acrid.
Gills: Adnate; close to moderately well-spaced; cream to pale ochraceous.
Stalk: Equal or tapered toward the base, becoming hollow; white to yellowish white with brown stains at the base.
Spore print: Pale orange-yellow.
Spores: 6–8.5 x 4.5–7 μm, elliptical, surface with isolated warts and a few connective lines, hyaline.
Occurrence: Usually in groups in woods, grassy woodland margins, parklands, and landscaped areas; mycorrhizal; early summer-fall; fairly common.
Edibility: Not edible.

Comments: This mushroom often grows in open grassy woods and along grassy roadsides. The fruitbodies are small to medium size, predominantly brownish, and have viscid caps with well-defined grooves on the margin. The name *amoenolens* means "pleasant smelling," which is perhaps a rather generous assessment.

Latin name: *Russula angustispora* Bills
Common name: None
Order: Russulales
Family: Russulaceae
Width of cap: ¾–3 in. (2–7.5 cm)
Length of the stalk: ⅝–1½ in. (1.5–4 cm)
Cap: Hemispheric with an incurved margin at first, becoming convex to flattened with a depressed center, eventually funnel-shaped; white at first, developing cream to yellowish or yellowish brown blotches or streaks, or these colors dominating, clay-colored with age; surface smooth to somewhat wrinkled, dull, minutely velvety, especially in the center, margin even; **flesh** white to pale buff, hard, brittle, odor not distinctive, taste mild.
Gills: Subdecurrent to decurrent, close to medium spaced, often forked or anastomosing; white to pale yellow, developing yellowish brown to brown stains with age.
Stalk: Equal or tapered toward the base; white or colored like the cap; surface dull, glabrous, with or without lengthwise wrinkles or irregularities.
Spore print: Yellow
Spores: 6.5–8.5 x 4.5–5 μm, elliptical to broadly elliptical, surface with isolated warts and some low connective lines, hyaline.
Occurrence: Usually in groups on the ground under Virginia pine (*Pinus virginiana*); mycorrhizal; late summer-fall; fairly common where Virginia pine is present.
Edibility: Unknown.

Comments: The stature and color of this mushroom is very similar to *Lactarius deceptivus* (p.88), which exudes white latex from broken tissue. The nearly identical *Russula brevipes* (not illustrated) often has a pale greenish band at the stalk apex and has larger spores (8–11 x 6.5–10 μm) that are white to cream in deposit. *Angustispora* means "having narrow spores."

Latin name: *Russula ballouii* Peck
Common name: Ballou's Russula
Order: Russulales
Family: Russulaceae
Width of the cap: 1–3 in. (2.5–7.5 cm)
Height of the stalk: 1–2 in. (2.5–5 cm)
Cap: Broadly convex, becoming flattened to depressed in the center; yellowish ochre to brownish orange; surface dry, with the cuticle breaking up into small scale-like patches and revealing a cream to yellowish ground color; **flesh** white, odor not distinctive, taste typically acrid.
Gills: Adnate, close to crowded; white to pale yellow.
Stalk: Cylindrical, equal; creamy white to colored like the cap, especially toward the base, surface smooth, breaking into scale-like patches at the base.
Spore print: White.
Spores: 7.5–9.5 x 7–8 µm, oval, surface with warts and ridges forming a partial reticulum, hyaline.
Occurrence: Solitary or in small groups on the ground in broad-leaved woods and grassy woodland margins; mycorrhizal; summer-fall; uncommon.
Edibility: Unknown.

Comments: Compare with *Russula compacta* (p.220), which can be similar in color, but its gills quickly stain brownish when rubbed. The specific epithet honors American mycologist W.H. Ballou (1857–1937).

Latin name: *Russula brunneola* Burlingham
Common name: None
Order: Russulales
Family: Russulaceae
Width of cap: 2–5 in. (5–12.5 cm)
Length of the stalk: 1¼–3 in. (3–7.5 cm)
Cap: Convex, becoming flattened to depressed in the center; sepia-brown to purple-brown; surface smooth, dry, often with a whitish bloom at first, margin becoming tuberculate-striate, pellicle peels about halfway from the margin to the disc; **flesh** brittle, odor not distinctive, taste mild.
Gills: Adnate, close to crowded, forked near the stalk, brittle; white to cream.
Stalk: Cylindrical, equal; white or tinged pale purplish brown.
Spore print: White.
Spores: 6–9.5 x 4–7 µm, elliptical, surface with isolated low warts, hyaline.
Occurrence: Usually in groups on the ground in conifer and mixed woods; mycorrhizal; summer-fall; common.
Edibility: Edible.

Comments: This mushroom is often found in high elevation spruce woods and under planted Norway spruce (*Picea abies*). *Brunneola* means "somewhat brownish."

Latin name: *Russula compacta* Frost
Common name: Firm Russula
Order: Russulales
Family: Russulaceae
Width of cap: 2½–6 in. (6.5–15 cm)
Length of the stalk: 1½–3 in. (4–7.5 cm)
Cap: Convex with an inrolled margin at first, becoming flattened to depressed in the center; sometimes whitish to pale buff at first, but soon becoming cinnamon-tan to reddish cinnamon; surface tacky when fresh, dull, often cracking in the center; **flesh** thick, firm, brittle, white, odor and taste fishy or metallic, somewhat unpleasant, especially when old.
Gills: Adnate, moderately close to crowded; white to buff, quickly staining brownish when rubbed or injured.
Stalk: Equal, firm; white or tinged the color of the cap, staining brown when handled or bruised; surface smooth.
Spore print: White.
Spores: 7–10 x 6–8 μm, broadly elliptical, surface with warts and ridges forming a partial reticulum, hyaline.
Occurrence: Solitary or more often in groups on the ground in broad-leaved and mixed woods; mycorrhizal; summer-fall; common.
Edibility: Not recommended. Edible according to some reports, but not popular.

Comments: This medium to large Russula often fruits in dry periods when other mushrooms are scarce. The firm flesh, compact stature, brown-staining gills, and peculiar odor are distinctive. Compare with *Russula ballouii* (p.218), which is similarly colored. *Compacta* means "compact" or "solid."

Latin name: *Russula crassotunicata* Singer
Common name: Rubber-skin Russula
Order: Russulales
Family: Russulaceae
Width of cap: 1–3 in. (2.5–7.5 cm)
Length of the stalk: 1–2½ in. (2.5–6.5 cm)
Cap: Cushion-shaped to convex at first, becoming broadly convex with a depressed center, margin even to slightly striate, incurved when young; white to pale yellow or pale orange-yellow, often with brownish stains; surface viscid when moist, becoming felted when dry, glabrous, usually developing cracks, especially in the center, cuticle thick, elastic, peelable for about two-thirds from the margin; **flesh** firm, brittle, white to yellowish, slowly staining brownish when exposed, odor not distinctive or mildly of coconut, taste acrid, often bitter.
Gills: Adnate to adnexed, moderately close to subdistant, forking near the stalk; white at first, becoming yellowish, bruising slowly brownish.
Stalk: Equal or enlarged toward the base; white to yellowish, bruising slowly brownish; surface dry, pruinose when young, becoming glabrous, smooth to somewhat wrinkled.
Spore print: White.
Spores: 9–11 x 6.8–9 μm, subglobose, surface with isolated warts, hyaline.
Occurrence: In small groups on the ground in conifer and mixed woods, especially with spruce; mycorrhizal; summer-fall; uncommon.
Edibility: Unknown.

Comments: The unusual elastic cap cuticle and cracking on the disc are distinctive features. *Crassotunicata* means "with a thick coating," in reference to the cap cuticle.

Latin name: *Russula dissimulans* R. Shaffer
Common name: Red and Black Russula
Order: Russulales
Family: Russulaceae
Width of cap: 2–5 in. (5–12.5 cm)
Length of the stalk: 1–3 in. (2.5–7.5 cm)
Cap: Convex to broadly convex with an incurved margin, becoming nearly flat with a depressed center, or somewhat funnel-shaped with age; dingy white at first, soon turning blackish brown; surface velvety, dull, dry, or slightly tacky when moist; **flesh** white, staining reddish, then reddish brown, and finally black, odor not distinctive, taste mild to acrid.
Gills: Adnate, moderately close to subdistant, lamellulae abundant; off white, becoming grayish with age.
Stalk: Equal; dull white, becoming blackish brown with age; surface smooth, bruising reddish then dark reddish brown, and finally black.
Spore print: White.
Spores: 7.5–11 x 6.5–9 μm, broadly elliptical, surface with low warts and ridges, often forming a partial reticulum, hyaline.
Occurrence: Solitary, or more typically gregarious on the ground in conifer and mixed woods; mycorrhizal; early summer-fall; common, sometimes abundant.
Edibility: Not edible.

Comments: This is one of the more common whitish Russulas that ages blackish brown, and stains slowly reddish then black when bruised. *Russula densifolia* (not illustrated) is nearly identical except that it has close to crowded gills. *Russula albonigra* (not illustrated) does not have a reddish phase but stains directly blackish when bruised. It may take 10 minutes or longer before the staining progression is complete. These Russulas are frequent hosts for parasitic mushrooms such as *Collybia tuberosa* (not illustrated) and *Collybia cookei* (not illustrated), which are small, thin-fleshed whitish mushrooms that have tuber or seed-like sclerotia at the base of the stalk. Look for them on old, blackened Russula fruitbodies. *Asterophora lycoperdoides* (p.490) and *Asterophora parasitica* (p.491) are also parasitic on old Russulas. *Dissimulans* means "dissimilar" in reference to the color variation of the fruitbodies.

Latin name: *Russula earlei* Peck
Common name: Beeswax Russula
Order: Russulales
Family: Russulaceae
Width of cap: 1½–4 in. (4–10 cm)
Length of the stalk: 1–2 in. (2.5–5 cm)
Cap: Convex to broadly convex; straw yellow to butterscotch tan; surface roughened, irregularly pitted to lumpy, viscid when wet, margin sometimes faintly striate with age; **flesh** whitish to pale yellow, odor not distinctive, taste mild or slightly acrid.
Gills: Adnate, widely-spaced; white to pale yellow; thick, waxy looking.
Stalk: Equal or tapering upward; white to yellowish; surface somewhat roughened.
Spore print: White.
Spores: 5.5–7 x 3.5–5.5 μm, subglobose, minutely warted, hyaline.
Occurrence: Solitary or in small groups on the ground in broad-leaved woods, especially with oak and beech; mycorrhizal; summer-early fall; occasional.
Edibility: Unknown.

Comments: This odd Russula is thought to be the most primitive in the genus. The entire mushroom looks as if it might be made of wax, and many who encounter it for the first time think they have found a species of wax cap (*Hygrophorus*). However, the amyloid spores and presence of globose cells (sphaerocysts) in the structural tissue of the fruitbody confirm its relationship with *Russula.* The name *earlei* honors American mycologist Franklin S. Earle (1856–1929).

Latin name: *Russula eccentrica* Peck
Common name: None
Order: Russulales
Family: Russulaceae
Width of cap: 1½–5 in. (4–12.5 cm)
Length of the stalk: 1–3 in. (2.5–7.5 cm)
Cap: Convex to broadly convex with an incurved margin when young, becoming flattened with age; white at first, soon grayish to grayish pink, or pinkish buff, eventually becoming grayish brown to dark brown; margin even or with coarse striations; surface viscid to dry, shiny to dull, smooth to slightly uneven or pitted, sometimes cracking; **flesh** firm but brittle, white to grayish pink, odor not distinctive or somewhat fruity, taste mild or bitter.
Gills: Adnate to subdecurrent, moderately well-spaced, often forked and intervenose; pale pink to vinaceous or pinkish cinnamon, sometimes exuding clear droplets when young.
Stalk: Cylindrical to slightly compressed, often eccentric, equal or tapering downward, becoming hollow with age; colored like the cap or paler; surface dry, glabrous or slightly pruinose.
Spore print: White.
Spores: 6.5–8.5 x 5–6 μm, elliptical, with low warts and fine lines forming a partial reticulum, hyaline.
Occurrence: Usually in groups on the ground in broad-leaved and mixed woods, especially with hemlock, beech, and oak; mycorrhizal; late spring-summer, uncommon.
Edibility: Unknown.

Comments: This mushroom is distinctive because of its pinkish buff gills, which are unusual for the genus. *Entoloma sinuatum* (p.277) is somewhat similar but has a sinuate gill attachment and pink spores. *Eccentrica* means "off-center," referring to the stalk attachment.

Latin name: *Russula flavisiccans* Bills
Common name: None
Order: Russulales
Family: Russulaceae
Width of cap: 1–4½ in. (2.5–11.5 cm)
Length of the stalk: ¾–2½ in. (2–6.5 cm)
Cap: Hemispherical to convex at first, becoming broadly convex to flattened or depressed with age, margin even or obscurely sulcate; pink to pinkish red, mottled with areas of grayish pink, violet, pale yellow, ochraceous buff, or brown; surface dull, minutely velvety or pruinose, somewhat viscid when wet, becoming cracked with age; **flesh** hard and firm when young, becoming more brittle with age; white to cream, odor not distinctive, taste sometimes mild but usually bitter or disagreeable.
Gills: Adnate to sinuate, moderately close, with some forking, especially near the stalk, brittle; white, becoming pale yellow with age, often developing yellowish brown stains.
Stalk: More or less equal or tapered at the base, solid and hard; white or white fused with pink; surface glabrous to minutely pruinose above, sometimes scaly with age.
Spore print: White to pale yellow.
Spores: 7–8.5 x 6.5–8 μm, globose to subglobose, with low warts and lines connecting to form a reticulum, hyaline.
Occurrence: Solitary or gregarious on the ground in broad-leaved woods, especially with oak, hickory, and beech; mycorrhizal; summer-fall; fairly common.
Edibility: Unknown.

Comments: This mushroom has been identified as *Russula lepida* (not illustrated), a similar European species having a cedarwood odor (especially noticeable at the base of the stalk). It is unclear whether these 2 entities are separate species. *Flavisiccans* means "drying yellowish."

Latin name: *Russula granulata* Peck
Common name: Granulated Russula
Order: Russulales
Family: Russulaceae
Width of cap: 1½–3 in. (4–7.5 cm)
Length of the stalk: 1¼–2½ in. (3–6.5 cm)
Cap: Convex, becoming nearly flat, often depressed in the center; margin incurved at first; yellow-brown to tawny; surface viscid when moist, with crustose patches and granules forming in the center as the cap expands, margin tuberculate-striate; **flesh** yellowish to white, odor slightly rancid, unpleasant, or not distinctive, taste slowly acrid.
Gills: Adnate, close, often forked near the stalk; whitish to pale yellow.
Stalk: Equal; white, often with brownish stains at the base; smooth.
Spore print: Cream or pale ochre.
Spores: 5.7–8 x 4.4–6 μm, elliptical, surface with low warts, hyaline.
Occurrence: Scattered or in groups on the ground in high elevation red spruce and northern hardwood forests; mycorrhizal; early summer-fall; common.
Edibility: Not edible.

Comments: This small- to medium-sized Russula is distinguished by the viscid cap with granular patches concentrated on the disc. *Russula fragrantissima* (not illustrated) is similar but lacks crustose patches on the cap, is generally more robust, and has a more pronounced rancid odor. *Russula laurocerasi* (p.227) also lacks crustose patches on the cap, and it has a fragrant almond-like odor. *Granulata* means "covered with granules."

Latin name: *Russula laurocerasi* Melzer
Common name: Almond-scented Russula
Order: Russulales
Family: Russulaceae
Width of cap: 1½–4½ in. (4–11.5 cm)
Length of the stalk: 1½–4 in. (4–10 cm)
Cap: Nearly round in the button stage, becoming convex with an incurved margin, then nearly flat or centrally depressed with age; yellowish to ochre-brown; surface glabrous, smooth, viscid when moist, margin tuberculate-striate; **flesh** white to yellowish, thick, firm, brittle, odor of bitter almonds or cherry bark, taste mild to somewhat acrid.
Gills: Adnate to adnexed, moderately close; creamy white to yellowish, sometimes developing brownish stains; brittle.
Stalk: More or less equal to slightly spindle-shaped; white to pale yellowish, often with brown stains; surface smooth to somewhat wrinkled.
Spore print: Creamy white to yellowish.
Spores: 7–10.5 x 7–9 μm, elliptical to oval, surface with high warts and prominent ridges that form a partial or complete reticulum, hyaline.
Occurrence: Solitary or in groups on the ground in broad-leaved or mixed woods; mycorrhizal; summer-fall; common.
Edibility: Not edible.

Comments: This common mushroom can usually be distinguished from similar species by its pleasant bitter almond odor. *Russula fragrantissima* (not illustrated) is nearly identical but has oily tasting flesh and a rancid odor. *Russula foetentula* (not illustrated) is also similar but lacks an almond odor and usually has reddish brown stains at the base of the stalk and sometimes on the cap. Compare with *Russula granulata* (p.226), which has crustose patches and granules in the center of its cap. *Laurocerasi* means "laurel cherry," relating to the odor that is sometimes likened to cherry bark.

Latin name: *Russula modesta Peck*
Common name: Modest Russula
Order: Russulales
Family: Russulaceae
Width of cap: 1½–3½ in. (4–9 cm)
Length of the stalk: 1–2½ in. (2.5–6.5 cm)
Cap: Convex at first, becoming flattened to depressed in the center, margin incurved at first; color variable from dull gray to dull grayish green or olivaceous buff to olivaceous brown, often dingy yellowish buff in the center; surface viscid when moist, soon dry and dull, usually with a whitish bloom; **flesh** white to grayish, brittle, taste mild.
Gills: Adnate to subdecurrent, close, intervenose, often forked, especially near the stalk; white to pale buff, often with brownish stains; fragile, brittle.
Stalk: More or less equal; white or sometimes yellowish brown at the base; surface smooth to somewhat wrinkled.
Spore print: Cream to pale yellow.
Spores: 6–7 x 4.5–6 μm, subglobose, surface with isolated warts and occasionally ridges that form a partial reticulum, hyaline.
Occurrence: Solitary to gregarious on the ground in broad-leaved woods, especially with beech, birch, and oak, also under red spruce and hemlock; mycorrhizal; summer-fall; common.
Edibility: Edible.

Comments: This rather drab mushroom is recognized by the overall grayish green tones of the cap, and the whitish bloom on fresh specimens. The striate margin, which is obscure on young buttons, becomes more conspicuous as the cap expands. *Russula virescens* (p.234) is similarly colored but has a cracked "mosaic" pattern on the cap. Compare also with *Russula redolens* (p.231), which has a darker blue-green cap and a strong fishy or parsley-like odor and unpleasant taste. *Modesta* means "modest."

Latin name: *Russula ochroleucoides* Kauffm.
Common name: Matt Yellow Russula
Order: Russulales
Family: Russulaceae
Width of cap: 1½–4½ in. (4–11.5 cm)
Length of the stalk: 1½–3 in. (4–7.5 cm)
Cap: Hemispherical to convex at first, becoming broadly convex to flattened or depressed in the center, margin even or obscurely striate; yellow to pale yellow or orange-yellow, sometimes brownish; surface dry, dull, velvety, often with a whitish bloom, smooth to occasionally concentrically wrinkled near the margin when young; **flesh** hard, brittle, white to buff, odor not distinctive, taste usually bitter or slightly acrid.
Gills: Adnate to adnexed, moderately close, with some forking, especially near the stalk; white, becoming cream to pale yellow with age, sometimes with yellowish brown spots or stains.
Stalk: Equal or tapered at the base, solid when young, spongy with age; white or tinged yellow, especially at the base; surface dry, dull, glabrous to minutely pruinose, smooth or with lengthwise wrinkles and irregularities.
Spore print: Pale yellowish white.
Spores: 7–8.5 x 5.5–7 µm, globose to subglobose, surface with warts, ridges, and fine lines that sometimes form a partial reticulum, hyaline.
Occurrence: Solitary or gregarious on the ground in broad-leaved and mixed woods, especially with oak and beech; mycorrhizal; early summer-fall; common.
Edibility: Unpalatable.

Comments: The cap of this handsome mushroom is predominately a rich yellow but at times tends toward brownish. *Russula flavida* (not illustrated) is generally smaller, with a deeper yellow to orange-yellow cap and stalk. *Ochroleucoides* means "resembling *Russula ochroleuca,*" which is a European species. *Ochroleuca* means "yellowish white."

Latin name: *Russula pulchra* Burlingham
Common name: None
Order: Russalales
Family: Russulaceae
Width of cap: 2–5 in. (5–12.5 cm)
Length of the stalk: 1–4 in. (2.5–10 cm)
Cap: Convex at first, becoming broadly convex to flattened or shallowly depressed with age; pinkish red to reddish orange or peach-colored, often with whitish or yellowish cream patches; surface dry, dull, pruinose, smooth but often developing cracks or fissures, margin striate; **flesh** white, odor and taste not distinctive.
Gills: Adnate, moderately well-spaced, forked near the stalk; white to cream.
Stalk: Equal; white or tinged pinkish; surface smooth, dry.
Spore print: Cream.
Spores: 6.5–9 x 5.5–7.5 μm, elliptical, surface with isolates warts, hyaline.
Occurrence: Solitary or in groups on the ground under broad-leaves, especially oak; mycorrhizal; early summer-fall; occasional to locally common.
Edibility: Edible.

Comments: The dull "matt" surface of the cap and frequently cracked cuticle help to distinguish this species from other reddish Russulas. *Pulchra* means "beautiful."

Latin name: *Russula redolens* Burlingham
Common name: Parsley-scented Russula
Order: Russulales
Family: Russulaceae
Width of cap: 1½–3 in. (4–7.5 cm)
Length of the stalk: 1½–2½ in. (4–6.5 cm)
Cap: Broadly convex to nearly flat with a central depression; grayish blue-green to grayish green, sometimes with purplish tones; surface viscid when moist, smooth to slightly wrinkled, margin striate, cap cuticle easily peeling at least halfway from the margin to the disc; **flesh** whitish, brittle, odor and taste "fishy," or like parsley, unpleasant.
Gills: Adnate to subdecurrent, crowded, forked near the stalk; white to pale yellow; fragile.
Stalk: Equal or tapered toward the base; white to dingy grayish where handled; surface dry, dull, smooth to somewhat wrinkled.
Spore print: White to cream.
Spores: 6–8 x 4.5–6 μm, elliptical, surface with isolated warts, hyaline.
Occurrence: Solitary or in small groups on the ground in woods, including high elevation red spruce and northern hardwood forests; mycorrhizal; summer-fall; occasional.
Edibility: Unknown.

Comments: This very fragile mushroom is distinctive for the unusual bluish green cap color, and peculiar odor of parsley or fish (especially noticeable when the flesh is crushed). *Russula modesta* (p.228) has a paler grayish green cap, and lacks a characteristic odor. *Redolens* means "emitting a scent."

Latin name: *Russula subtilis* Burlingham
Common name: None
Order: Russulales
Family: Russulaceae
Width of cap: ¾–1¾ in. (2–4.5 cm)
Length of the stalk: ¾–1¾ in. (2–4.5 cm)
Cap: Convex to broadly convex, often depressed in the center; pinkish violet to vinaceous, darker in the center; surface granular to finely scaly toward the margin; **flesh** brittle, white, odor not distinctive, taste slowly acrid.
Gills: Adnate, close to moderately well-spaced.
Stalk: Equal; whitish or flushed with the cap color.
Spore print: White.
Spores: 6–7.7 x 4.5–6.3 μm, subglobose, surface with warts and lines forming a partial reticulum, hyaline.
Occurrence: Usually in loosely scattered groups on the ground in broad-leaved and mixed woods; mycorrhizal; summer-fall; occasional.
Edibility: Unknown.

Comments: This small Russula has a rather unremarkable appearance, but is quite pleasing on closer inspection. *Subtilis* means "fine" or "delicate."

Latin name: *Russula variata* Banning
Common name: Variable Russula
Order: Russulales
Family: Russulaceae
Width of cap: 2–5 in. (5–12.5 cm)
Length of the stalk: 1–3 in. (2.5–7.5 cm)
Cap: Convex to flat, becoming depressed in the center or somewhat funnel-shaped with age; color variable from green to yellowish green, lavender, reddish purple, olive, or a mixture of these colors; surface viscid when moist, glabrous, smooth, margin sometimes cracking; **flesh** white, firm, brittle, odor not distinctive, taste slowly peppery to sharply acrid.
Gills: Adnate to sightly decurrent, crowded, repeatedly forked, flexible; white to cream.
Stalk: More or less equal or tapering downward; white; surface smooth, dull, sometimes with lengthwise wrinkles.
Spore print: White.
Spores: 7–11 x 5.5–9 μm, ovoid to subglobose, surface with isolated warts, hyaline.
Occurrence: Solitary to gregarious on the ground in broad-leaved and mixed woods, especially with oak; mycorrhizal; summer; common.
Edibility: Edible. The acrid taste of the flesh usually dissipates in cooking.

Comments: Although quite variable in color, this common summer mushroom is recognizable by the crowded, forking gills that are pliable and "greasy" rather than brittle and fragile like most others in the genus. *Russula cyanoxantha* (not illustrated) is similar but has mild tasting flesh. *Russula aeruginea* (not illustrated) has a green to yellowish green cap with a slightly grooved margin and cream-colored spores. *Russula virescens* (p.234) has a green, "quilted" cap, and gills that easily crumble when rubbed. *Variata* means "variable."

Latin name: *Russula virescens* (Schaeff.) Fr.
Common name: Green Quilt Russula
Order: Russulales
Family: Russulaceae
Width of cap: 2–4½ in. (5–11.5 cm)
Length of the stalk: 1¼–2½ in. (3–6.5 cm)
Cap: Nearly globose at first, soon becoming convex to broadly convex with a depressed center; grayish green to dull bluish green, sometimes with yellowish green to ochraceous areas; surface dry, dull to velvety, soon cracking into quilt-like patches, margin even or sometimes slightly grooved; **flesh** white, brittle, odor and taste not distinctive.
Gills: Adnate to adnexed, or almost free from the stalk, close; white, fragile and brittle.
Stalk: Equal; white; surface smooth.
Spore print: White.
Spores: 6.2–8.5 x 5–6.5 μm, broadly elliptical, surface with isolated low warts and fine connecting lines that form a partial reticulum, hyaline.
Occurrence: Solitary or in groups on the ground in woods, especially with oak, beech, and pine; mycorrhizal; summer-early fall; occasional.
Edibility: Edible and good.

Comments: The tender flesh of the Green Quilt Russula is attractive to a variety of insects, making it difficult to find unblemished specimens. *Russula aeruginea* (not illustrated) differs in having a smooth, non-cracking grayish olive to yellowish green cap with a grooved margin. *Russula crustosa* (not illustrated) also has a greenish "quilted" cap, but with more yellow-brown to orangish tones, and cream-colored spores. *Russula subgraminicolor* (not illustrated) has a smooth bluish green cap with a paler cap margin. Compare with *Russula variata* (p.233), which can have a green cap with cracks on the margin. *Virescens* means "becoming green."

Medium to Large Mushrooms with White to Pale Pinkish Buff Spores That Grow on the Ground, Have Fibrous Flesh, and Have Gills That Are Free or Variously Attached but Not Strongly Decurrent.

This section includes terrestrial mushrooms that exhibit a broad range of macroscopic features, but all have a ringless stalk that is typically more than ¼-inch (6 mm) thick at the apex. It is a catch-all group united by size, spore color, and the lack of features used to categorize the other groups of gilled mushrooms. Additional light-spored mushrooms having a marginal stalk thickness in the ¼-inch range are found in Section 1-H(a). For similar medium to large terrestrial mushrooms having dark-spores, see Section 1-K(b).

Latin name: *Amanita alba* Pers.
Synonym: *Amanita vaginata* var. *alba* Gillet
Common name: White Grisette
Order: Agaricales
Family: Amanitaceae
Width of cap: 1½–3 in. (4–7.5 cm)
Length of the stalk: 2½–5 in. (6.5–12.5 cm)
Cap: Oval at first, becoming bell-shaped to convex, then flat or depressed in the center, with or without a low umbo; white; surface glabrous, smooth with long striations on the margin; **flesh** white, odor and taste not distinctive.
Gills: Free from the stalk, close; white.
Stalk: Slender, nearly equal or tapered upward from an enlarged base that is seated within a membranous white, saccate volva; surface smooth, white.
Spore print: White.
Spores: 9.5–12 x 9.5–11.5 μm, globose to subglobose, smooth, hyaline, non-amyloid.
Occurrence: Solitary or in small groups on the ground in mixed woods; mycorrhizal; early summer-fall; uncommon.
Edibility: Edible but not recommended. Dangerous.

Comments: Careful attention to the defining features of this all-white mushroom will reduce the possibility of confusion with the dangerously poisonous *Amanita virosa* (p.62) and its relatives, which have a skirt-like ring on the stalk and a smooth, unlined cap margin. The similarity of these mushrooms is simply too close to risk eating the White Grisette. A mistake could be fatal. *Alba* means "white."

Latin name: *Amanita ceciliae* (Berk. & Broome) Bas
Common names: Strangulated Amanita; Cecilia's Amanita
Order: Agaricales
Family: Amanitaceae
Width of cap: 2–4½ in. (5–11.5 cm)
Length of the stalk: 2½–6 in. (6.5–15 cm)
Cap: Convex to flat, with or without a low umbo; grayish brown to bronze, darker in the center; surface smooth beneath scattered gray cottony patches, slightly viscid when moist, margin striate; **flesh** white, odor and taste not distinctive.
Gills: Free from the stalk, close to crowded; whitish.
Stalk: Equal or tapering upward, becoming hollow; white to grayish; surface smooth to slightly scurfy, base girdled with irregular gray cottony bands or patches of the Volva.
Spore print: White.
Spores: 10.2–11.7 μm, globose, smooth, hyaline, non-amyloid.
Occurrence: Solitary or in scattered groups on the ground in conifer woods, especially with pine, also in broad-leaved woods; mycorrhizal; spring-fall; common.
Edibility: Unknown.

Comments: The fragile volva usually crumbles, leaving grayish fragments in the soil near the stalk base. *Amanita sinicoflava* (p.240) is similar but has a sack-like volva with a free margin, and does not have scattered warts on the cap, although it sometimes will have a single large patch. Compare with *Amanita porphyria* (p.59), which is similarly colored but has a skirt-like ring on the stalk. This mushroom was named by the nineteenth-century British mycologist Miles Joseph Berkeley (1803–1889) in honor of his wife, Cecilia. The common name "Strangulated Amanita" refers to the tightly clasping volva.

Latin name: *Amanita farinosa* Schwein.
Common names: Powdery Amanita, Powder-cap Amanita
Order: Agaricales
Family: Amanitaceae
Width of cap: 1–2½ in. (2.5–6.5 cm)
Length of the stalk: 1–3 in. (2.5–7.5 cm)
Cap: Bell-shaped at first, becoming broadly convex to flat or depressed in the center; grayish to grayish brown beneath a dense coating of gray powder; surface smooth, margin striate; **flesh** thin, fragile, odor and taste not distinctive.
Gills: Free from the stalk, close; white.
Stalk: Slender or sometimes stocky, equal or tapering upward from a small rounded bulb at the base; white to pale gray; surface smooth to powdery.
Spore print: White.
Spores: 6.3–9.4 x 4.5–8 μm, subglobose to elliptical, smooth, hyaline, non-amyloid.
Occurrence: Solitary or in small groups on the ground in broad-leaved and conifer woods, and woodland margins; mycorrhizal; early summer-fall; occasional.
Edibility: Unknown.

Comments: The striations on the cap margin may be obscured in the button stage by the dense coating of gray powder (universal veil). Small, squatty specimens might be mistaken for a species of *Russula,* especially if the powdery coating is washed away by rainfall. Russulas, however, do not have free gills. *Farinosa* means "having a mealy surface."

Latin name: *Amanita parcivolvata* (Peck) Gilbert
Common names: False Fly Agaric, False Caesar's Mushroom
Order: Agaricales
Family: Amanitaceae
Width of cap: 1½–4 in. (4–10 cm)
Length of the stalk: 2–4½ in. (5–11.5 cm)
Cap: Bell-shaped to convex at first, becoming flattened with age; red to orange-red, paler toward the margin; surface smooth, shiny, viscid when moist, covered at first with scattered yellowish to cream-colored warts or patches, margin striate; **flesh** white to yellowish, odor not distinctive.
Gills: Free from the stalk, close to crowded; yellowish; edges fringed.
Stalk: Equal to slightly tapering upward from a swollen basal bulb; pale yellow; surface powdery to scurfy; volva at base fragmented and indistinct.
Spore print: White.
Spores: 10–14 x 6.5–8 μm, elliptical, smooth, hyaline, non-amyloid.
Occurrence: Solitary to loosely scattered on the ground in broad-leaved and mixed woods, and under trees in lawns, especially oak; mycorrhizal; summer-fall; common.
Edibility: Poisonous.

Comments: This mushroom is often misidentified as the Fly Agaric (see *Amanita muscaria* var. *formosa* [p.56]). However, the common Fly Agaric that occurs in eastern North America is typically yellow-orange, not red, and it has white gills and a prominent skirt-like ring on the stalk. Compare with *Amanita jacksonii* (p.52), which has a ring on the stalk and a substantial white saccate volva. *Parcivolvata* means "having a frugal volva."

Latin name: *Amanita sinicoflava* Tulloss
Common name: None
Order: Agaricales
Family: Amanitaceae
Width of cap: 1½–3 in. (4–7.5 cm)
Length of the stalk: 2½–5 in. (6.5–12.5 cm)
Cap: Bell-shaped to convex when young, becoming flattened with an umbo with age; yellow-ochre to olive-tan; surface smooth, sometimes with a thick patch of cottony gray volva material adhering, margin striate; **flesh** white, odor and taste not distinctive.
Gills: Free or barely reaching the stalk, close; white to cream.
Stalk: Equal or tapering upward, hollow, base seated in a fragile, sack-like grayish volva with a free margin; pale buff; surface scurfy.
Spore print: White.
Spores: 9–12 x 8–11.5 μm, subglobose, smooth, hyaline, non-amyloid.
Occurrence: Usually solitary on the ground in mixed woods, especially beneath conifers; mycorrhizal; early summer-fall; uncommon.
Edibility: Unknown. Possibly poisonous.

Comments: This recently described species may be more common than reports indicate, but it has probably passed as a variation of *Amanita fulva* (not illustrated), which has a more orange-brown cap and a membranous white volva. *Amanita ceciliae* (p.237) has scattered gray warts on its cap and clasping gray cottony bands at the base of the stalk. Compare with *Amanita umbrinolutea* (p.241), which has a large sack-like white volva at the stalk base. *Sinicoflava* means "Chinese yellow."

Latin name: *Amanita umbrinolutea* (Secr. ex Gillet) Bertillon
Common name: None
Order: Agaricales
Family: Amanitaceae
Width of cap: 2½–5 in. (6.5–12.5 cm)
Length of the stalk: 4–7 in. (10–18 cm)
Cap: Bell-shaped to convex with an umbo, becoming flattened with age; bister-brown to bronze and yellowish brown in broad concentric bands, darkest in the center, paler to nearly whitish toward the margin; surface glabrous, smooth, margin striate; **flesh** white, odor and taste not distinctive.
Gills: Free from the stalk, crowded; white.
Stalk: More or less equal to tapering upward, seated in a large, white, sack-like membranous volva at the base; pallid to brownish; surface smooth to somewhat powdery or scurfy.
Spore print: White.
Spores: 11–16 x 9.5–13 μm, subglobose, smooth, hyaline, non-amyloid.
Occurrence: Solitary or in groups on the ground in broad-leaved and mixed woods; mycorrhizal; summer-fall; uncommon.
Edibility: Unknown.

Comments: This large, handsome species in not well-known in North America. It was first described from Europe, where it is considered to be edible. However, more study is needed to determine whether the American version is actually the same entity. It is sometimes possible to find undeveloped "eggs" in the vicinity of mature fruitbodies. If these are cut lengthwise, the structure of a developing mushroom can clearly be seen within (as shown in the photo). *Amanita fulva* (not illustrated) is similar but has an orange-brown cap and smaller, globose spores (9.7–12.5 μm). *Amanita vaginata* (not illustrated) is also similar but has a more uniformly colored gray or brownish gray cap. The name *umbrinolutea* is a combination of two Latin words meaning "shade giving" and "yellowish."

Latin name: *Clitocybe irina* (Fr.) H.E. Bigelow & A.H. Sm.
Synonym: *Lepista irina* (Fr.) H.E. Bigelow
Common name: False Blewit
Order: Agaricales
Family: Tricholomataceae
Width of cap: 2–5 in. (5–12.5 cm)
Length of the stalk: 1½–3 in. (4–7.5 cm)
Cap: Convex to broadly convex, becoming flattened, with or without a low umbo, margin inrolled at first, often wavy with age; whitish to pale buff or pinkish buff, darker in the center; surface smooth to somewhat wrinkled, dull, with or without water-spots, margin sometimes slightly grooved; **flesh** thick, whitish, odor aromatic or not distinctive, taste not distinctive.
Gills: Adnate to notched, crowded; whitish to cream-colored, becoming pale pinkish gray, easily separated from the cap.
Stalk: Equal to enlarged at the base, solid at first, becoming more or less hollow; colored like the cap, becoming dingy with age, staining pinkish brown where bruised; surface smooth to fibrillose.
Spore print: Pinkish buff.
Spores: 7–10 x 4–5 μm, elliptical, smooth to minutely roughened, hyaline, non-amyloid.
Occurrence: Gregarious, sometimes growing in clusters and fairy rings on the ground in broad-leaved and conifer woods, woodland margins, grassy places, or on bare soil; saprobic; late summer-fall; fairly common.
Edibility: Edible but not recommended (see comments below).

Comments: Some individuals have reported gastrointestinal upsets after eating this mushroom. Also, care must be taken not to confuse it with poisonous species such as *Entoloma sinuatum* (p.277), which has a grayish to pale brown cap, gills that are notched at the stalk, and salmon-pink spores. Compare with *Clitocybe subconnexa* (p.246), which is similar but typically grows in dense clusters and has smaller spores. *Clitocybe nuda* (p.244) can also resemble *Clitocybe irina.* It is usually more robust, has violet or pinkish lilac tones on the cap, and fruits late in the season. Compare also with *Entoloma abortivum* (p.133), *Clitocybe tarda* (p.168), and *Clitopilus prunulus* (p.131). The name *irina* refers to "iris," perhaps applied to this species because of the "perfumy" odor of some collections.

Latin name: *Clitocybe nuda* (Bull.) H.E. Bigelow & A.H. Sm.
Synonym: *Lepista nuda* (Bull.) Cooke
Common names: Blewit; Wood Blewit
Order: Agaricales
Family: Tricholomataceae
Width of cap: 2–5 in. (5–12.5 cm)
Length of the stalk: 1½–3 in. (4–7.5 cm)
Cap: Convex at first, becoming flattened or somewhat depressed in the center, sometimes with a low umbo; margin inrolled at first, later expanding and becoming wavy; violet to grayish violet at first, becoming duller violet-brown to pinkish cinnamon, pinkish tan, or pale pinkish buff with age; surface glabrous, smooth, hygrophanous, sometimes faintly striate when moist; **flesh** thick, firm at first, pale violet or purplish marbled, odor slightly fruity or not distinctive, taste mild to slightly bitter.
Gills: Adnate to sinuate or subdecurrent, close to crowded; purple to pale violet, becoming brownish with age.
Stalk: More or less equal to club-shaped, or bulbous at the base, solid; grayish to pale purplish, often bruising darker purplish, with white-to lilac-tinged cottony mycelium at the base; surface fibrillose to scurfy.
Spore print: Pinkish buff to pinkish tan.
Spores: 6–8 x 3–5 μm, elliptical, minutely roughened, hyaline, non-amyloid.
Occurrence: Occasionally solitary but usually gregarious, often in dense clusters, sometimes in rings or arcs on soil, leaf litter, and humus in woods, residential areas, gardens, compost, mulch beds; saprobic; late summer-fall, or rarely in spring; common.
Edibility: Edible but must be well-cooked. Highly regarded by some, but others think it is overrated.

Comments: This stout and squatty mushroom is usually found late in the year, especially where there is an accumulation of decaying leaves. Although edible, the taste varies and some collections are unpalatable. Firm, young specimens are best. Certain *Cortinarius* species, including some that may be poisonous, are very similar, but they have rusty brown spores and a web-like partial veil that leaves a rusty brown fibrillose ring zone on the stalk. Compare with *Cortinarius pyriodorus* (p.274), which has dingy yellow to yellowish brown flesh and a pungent odor. Also compare with *Clitocybe irina* (p.242), which lacks purplish tones on the cap. *Nuda* means "naked" or "bare," referring to the smooth cap or cap margin. The common name Blewit is probably a corruption of "Blue Hat," which is one of the British names for this mushroom.

Latin name: *Clitocybe subconnexa* Murrill
Synonym: *Clitopilus caespitosus* Peck
Common name: Clustered Clitocybe
Order: Agaricales
Family: Tricholomataceae
Width of cap: 1¼–3½ in. (3–9 cm)
Length of the stalk: 1–2½ in. (2.5–6.5 cm)
Cap: Convex with an inrolled margin at first, becoming flattened to depressed in the center, sometimes with a flaring margin with age; white to buff or pale pinkish brown; surface dry, smooth, satiny, white; **flesh** white to pinkish, brittle, odor mildly fragrant or not distinctive, taste mild to bitterish in older specimens.
Gills: Adnate to short-decurrent, crowded, occasionally forked; whitish to pale pinkish buff, readily separable from the cap.
Stalk: More or less equal or tapering upward; colored like the cap; surface smooth to silky-fibrillose, or somewhat scurfy at the apex, base cottony.
Spore print: Pinkish buff.
Spores: 4.5–6 x 3–3.5 µm, elliptical, minutely roughened, hyaline, non-amyloid.
Occurrence: Typically in caespitose clusters, sometimes in fairy rings, on the ground under broad-leaved trees; saprobic; summer-fall; occasional.
Edibility: Edible when young but not recommended.

Comments: As an edible this mushroom is mediocre at best, and since misidentifications are likely, it cannot be recommended. The stocky stature and caespitose growth are good field characters. The similar *Leucopaxillus albissimus* (not illustrated) grows beneath conifer trees, has a white to pale tan cap, a white stalk with a dense mat of mycelium at the base, short decurrent gills that often run down the stalk as lines, bitter tasting flesh, and white amyloid spores. Compare with *Clitocybe irina* (p.242). *Subconnexa* means "almost connected," in reference to the clustered manner of growth.

Latin name: *Hygrocybe punicea* (Fr.) P. Kumm.
Synonym: *Hygrophorus puniceus* Fr.
Common name: Crimson Waxy Cap
Order: Agaricales
Family: Tricholomataceae
Width of cap: 1½–3 in. (4–7.5 cm)
Length of the stalk: 1½–3 in. (4–7.5 cm)
Cap: Broadly conical to bell-shaped, becoming flattened with an umbo, margin often uplifted with age; crimson to blood red, at times with a narrow yellow margin, fading and often developing yellowish orange patches with age; surface smooth, lubricous, more or less streaked, somewhat viscid when moist, shiny; **flesh** yellowish orange, odor and taste not distinctive.
Gills: Adnexed to notched, often with a decurrent tooth, moderately well-spaced, broad, thick, waxy; pale yellow to yellowish orange with a yellow gill edge.
Stalk: Cylindrical to furrowed or compressed, more or less equal, becoming hollow; yellowish orange, usually white at the base; surface smooth, sometimes streaked.
Spore print: White.
Spores: 8–11 x 5–6 μm, oblong-elliptical, smooth, hyaline, non-amyloid.
Occurrence: Usually in small groups on the ground in mixed woods; saprobic; summer-fall; occasional.
Edibility: Not recommended. Probably edible but reports vary.

Comments: This colorful wax cap is larger and more robust than most others in the genus *Hygrocybe.* It is often confused with the smaller *Hygrocybe coccinea* (not illustrated), which has adnate gills and a red stalk. Compare with *Hygrocybe conica* (p.173), which can sometimes approach the Crimson Wax Cap in size and color, but turns black with age or when handled. *Punicea* means "crimson" or "blood red."

Latin name: *Hygrophorus tennesseensis* A.H. Sm. & Hesler
Common name: Tennessee Waxy Cap
Order: Agaricales
Family: Tricholomataceae
Width of cap: 2–4 in. (5–10 cm)
Length of the stalk: 1½–4 in. (4–10 cm)
Cap: Convex with an incurved margin at first, becoming flattened or depressed in the center; brown to tawny-brown in the center, paler to nearly white toward the margin; surface glabrous, smooth, viscid to glutinous; **flesh** thick in the center, thin at the margin, firm, white, odor not distinctive or of raw potatoes, taste bitter.
Gills: Adnate to subdecurrent, moderately well-spaced; white to cream.
Stalk: Equal or tapering downward, solid; white; surface smooth, more or less fibrillose-streaked, often scurfy at the apex.
Spore print: White.
Spores: 8–10 x 4–5 μm, elliptical, smooth, hyaline, non-amyloid.
Occurrence: Usually in clusters on the ground in woods under conifers; mycorrhizal; late summer-fall; occasional to locally common.
Edibility: Not edible.

Comments: The waxy, clean-looking gills and glutinous cap when young help to distinguish this mushroom from several brownish species of *Tricholoma* that are similar in stature. Compare with *Hebeloma crustuliniforme* (p.279), which has pale brown gills and a radish-like odor. *Tennesseensis* refers to the state of Tennessee, from which this mushroom was first described.

Latin name: *Laccaria ochropurpurea* (Berk.) Peck
Common name: Purple-gilled Laccaria
Order: Agaricales
Family: Hydnangiaceae
Width of cap: 1½–6 in. (4–15 cm)
Length of the stalk: 2–6 in. (5–15 cm)
Cap: Convex with an incurved margin at first, becoming flattened to depressed in the center, margin uplifted and wavy with age; purple to purple-brown, fading to grayish or dull white; surface smooth to somewhat fibrous-scaly, hygrophanous; **flesh** firm, white, odor and taste not distinctive.
Gills: Adnate to short-decurrent, close to moderately well-spaced, thick; pale to dark purple and remaining purple in contrast to the fading cap.
Stalk: Nearly equal to spindle-shaped, or tapering upward from a swollen base, solid, fibrous; more or less colored like the cap; surface fibrous-roughened, scurfy, sometimes striate.
Spore print: White to pale lilac.
Spores: 6.5–11 x 6.5–9.5 μm, globose to subglobose, spiny, hyaline, non-amyloid.
Occurrence: Solitary or more often in groups on the ground in broad-leaved and mixed woods, especially with oak. Also in open grassy woods and woodland margins; mycorrhizal; summer-fall; common.
Edibility: Edible but not highly regarded.

Comments: The Purple-gilled Laccaria is the largest and most easily recognized species in the genus. The thick, waxy-looking gills are reminiscent of *Hygrophorus,* however members of that genus have sharp-edged gills and smooth white spores. *Cortinarius* species have a web-like partial veil covering the gills when young, and rusty brown spores. The fruitbodies of *Laccaria* species resist decay and persist longer than those of most fleshy mushrooms. *Ochropurpurea* means "ochre and purple."

Latin name: *Limacella illinita* var. *illinita* (Fr.) Maire
Common name: White Slime Mushroom
Order: Agaricales
Family: Pluteaceae
Width of cap: 1–2½ in. (2.5–6.5 cm)
Length of the stalk: 1½–3 in. (4–7.5 cm)
Cap: Hemispherical to nearly globose at first, becoming bell-shaped to broadly convex, or nearly flat with age, usually with a slight, broad umbo; white to cream or pale brown; surface smooth, viscid to glutinous; **flesh** thin, white, odor and taste not distinctive.
Gills: Free from the stalk, or barely reaching the stalk, close; white.
Stalk: More or less equal or slightly tapering upward; colored like the cap; surface viscid to glutinous, slippery.
Spore print: White.
Spores: 4.5–5.5 x 5–6.3 μm, globose to broadly elliptical, smooth, hyaline, non-amyloid.
Occurrence: Solitary or in small groups on the ground in conifer and mixed woods and along forest roads, also reported to occur in grassy places and on bare ground; saprobic, possibly also mycorrhizal; summer-fall; uncommon.
Edibility: Unknown.

Comments: When fresh, the glutinous coating on this distinctive mushroom is thick and slimy, sometimes dripping from the cap margin. The fruitbody is so slippery that it is nearly impossible to pick. *Limacella illinita* var. *argillacea* (not illustrated) is identical except that the young caps are dull brown. The all-white *Hygrophorus eburneus* (not illustrated) is also glutinous and slimy when fresh, but it has waxy gills that are clearly attached to the stalk. *Illinita* means "smeared." The genus name *Limacella* is derived from the Latin word limi, which means "slime."

Latin name: *Lyophyllum decastes* (Fr.) Singer
Synonym: *Clitocybe multiceps* (Peck) Sacc.
Common name: Fried Chicken Mushroom
Order: Agaricales
Family: Tricholomataceae
Width of cap: 1½–4 in. (4–10 cm)
Length of the stalk: 1½–3½ in. (4–9 cm)
Cap: Convex with an incurved margin at first, becoming flattened with age; color variable from gray to grayish brown or yellowish brown; surface smooth, silky, dry to lubricous when moist; **flesh** thick in the center, thin at the margin, firm, white, odor and taste not distinctive.
Gills: Adnate to sinuate or subdecurrent, close to crowded; whitish to buff.
Stalk: More or less equal or swollen in the middle, solid, tough, fibrous; white to pale grayish brown, or tinged with the color of the cap; surface smooth.
Spore print: White.
Spores: 4–5.5 x 4–5 µm, globose to subglobose, smooth, hyaline, non-amyloid.
Occurrence: Usually in clusters on soil in open grassy woods, disturbed ground, roadsides, lawns, cultivated areas, and woodland margins; saprobic, summer-fall; common.
Edibility: Edible with caution.

Comments: *Lyophyllum decastes* represents a complex of closely related species that more or less fit the description above. Although edible for most persons, there have been reports of gastrointestinal upsets from eating members of this group. The flavor doesn't taste anything like fried chicken but the cooked flesh is a bit fibrous and firm, somewhat like chicken. It is essential to confirm the white spore color so as not to confuse this edible mushroom with similar poisonous species in the pink-spored genus *Entoloma. Clitocybe subconnexa* (p.246) is similar in stature and growth habit but has a white cap and pinkish buff spores. *Decastes* means "in groups of tens," referring to the clustered manner of growth.

Latin name: *Marasmius nigrodiscus* (Peck) Halling
Synonyms: *Gymnopus tenuifolius* Murrill; *Collybia nigrodisca* Peck
Common name: Black-eyed Marasmius
Order: Agaricales
Family: Marasmiaceae
Width of cap: 1¼–3 in. (3–7.5 cm)
Length of the stalk: 2–4 in. (5–10 cm)
Cap: Convex to nearly flat, more or less umbonate, margin incurved at first, becoming uplifted with age; cream to pale yellow, pale tan, or yellow-brown, darker on the disc; surface glabrous, smooth or wrinkled, dry, often with short striations on the margin; **flesh** thin, whitish, odor of bitter almond or not distinctive, taste mild to bitter.
Gills: Adnate to adnexed or sinuate, close to moderately well-spaced; white to dull yellow, sometimes with smoky gray stains.
Stalk: Nearly equal, or tapering upward, sometimes twisted, fibrous; white to more or less colored like the cap; surface dry, often with longitudinal grooves.
Spore print: White to pale buff.
Spores: 7–9 x 3–5 μm, elliptical to teardrop-shaped, smooth, hyaline, non-amyloid.
Occurrence: Solitary or more often in scattered groups or clusters on the ground in broad-leaved and mixed woods, also reported to occur in conifer woods; saprobic; summer-early fall; occasional.
Edibility: Reported to be edible.

Comments: This species is unusually large for a Marasmius. Compare with *Marasmius oreades* (p.182), which grows in grasslands. *Nigrodiscus* means "black centered."

Latin name: *Rhodocollybia maculata* (Alb. & Schwein.) Singer
Synonym: *Collybia maculata* (Alb. & Schwein.) Quél.
Common name: Spotted Collybia
Order: Agaricales
Family: Marasmiaceae
Width of cap: 1½–4 in. (4–10 cm)
Length of the stalk: 2–4 in. (5–10 cm)
Cap: Convex with an incurved margin at first, becoming flattened to depressed in the center, with or without a low, broad umbo; white to cream, soon developing rust-brown spots and stains, especially on the disc; surface glabrous, smooth; **flesh** thick, white, tough, odor somewhat sweetish, taste bitter.
Gills: Adnexed to sinuate or free from the stalk, close to very crowded; whitish, becoming spotted rust brown.
Stalk: Slender, equal, sometimes slightly rooting, fibrous, tough; colored like the cap and also developing rust-brown spots and streaks; surface smooth to somewhat grooved.
Spore print: Pinkish buff.
Spores: 5.5–7 x 5–6 μm, globose to subglobose, smooth, hyaline, some dextrinoid.
Occurrence: Usually gregarious on humus and rich soil in conifer and mixed woods, sometimes growing in fairy rings or arcs; saprobic; spring-fall; common.
Edibility: Non-poisonous but bitter.

Comments: The stark white caps of this mushroom first appear in early June, and it is one of the earliest gilled mushrooms substantial enough to tempt the pothunter. The flesh, however, is bitter and unpalatable. *Tricholoma inamoenum* (p.258) is similar but does not become spotted, and it has a distinctive coal-tar odor. *Maculata* means "spotted."

Latin name: *Tricholoma aurantium* (Schaeff.) Ricken
Common name: Golden Orange Tricholoma
Order: Agaricales
Family: Tricholomataceae
Width of cap: 1½–5 in. (4–12.5 cm)
Length of the stalk: 1½–3½ in. (4–9 cm)
Cap: Convex, becoming nearly flat, often with a low, broad umbo; yellow-orange, to orange or orange-brown, sometimes with olive tones; surface viscid when moist, shiny, breaking into appressed scales or patches; **flesh** thick in the center, white, odor and taste farinaceous, unpleasant.
Gills: Adnexed to sinuate, or subdecurrent, close to crowded; white to cream, often developing rusty brown stains with age.
Stalk: Equal or swollen in the middle, solid at first, becoming hollow; colored like the cap from the base upward to a clearly defined zone near the apex, white above; surface breaking into scales or patches, often forming irregular bands and revealing a white ground color.
Spore print: White.
Spores: 4–5 x 3–3.5 μm, subglobose to broadly elliptical, smooth, hyaline, non-amyloid.
Occurrence: Usually gregarious, often clustered, in conifer and mixed woods, and woodland borders, sometimes in rings; mycorrhizal; summer-fall; fairly common.
Edibility: Not edible.

Comments: When fresh and young this is an attractive mushroom, especially when the stalk develops a "snake-skin" pattern of color zones. It is atypical for the genus because there is a hint of a partial veil. Although the veil is merely suggested and insubstantial, its presence has caused some mycologists to place this species in the genus *Armillaria. Tricholoma pessundatum* (not illustrated) is similar but grows under conifers in the fall and has a smooth reddish brown cap and stalk. *Aurantium* means "orange."

Latin name: *Tricholoma flavovirens* (Pers.) S. Lundell
Synonym: *Tricholoma equestre* (L.) Quél.
Common names: Canary Tricholoma; Man on Horseback
Order: Agaricales
Family: Tricholomataceae
Width of cap: 1½–4½ in. (4–11.5 cm)
Length of the stalk: 1–2½ in. (2.5–6.5 cm)
Cap: Convex, often with a slight, broad umbo, becoming more or less flattened with age, margin incurved at first, later uplifted and flaring; yellow to greenish yellow or yellowish brown, often reddish brown toward the center; surface smooth to lightly cracked, viscid when moist, somewhat streaked on the disc; **flesh** thick, firm; white to pale yellow, odor and taste not distinctive.
Gills: Adnexed to sinuate, or nearly free from the stalk, close; pale yellow to sulphur yellow.
Stalk: More or less equal, or with an enlarged to nearly bulbous base, solid when young; white to yellow or brownish; surface smooth, dry.
Spore print: White.
Spores: 6–7.5 x 4–5 μm, elliptical, smooth, hyaline, non-amyloid.
Occurrence: Usually gregarious, often clustered on the ground under pines, sometimes in rings; mycorrhizal; fall-early winter; common.
Edibility: Not recommended (see comments below).

Continued on next page

Comments: The Canary Tricholoma typically fruits late in the year, often in company with *Tricholoma portentosum* (p.259). The Canary Tricholoma (*Tricholoma flavovirens*) has been enjoyed as a safe, delicious edible mushroom for many years in countries around the world. However, a disturbing recent report from France, which was published in the *New England Journal of Medicine,* has implicated *Tricholoma equestre* in 12 cases of serious poisoning, including 3 deaths. The individuals involved suffered from a muscle-destroying illness called rhabdomyolysis. Many mycologists believe *Tricholoma equestre* is a synonym of *Tricholoma flavovirens.* Because of the long history of use of this mushroom for food, one might question the veracity of the report. However, until further information either supports or refutes the findings, *Tricholoma flavovirens* should be avoided for eating.

Tricholoma leucophyllum (not illustrated) is nearly identical except that it has white gills. Compare with *Tricholoma sejunctum* (p.262), which has a greenish yellow cap with blackish radiating fibrils on the disc, and white gills that are sometimes tinged yellowish with age. *Flavovirens* means "yellow green."

Latin name: *Tricholoma fumosiluteum* (Peck) Sacc.
Common name: Smoky Yellow Tricholoma
Order: Agaricales
Family: Tricholomataceae
Width of cap: 1½–3 in. (4–7.5 cm)
Length of the stalk: 2–5 in. (5–12.5 cm)
Cap: Bell-shaped to convex, often with a low umbo; smoky yellow at first, especially on the disc, brighter yellow when expanded; surface glabrous, moist, sometimes with water spots; **flesh** white to somewhat yellowish beneath the cuticle, odor and taste not distinctive or farinaceous.
Gills: Adnexed to sinuate, close to moderately well-spaced; white to pale yellow; edges becoming eroded with age.
Stalk: More or less equal or tapering toward the base, elongated when growing in mossy areas; smooth, glabrous; white to yellowish.
Spore print: White
Spores: 6–7 x 4.5–5 μm, elliptical to subglobose, smooth, hyaline, non-amyloid.
Occurrence: Solitary or in small groups on the ground, often growing in moss or with the liverwort *Bazzania* under conifers, especially red spruce (*Picea rubens*), also reported to occur in broad-leaved woods; mycorrhizal; fall; uncommon.
Edibility: Unknown.

Comments: This mushroom is more common in the northeastern United States, but also occurs at high elevations in the central Appalachians. *Fumosiluteum* means "smoky yellow."

Latin name: *Tricholoma inamoenum* (Fr.) Gillet
Common name: Stinky White Tricholoma
Order: Agaricales
Family: Tricholomataceae
Width of cap: 1–2 in. (2.5–5 cm)
Length of the stalk: 1¼–2½ in. (3–6.5 cm)
Cap: Broadly conical to convex, becoming nearly flat, with or without a low umbo; white to cream, often tinged dull yellow to pale brownish on the disc; surface smooth; **flesh** white, thick at the center, odor strongly of coal tar or burned rubber.
Gills: Adnate to notched, close to moderately well-spaced; white to cream.
Stalk: Equal to somewhat enlarged at the base, solid; white, sometimes brownish near the base; surface smooth to scurfy at the apex.
Spore print: White.
Spores: 10–12 x 6.5–8 μm, elliptical, smooth, hyaline, non-amyloid.
Occurrence: Usually in small groups on the ground under conifers, especially Norway spruce (*Picea abies*); mycorrhizal; summer-fall; occasional.
Edibility: Poisonous.

Comments: This small *Tricholoma* is recognized by its overall whitish color and strong, unpleasant odor. *Tricholoma resplendens* (not illustrated) is larger, has a viscid white cap, and lacks a distinctive odor. *Hygrophorus eburneus* (not illustrated) has a viscid to glutinous white cap and stalk, and waxy gills. *Inamoenum* means "unpleasant" in reference to the odor of the flesh.

Latin name: *Tricholoma portentosum* (Fr.) Quél.
Common names: Marvelous Tricholoma; Streaked Tricholoma
Order: Agaricales
Family: Tricholomataceae
Width of cap: 1½–4½ in. (4–11.5 cm)
Length of the stalk: 1½–4 in. (4–10 cm)
Cap: Broadly conical to convex, becoming flattened to depressed in the center, with or without an obtuse umbo, margin incurved at first; gray to dark brownish gray, streaked with blackish fibrils radiating from the nearly black disc, often tinged yellowish on the margin; surface smooth, viscid and sticky when moist, shiny; **flesh** white, tinged or slowly staining lemon yellow, odor and taste not distinctive.
Gills: Sinuate, close to moderately well-spaced; white to pale gray, often tinged lemon yellow near the cap margin; gill edges uneven.
Stalk: Equal or tapering toward the base, solid; white to tinged yellow; surface smooth to lightly fibrillose streaked; flesh staining yellow when cut, especially at the base.
Spore print: White.
Spores: 5–7 x 3–5 μm, elliptical, smooth, hyaline, non-amyloid.
Occurrence: Gregarious, often in clusters on the ground beneath pines; mycorrhizal; fall-early winter; locally common.
Edibility: Edible and very good.

Continued on next page

Comments: This late season edible mushroom is often found in company with *Tricholoma flavovirens* (p.255). Soil and debris adhere to the viscid cap, making it difficult to clean, but peeling the cuticle eliminates most of the problem. There are several gray to gray-brown species of *Tricholoma,* not all of which are safe edibles. The most reliable field mark that distinguishes *Tricholoma portentosum* is the lemon yellow tint or staining of the flesh, which is especially noticeable at the cut stalk base. *Tricholoma virgatum* (not illustrated) lacks yellowish tints, has a sharply pointed umbo on an ash-gray cap, and has bitter, acrid tasting flesh. *Tricholoma pullum* (not illustrated) is similar but grows under broad-leaved trees and has acrid tasting flesh. *Tricholoma sejunctum* (p.262) has blackish streaks over a greenish yellow to olive cap, and bitter tasting flesh. *Tricholoma saponaceum* (p.261) lacks radial streaks on the cap, has pinkish flesh at the base of the stalk, and has a soap-like odor. *Tricholoma terreum* (p.263) has a more or less scaly cap. Also compare with *Megacollybia platyphylla* (p.209) and *Entoloma madidum* (p.276). The specific epithet is derived from the Latin word portentum, which means "marvelous."

Latin name: *Tricholoma saponaceum* (Fr.) Quél.
Common name: Soapy Tricholoma
Order: Agaricales
Family: Tricholomataceae
Width of cap: 1½–4 in. (4–10 cm)
Length of the stalk: 1½–3½ in. (4–9 cm)
Cap: Convex with an incurved margin at first, becoming broadly convex to flattened, with or without a low, broad umbo; color variable from olive-gray, brownish gray, grayish green, greenish brown, to yellowish olive, or brown; surface lubricous, smooth, often cracking into scales; **flesh** thick, white, sometimes slowly staining pinkish, odor fragrant like soap, taste unpleasant.
Gills: Adnate to adnexed or sinuate, moderately well-spaced; white or sometimes with a pinkish or greenish tinge.
Stalk: Nearly equal to swollen in the middle, or tapering downward, base often rooting, solid; white or colored more or less like the cap, often pinkish to pinkish orange at the base.
Spore print: White.
Spores: 5–7 x 3.5–4 μm, elliptical, smooth, hyaline, non-amyloid.
Occurrence: Solitary or in groups, often caespitose, on the ground in broad-leaved and conifer woods; saprobic or mycorrhizal; summer-fall; occasional.
Edibility: Not edible, possibly poisonous.

Comments: The Soapy Tricholoma has a dull, dark cap, usually with olive or grayish green tones predominating. However, because it is variable in color it can be troublesome to identify. The distinguishing features are the soapy odor and pinkish color of the flesh at the base of the stalk. *Saponaceum* means "soapy."

Latin name: *Tricholoma sejunctum* (Sowerby) Quél.
Common name: Separated Tricholoma
Order: Agaricales
Family: Tricholomataceae
Width of cap: 1½–3½ in. (4–9 cm)
Length of the stalk: 2–3½ in. (5–9 cm)
Cap: Broadly conical to bell-shaped or convex at first, becoming broadly convex with an umbo; greenish yellow to grayish brown with dark brown or blackish radial fibrils concentrated on the disc; surface somewhat viscid when wet, glabrous or rarely developing small scales on the disc, streaked with fibrils; **flesh** white to slightly yellowish, odor farinaceous or not distinctive, taste bitter or astringent.
Gills: Adnate to sinuate, close to moderately well-spaced, fairly broad, margin eroded with age; white to cream or yellowish, especially near the margin.
Stalk: More or less equal, fibrous, tough, solid or becoming hollow; white or tinged yellow; surface smooth, silky, at times slightly scurfy at the apex.
Spore print: White.
Spores: 6–7 x 3.5–5.5 μm, subglobose, smooth, hyaline, non-amyloid.
Occurrence: Solitary or gregarious on the ground in conifer and broad-leaved woods, especially with spruce and pine; mycorrhizal; summer-fall; common.
Edibility: Not edible, sometimes reported to be poisonous.

Comments: *Tricholoma intermedium* (not illustrated) is similar but has an unstreaked cap, and the gills and stalk lack yellowish tones. *Tricholoma flavovirens* (p.255) has yellow gills and is found under pines in the fall. *Tricholoma portentosum* (p.259) has a dark gray to nearly black cap. *Sejunctum* means "separated," referring either to the gills that sometimes pull away from the stalk with age, or to the cap margin that often splits with age.

Latin name: *Tricholoma terreum* (Schaeff.) Quél.
Common name: Earth-gray Tricholoma
Order: Agaricales
Family: Tricholomataceae
Width of cap: 1–2½ in. (2.5–6.5 cm)
Length of the stalk: 1–2 in. (2.5–5 cm)
Cap: Convex, becoming flattened, usually with a low umbo, margin uplifted with age; dark gray to ash gray or brownish gray; surface dry, fibrillose to tomentose, becoming more or less scaly; **flesh** thin, fragile, white to pale gray, odor and taste farinaceous or not distinctive.
Gills: Adnexed to sinuate, moderately close; white to grayish.
Stalk: Equal, fragile, easily splitting lengthwise; white or tinged grayish; surface smooth to fibrillose.
Spore print: White.
Spores: 5–7.5 x 4–5 μm, elliptical, smooth, hyaline, non-amyloid.
Occurrence: Gregarious, often in clusters on the ground under pines and other conifers; mycorrhizal; fall; common.
Edibility: Edible with caution.

Comments: This dark *Tricholoma* appears late in the season and is often present after the first frost. Although eaten in Europe and sold in the markets there, it can be confused with other gray, scaly Tricholomas of dubious edibility. The similar *Tricholoma pardinum* (not illustrated) is known to be poisonous. It is a northern species that may extend down into the higher elevations of the central Appalachians. It is generally paler and more robust, has closer gills, thicker flesh, and larger spores (8–10 x 5.5–6.5 μm). Another northern species, *Tricholoma myomyces* (not illustrated), is nearly identical to *Tricholoma terreum,* but has a fleeting web-like partial veil that is only visible on very young specimens. *Tricholoma orirubens* (not illustrated) grows in broad-leaved woods and has a salmon to reddish tinge on the gills and bruised flesh. The name *terreum* is derived from the Latin word *terra,* which means "earth."

Latin name: *Tricholoma vaccinum* (Pers.) Fr.
Common name: Russet Scaly Tricholoma
Order: Agaricales
Family: Tricholomataceae
Width of cap: 1–2½ in. (2.5–6.5 cm)
Length of the stalk: 1½–3 in. (4–7.5 cm)
Cap: Broadly conical, becoming convex to flattened, usually with a broad, low umbo, margin incurved at first; color variable from reddish cinnamon, brownish orange, to tan, often darker on the disc; surface dry, fibrillose to scaly, margin shaggy with veil remnants; **flesh** white to pinkish buff, slowly staining pale cinnamon where injured or with age, odor farinaceous or not distinctive, taste bitter.
Gills: Adnate to sinuate, close; dingy whitish, often with pale cinnamon to reddish brown stains.
Stalk: More or less equal, becoming hollow; dingy white to colored like the cap, paler at the apex, often with reddish brown stains; surface dry, smooth to fibrillose-scaly.
Spore print: White.
Spores: 6–7.5 x 4–5 μm, broadly elliptical, smooth, hyaline, non-amyloid.
Occurrence: Usually in groups or clusters on the ground or in moss under conifers, especially with pine and spruce; mycorrhizal; late summer-fall; occasional.
Edibility: Not recommended.

Comments: Although this small- to medium-sized *Tricholoma* has a thin, fibrillose partial veil, it does not leave a ring on the stalk. The somewhat similar *Tricholoma imbricatum* (not illustrated) has a broadly convex, slightly scaly, reddish brown cap with a smooth margin. *Tricholoma vaccinum* is reported to be edible in Europe, but there is no reliable information regarding the edibility of North American collections. The specific epithet is derived from the Latin word *vaccinus,* which means "cow-colored."

Latin name: *Xerula furfuracea* (Peck) Redhead, Ginns, & Shoemaker
Synonym: *Oudemansiella radicata* (Relhan) Singer
Common names: Rooted Agaric; Rooted Oudemansiella; Rooting Collybia
Order: Agaricales
Family: Marasmiaceae
Width of cap: 1½–6 in. (4–15 cm)
Length of the stalk: 3–10 in. (8–25 cm) above ground, below-ground portion often equally as long.
Cap: Broadly convex, becoming flattened to depressed in the center with a prominent umbo; light brown to grayish brown or yellow-brown; surface minutely velvety at first, soon glabrous and viscid or lubricous, radially wrinkled on the disc, margin even or faintly striate; **flesh** thin, white, odor and taste faintly metallic or not distinctive.
Gills: Adnate to sinuate, well-spaced, with numerous short gills; white.
Stalk: Slender, cylindrical, above-ground portion tapering gradually toward the apex; more or less colored like the cap but usually paler, especially at the apex; surface grainy to somewhat fibrillose or scaly, cortex often separating into a variegated "snake-skin" pattern; texture fibrous but easily breaking.
Spore print: White.
Spores: 14–17 x 9.5–12 μm, broadly elliptical, smooth, hyaline, non-amyloid.
Occurrence: Solitary or in groups on the ground in woods, semi-open areas, woodland margins, and urban areas, especially around or sometimes on decaying stumps and at the base of standing dead trunks of broad-leaved trees; saprobic on decaying wood (usually buried); spring-fall; common.
Edibility: The fleshy caps are edible but mediocre.

Comments: The gills of this common, easily recognized mushroom have a stark, clean appearance. The surprisingly long underground portion of the stalk can be difficult to extract without breaking it off. Compare with *Xerula megalospora* (p.188), which is generally smaller and paler in color. *Furfuracea* means "scurfy" or "mealy."

Medium to Large Mushrooms That Grow on the Ground, with Gills That Are Free from the Stalk or Variously Attached but Not Decurrent, and with Pinkish, Brown, or Black Spores

What the mushrooms in this section have in common are size, spore color, terrestrial growth habitat, and the lack of features used to categorize the other groups of gilled mushrooms. They typically have a stalk that is more than ¼ in. (6 mm) thick at the apex. For similar dark-spored mushrooms having a marginal stalk thickness in the ¼-inch range, see also Section 1-H(b). For similar medium to large terrestrial mushrooms that have light colored spores, see Section 1-K(a).

Latin name: *Agrocybe molesta* (Lasch) Singer
Synonym: *Agrocybe dura* (Bolton) Singer
Common name: Cracked-cap Agrocybe
Order: Agaricales
Family: Bolbitiaceae
Width of cap: 1½–3 in. (4–7.5 cm)
Length of the stalk: 1½–3½ in. (4–9 cm)
Cap: Convex, becoming flattened, sometimes with an obscure umbo; white to cream, or pale tan on the disc; surface dry to slightly tacky when moist, smooth, usually with a cracked cuticle in dry weather; **flesh** thick, white, odor not distinctive, taste not distinctive or slightly bitter.
Gills: Adnate, close; white at first, becoming grayish brown to dark brown; covered at first with a thin cottony-fibrillose partial veil.
Stalk: Equal or tapering toward the base, which is sometimes slightly enlarged, fibrous, hard; colored like the cap; surface smooth or slightly scurfy near the apex, partial veil sometimes leaving a thin evanescent ring on the upper stalk.
Spore print: Dark brown.
Spores: 10–14 x 6.5–8 μm, elliptical with an apical germ pore, smooth, non-amyloid.
Occurrence: Usually gregarious, sometimes in clusters on cultivated ground or in grassy places, lawns, pastures, gardens, mulch beds; saprobic; spring-early summer; occasional.
Edibility: Edible.

Comments: The Cracked-cap Agrocybe often appears in May and June after rainfall. *Agrocybe praecox* (not illustrated) has a browner cap that does not typically become cracked, and a conspicuous membranous ring on the stalk. *Agaricus* species have gills that are free from the stalk, and produce chocolate-brown spores. *Molesta* means "disturbed," perhaps in reference to the habitats where this species grows.

Latin name: *Coprinus comatus* Fr.
Common name: Shaggy Mane; Lawyer's Wig
Order: Agaricales
Family: Coprinaceae
Length of cap: 2–6 in. (5–15 cm)
Length of the stalk: 2½–7 in. (6.5–18 cm)
Cap: Oval to bluntly cylindrical with an incurved margin at first, becoming somewhat bell-shaped as the margin expands at maturity; surface shaggy with coarse scales; white to dingy white, disc yellowish brown, scales soon becoming yellowish brown to reddish brown, or blackish with age; **flesh** thin, white, odor and taste not distinctive.
Gills: Nearly free from the stalk, densely crowded; white at first, progressively becoming pinkish gray then black from the lower margin upward as the spores mature, finally dissolving into a black, inky fluid.
Stalk: Cylindrical, nearly equal or tapering slightly upward from a somewhat bulbous base, hollow; white; surface smooth to fibrillose; brittle, easily splitting; sometimes with an ephemeral ring near the base.
Spore print: Black.
Spores: 10–14 x 5.5–8 μm elliptical with an apical germ pore, smooth, purple-brown.
Occurrence: Gregarious on the ground in grassy places, lawns, pastures, or on hard-packed soil in waste areas and on roadsides; saprobic; spring-fall; common.
Edibility: Edible when young before the gills begin to deliquesce.

Comments: This is one of the easiest edible mushrooms for beginners to recognize. It can often be found in large quantities in late summer and fall, sometimes with hundreds of fruitbodies within a small area. The cap never fully opens as with most gilled mushrooms, but remains more or less bullet-shaped and goes through a "melt-down" (melt up!) process from the base of the cap upward as the spores are progressively released at the lower margin, and the "spent" supportive tissue dissolves in to a black, inky fluid. This remarkable spore dispersal mechanism is highly efficient and may represent advanced evolution. Once underway, the dissolution of the cap and gills is rapid, and after several hours little remains of the fruitbody other than a tall stalk with a ragged, dripping disc on top. Because Shaggy Manes are highly perishable, they should be refrigerated soon after gathering, and consumed within a day or two. Specimens growing in contaminated sites, such as along busy roadsides or in chemically treated lawns should not be eaten. Compare with *Coprinus variegatus* (p.206). *Comatus* means "having long hair."

Latin name: *Cortinarius armillatus* (Alb. & Schwein.) Fr.
Common names: Red-banded Cortinarius; Bracelet Cortinarius
Order: Agaricales
Family: Cortinariaceae
Width of cap: 2–4 in. (5–10 cm)
Length of the stalk: 3–6 in. (7.5–15 cm)
Cap: Hemispherical to bell-shaped at first, becoming broadly convex to nearly flat with a broad umbo; burnt-orange to tawny brown or brick red; smooth at first but soon becoming radially fibrillose to somewhat scaly; **flesh** dingy white to brownish, odor radish-like or not distinctive, taste mild to slightly bitter.
Gills: Adnate to adnexed, moderately well-spaced; pale brown to rusty brown, covered at first with a whitish web-like partial veil.
Stalk: Tapering upward from a swollen or bulbous base, solid; whitish to brownish with white mycelium at the base; surface with 1 or more orange-brown to reddish bands distributed along the length, and a thin web-like rusty brown fibrillose zone on the upper stalk.
Spore print: Rusty brown.
Spores: 7–12 x 5–7 μm, broadly elliptical, finely warted, pale brown, non-amyloid.
Occurrence: Usually gregarious on the ground in wet woods, especially along rivers and streams under birch, hemlock, and rhododendron; mycorrhizal with birch; late summer-fall; common.
Edibility: Edible but mediocre.

Comments: The distinctive reddish bands on the stalk and late season appearance make this is an easily recognized species. *Armillatus* means "ringed" or "bracelet-like."

Latin name: *Cortinarius collinitus* (Pers.) Fr.
Common name: Slimy-banded Cort
Order: Agaricales
Family: Cortinariaceae
Width of cap: 1½–4 in. (4–10 cm)
Length of the stalk: 2½–4½ in. (6.5–11.5 cm)
Cap: Hemispheric at first, becoming convex to flattened, or sunken in the center with an uplifted margin; color variable from yellow-brown to orange-brown, purplish, or cinnamon; surface viscid to glutinous, shiny, margin sometimes faintly striate; **flesh** white to grayish, odor and taste not distinctive.
Gills: Adnate to notched, close; white to grayish tan, becoming cinnamon-brown to rust-brown, sometimes with violaceous tones; covered at first with a web-like partial veil (cortina) and a glutinous universal veil.
Stalk: More or less equal or tapering downward, somewhat rooting, solid; whitish, often with areas or bands of yellowish to purplish gluten; surface slimy, usually with a ring zone of rusty brown fibrils (collapsed cortina) near the apex.
Spore print: Rusty brown.
Spores: 12–15 x 6–8 μm, elliptical, surface minutely roughened, non-amyloid.
Occurrence: Solitary or in small groups on the ground in broad-leaved and conifer woods; mycorrhizal; late summer-fall; occasional.
Edibility: Not recommended. Reported to be edible, but easily confused with similar species of unknown edibility.

Comments: This handsome mushroom is either highly variable in color, or it may represent a complex of closely related species. At first the entire fruitbody is encased in a transparent glutinous universal veil that often has a yellowish to purplish hue. As the stalk expands, the gluten tends to separate into irregular zones or bands. *Collinitus* means “smeared” or “greased.”

Latin name: *Cortinarius corrugatus* Peck
Common names: Corrugated Cortinarius; Wrinkled Cortinarius
Order: Agaricales
Family: Cortinariaceae
Width of cap: 1½–4 in. (4–10 cm)
Length of the stalk: 2–5 in. (5–12.5 cm)
Cap: Bell-shaped to convex at first, becoming broadly convex to flattened with a low, broad umbo, margin uplifted with age; tawny ochre to orange-brown or reddish brown; surface viscid, shiny, wrinkled, margin fringed with remnants of a web-like cortina when young; **flesh** white to buff, odor and taste not distinctive.
Gills: Adnate to adnexed, close; violet at first, becoming cinnamon-brown.
Stalk: More or less equal, usually with a small bulb at the base; colored like the cap, especially toward the base, paler above; surface scurfy, viscid on the lower portion, web-like partial veil sometimes leaving a faint band of rusty brown fibrils near the apex.
Spore print: Rusty brown.
Spores: 10–14 x 7–9 μm, elliptical, minutely roughened, pale brown, non-amyloid.
Occurrence: Solitary or in groups on the ground in broad-leaved and mixed woods, especially with oak, or under hemlock and rhododendron; mycorrhizal; summer-fall; fairly common.
Edibility: Not edible.

Comments: The prominently wrinkled, viscid cap, and purplish gills when young are distinctive features. Compare with *Psathyrella delineata* (p.213), which grows on or near decaying wood, and has purple-brown gills. *Corrugatus* means "having wrinkles" or "corrugated."

Latin name: *Cortinarius distans* Peck
Common name: None
Order: Agaricales
Family: Cortinariaceae
Width of cap: 1–2½ in. (2.5–6.5 cm)
Length of the stalk: 1½–3 in. (4–7.5 cm)
Cap: Bell-shaped to broadly bell-shaped or convex with a distinct umbo, margin incurved at first; cinnamon-brown, fading to tan; surface hygrophanous, glabrous or slightly scurfy; **flesh** thin, fragile, brownish, odor and taste not distinctive.
Gills: Adnate to notched, widely spaced, broad; light brown becoming cinnamon-brown; covered at first with a white web-like partial veil.
Stalk: Nearly equal to club-shaped; colored like the cap or paler; surface fibrillose-streaked; partial veil sometimes leaving a thin fibrillose zone on the upper stalk.
Spore print: Rusty brown.
Spores: 6–8 x 5–6 μm, oval, surface minutely roughened, pale brown, non-amyloid.
Occurrence: In groups and small caespitose clusters on the ground in broad-leaved woods, especially with oak; mycorrhizal; late spring-fall; occasional.
Edibility: Unknown.

Comments: This is one of the earlier fruiting species of *Cortinarius,* often first appearing in middle June. The widely spaced, broad gills are distinctive. *Cortinarius* is the largest genus of gilled mushrooms, with as many as 800 species estimated to occur in North America. Many are difficult to identify, and since the genus includes several poisonous species, experimenting with them for food is not recommended. *Distans* means "widely separated," in reference to the gill spacing.

Latin name: *Cortinarius pyriodorus* Peck
Common name: Lilac Conifer Cortinarius
Order: Agaricales
Family: Cortinariaceae
Width of cap: 2–5 in. (5–12.5 cm)
Length of the stalk: 1½–5 in. (4–12.5 cm)
Cap: Hemispherical to nearly globose at first, becoming convex to broadly convex or nearly flat with age, often with a broad umbo; lilac to silvery lilac, paler with age; surface dry, covered with fine whitish fibrils when young, smooth; **flesh** cinnamon-brown with whitish to buff marbling, odor unpleasant, like overripe pears, taste bitterish.
Gills: Adnate to notched, subdistant; cinnamon-brown, covered at first with a whitish web-like partial veil.
Stalk: Enlarged to bulbous at the base, solid; colored like the cap beneath a coating of whitish fibrils; surface dry; partial veil leaving a zone of appressed fibrils on the middle to upper stalk.
Spore print: Rust-brown.
Spores: 7–10 x 5–6 μm, elliptical, finely warted, pale brown.
Occurrence: Usually in groups, sometimes in clusters on the ground under conifers, especially Norway spruce (*Picea abies*); mycorrhizal; late summer-fall; fairly common.
Edibility: Unknown.

Comments: This robust mushroom first appears in late summer. The strong unpleasant odor, cinnamon-brown marbled flesh, and association with conifers are distinctive features. Some mycologists regard this species to be a variety of *Cortinarius traganus* (not illustrated), which is very similar and has a strong odor that is often described as "goaty." *Pyriodorus* means "having an odor of pears."

Latin name: *Entoloma luridum* Hesler
Common name: Yellow-gilled Entoloma
Order: Agaricales
Family: Entolomataceae
Width of cap: 1½–4 in. (4–10 cm)
Length of the stalk: 1½–3 in. (4–7.5 cm)
Cap: Bell-shaped at first, becoming convex, usually with a broad umbo, margin upturned and wavy with age; whitish to pale grayish or pale yellow; surface glabrous, smooth or wrinkled, somewhat viscid when moist; margin striate when wet, even when dry; **flesh** white, thick in the center, thin at the margin, odor not distinctive, taste acidic.
Gills: Adnexed to notched, broad, close, edges eroded; yellow at first, becoming buff then pinkish with age.
Stalk: Equal or tapering upward from a swollen base, hollow; whitish; surface smooth.
Spore print: Pinkish cinnamon.
Spores: 8–10 x 7–8 μm, subglobose, angular.
Occurrence: Solitary or in small groups on the ground in mixed woods, especially with oak; saprobic; summer-fall; occasional.
Edibility: Unknown.

Comments: Until the pinkish spore color is determined, this mushroom can easily be mistaken for a *Tricholoma. Entoloma grande* (not illustrated) is similar but larger, with caps up to 7 in. (18 cm) wide, and whitish gills that turn pink at maturity. *Luridum* means "dingy yellow."

Latin name: *Entoloma madidum* (Fr.) Gillet
Synonym: *Entoloma bloxamii* (Berk.) Sacc.
Common name: Midnight Blue Entoloma
Order: Agaricales
Family: Entolomataceae
Width of cap: 2–4½ in. (5–11.5 cm)
Length of the stalk: 2–4 in. (5–10 cm)
Cap: Broadly conical to convex, with or without an umbo; dark bluish gray or bluish violet to nearly black; surface viscid to lubricous, smooth to wrinkled, fibrillose streaked; **flesh** thick, white, odor and taste farinaceous or not distinctive.
Gills: Adnexed to notched, close to crowded, edges eroded; white to pale bluish gray, becoming pinkish cinnamon as the spores mature.
Stalk: More or less equal to tapering downward, solid; colored like the cap but usually paler, whitish at the base; surface fibrous, lined or streaked lengthwise.
Spore print: Pinkish cinnamon.
Spores: 6.5–8.5 x 6–8 μm, elliptical to subglobose, angular, smooth, non-amyloid.
Occurrence: Solitary or in small groups on the ground in broad-leaved and mixed woods; saprobic; late summer-fall; uncommon.
Edibility: Unknown. Many related species are poisonous.

Comments: The Midnight Blue Entoloma has a general aspect similar to Russula and at first glance might be mistaken as such. Compare with *Tricholoma portentosum* (p.259), which grows under pines in late fall and has white spores. *Madidum* means "wet," probably referring to the lubricous appearance of the cap.

Latin name: *Entoloma sinuatum* Fr.
Synonym: *Entoloma lividum* (Bull.) Fr. (according to some authors)
Common names: Livid Entoloma; Lead Poisoner
Order: Agaricales
Family: Entolomataceae
Width of cap: 2–5 in. (5–12.5 cm)
Length of the stalk: 1¾–4½ in. (5–11.5 cm)
Cap: Bell-shaped to convex at first, often with a low umbo, becoming flattened with a wavy margin with age; lead gray to brown, fading paler with age; surface glabrous, smooth, silky, radially fibrillose; **flesh** whitish or brownish, odor and taste mealy or "like cucumber."
Gills: Sinuate, close to moderately well-spaced, gill edges crenate; whitish to pale yellowish at first, becoming dull pink as the spores mature.
Stalk: Equal or enlarged at the base, sometimes twisted; white to pale grayish brown; surface silky, longitudinally fibrillose.
Spore print: Pink.
Spores: 7–10 x 7–9 μm, globose, angular, smooth.
Occurrence: Solitary or more often in groups on the ground in broad-leaved woods, especially with oak; saprobic; summer-fall; occasional to locally common.
Edibility: Poisonous.

Comments: There are a number of similar Entolomas that are difficult to distinguish in the field. All in the genus should be regarded as poisonous except for *Entoloma abortivum* (p.133). *Pluteus atricapillus* (p.211) is similar but it grows on decaying wood and has gills that are free from the stalk. *Clitocybe nuda* (p.244) has violaceous tones on the cap and pale pinkish buff spores. *Tricholoma* species have white spores. *Sinuatum* means "wavy."

Latin name: *Entoloma vernum* S. Lundell
Synonym: *Nolanea verna* (S. Lundell) Kotl. & Pouzar
Common name: Early Spring Entoloma
Order: Agaricales
Family: Entolomataceae
Width of cap: 1–2½ in. (2.5–6.5 cm)
Length of the stalk: 1½–3½ in. (4–9 cm)
Cap: Conical with an acute nipple-like umbo, becoming bell-shaped to broadly convex, margin incurved at first; dark brown to grayish brown, fading to pale brown or tan; surface glabrous, smooth to slightly wrinkled, satiny, often streaked, hygrophanous; **flesh** thin, fragile, watery brown, odor and taste not distinctive.
Gills: Adnexed to nearly free from the stalk, moderately close; whitish at first, becoming pink to dull rose as the spores mature.
Stalk: Equal or slightly tapering upward; gray to grayish brown with white mycelium at the base; surface twisted-striate.
Spore print: Salmon-pink.
Spores: 8–11 x 7–8 μm, elliptical, angular, smooth, non-amyloid.
Occurrence: Usually gregarious, sometimes in clusters on the ground in woods, disturbed ground and along roadsides; saprobic; spring; common.
Edibility: Probably poisonous. Reports vary but related species are poisonous.

Comments: This is one of the first gilled mushrooms to appear in spring and is a good indicator that conditions are right for hunting morels. By late spring or early summer it is gradually replaced by the very similar but larger *Entoloma strictius* (not illustrated). *Entoloma clypeatum* (not illustrated), which is also similar, has a brown to grayish brown cap with an obtuse umbo and a white stalk and grows in grassy areas near briars, cherry, apple, and other members of the Rose family (Roseaceae). Compare with *Pluteus atricapillus* (p.211), which also appears in early spring but grows on decaying wood. *Vernum* means "vernal" or "of the spring."

Latin name: *Hebeloma crustuliniforme* (Bull.) Quél.
Common name: Poison Pie
Order: Agaricales
Family: Bolbitiaceae
Width of cap: 1½–3 in. (4–7.5 cm)
Length of the stalk: 1½–3 in. (4–7.5 cm)
Cap: Convex becoming flattened with an umbo, margin incurved at first, upturned with age; whitish to light brown, darker yellow-brown to reddish brown in the center; surface viscid and sticky when moist; **flesh** white, odor of radishes, taste bitter.
Gills: Adnexed to sinuate, crowded; whitish at first, becoming gray-brown, gill edges whitish, fringed, sometimes beaded with moisture that leaves brownish spots when dried.
Stalk: More or less equal to bulbous at the base; white to brownish, with white mycelium at the base; surface pruinose to scurfy, especially near the apex.
Spore print: Dull cinnamon-brown.
Spores: 9–13 x 5.5–7 μm, elliptical, minutely roughened, pale brown, dextrinoid.
Occurrence: Usually in groups on the ground in broad-leaved and conifer woods, parklands, residential areas, sometimes in fairy rings or arcs; mycorrhizal; late summer-fall; occasional.
Edibility: Poisonous.

Comments: *Hebeloma sinapizans* (not illustrated) is similar but has a darker cap, lacks brownish spots on the gills, and usually has a tooth-like cone of tissue that projects into the hollowed stalk at the apex. The name *crustuliniforme* is derived from the Latin word *crustulum,* which means "a small cake, or biscuit," in reference to the color and shape of the fresh caps.

Austroboletus betula, p.298

Stalked Mushrooms with Pores on the Underside of the Cap

This group includes the boletes and stalked polypores. Many of these mushrooms are of considerable interest to the mycophagist. They also play an important role in the ecology of the forest. With few exceptions the boletes form mutually beneficial mycorrhizal associations with trees. Although a few stalked polypores form mycorrhiza, most are wood-decaying saprobes or parasites. Most polypores are leathery or woody and inedible, but there are notable exceptions. Stalkless polypores (Bracket Fungi) are grouped separately in Section 3.

Key to the Groups of Boletes and Stalked Polypores

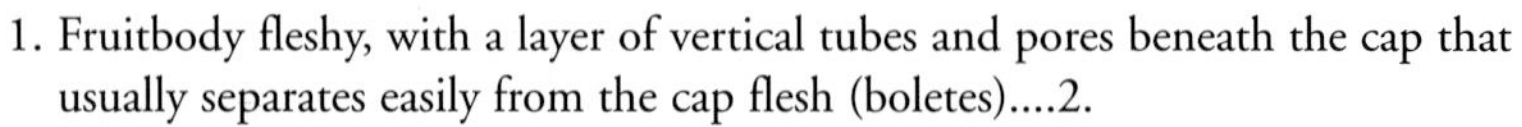

1. Fruitbody fleshy, with a layer of vertical tubes and pores beneath the cap that usually separates easily from the cap flesh (boletes)....2.
1. Fruitbody leathery, woody, fibrous, or fleshy only when young, with a shallow tube and/or pore layer that does not easily separate from the cap flesh (stalked polypores)....**Section 2-F** (p.343)

2. Stalk with a membranous, fibrous, or glutinous ring on the upper or mid-portion (partial veil)....**Section 2-A** (p.283).
2. Stalk lacking a ring....3.

3. Pore surface orange to red, or some shade of brown....**Section 2-B** (p.289).
3. Pore surface white, yellow, pinkish, or buff....4.

4. Stalk with a well-defined net-like pattern (reticulation), at least on the upper portion.....**Section 2-C** (p.297).
4. Stalk smooth or variously ornamented, but lacking reticulation, or indistinctly reticulate at the apex only....5.

5. Pore surface bruising bluish, blackish blue, or bluish green.....**Section 2-D** (p.314).
5. Pore surface unchanging when bruised, or bruising colors other than bluish, blackish blue, or bluish green.....**Section 2-E** (p.323).

Boletes with a Ring on the Stalk

The ring that is present on the stalk of a small number of boletes is formed from a partial veil, which in young specimens extends from the cap margin to the stalk and covers the immature pore surface. When the spores are mature, the partial veil separates from the cap margin and collapses on the stalk, and remains as a membranous or cottony-fibrillose ring. *Strobilomyces floccopus* (p.308) also has a partial veil, but it leaves a ring that merges with—and is often indistinguishable from—its shaggy stalk.

Latin name: *Pulveroboletus ravenelii* (Berk. & M.A. Curtis) Murrill
Synonym: *Boletus ravenelii* Berk. & M.A. Curtis
Common names: Ravenel's Bolete; Powdery Sulphur Bolete
Order: Boletales
Family: Boletaceae
Width of cap: 1–4 in. (2.5–10 cm)
Length of stalk: 1½–4 in. (4–10 cm)
Cap: Convex to nearly flat, margin incurved at first; bright yellow, soon becoming reddish orange to reddish brown on the disc; surface dry and powdery to matted and tacky or glabrous when moist, margin often fringed with partial veil remnants; **flesh** thick, white to yellow, staining pale blue then brownish when cut or injured, odor not distinctive or like hickory leaves, taste acidic.
Pore surface: Bright yellow becoming greenish yellow to greenish olive, bruising greenish blue; pores small, 1–3 per mm; covered at first with a yellow, cottony partial veil.
Stalk: More or less equal, solid; bright yellow; surface floccose to fibrillose-matted from the base up to a delicate cottony ring near the apex, smooth above; ring fragile, sometimes integrated with the floccose surface of the stalk and inconspicuous.
Spore print: Olive-brown to olive-gray.
Spores: 8–10.5 x 4–5 μm, elliptical to oval, smooth, pale brown.
Occurrence: Solitary or in small groups on the ground under conifers in woods, especially with pine, hemlock, and rhododendron; mycorrhizal; summer-fall; occasional.
Edibility: Edible.

Comments: This bolete is easy to recognize when young by the floccose-powdery coating on the fruitbody, and the delicate cottony ring that remains on the upper stalk, at least initially after the partial veil ruptures. The specific epithet honors American botanist-mycologist Henry William Ravenel (1814–1887).

Latin name: *Suillus grevillei* (Klotzsch) Singer
Synonym: *Suillus proximus* A.H. Sm. & Thiers
Common names: Tamarack Jack; Larch Suillus
Order: Boletales
Family: Suillaceae
Width of cap: 2–4 in. (5–10 cm)
Length of stalk: 1½–4 in. (4–10 cm)
Cap: Convex to nearly flat, with or without a low, broad umbo; color variable from yellow to orange-brown or dark reddish chestnut; surface smooth, viscid and sticky, shiny when dry, margin sometimes with hanging fragments of partial veil; **flesh** pale yellow to orange-yellow, odor and taste not distinctive or somewhat metallic.
Pore surface: Yellow, becoming ochraceous to olivaceous in age, staining cinnamon-brown to reddish brown when bruised; pores angular, small, 1–3 per mm.
Stalk: More or less equal or swollen near the base, solid; colored like the cap or paler, yellow above the ring, usually white at the base; surface scurfy to somewhat granular; partial veil white, or white blended with yellow, ring flaring upward, cottony-membranous to somewhat glutinous; flesh sometimes developing bright green staining at the base when exposed.
Spore print: Olive-brown to cinnamon.
Spores: 8–10 x 2.5–3.5 μm, elliptical to spindle-shaped, smooth, pale straw-colored to hyaline, non-amyloid.
Occurrence: Solitary or more often in groups on the ground under larch (*Larix*); mycorrhizal; summer-fall; locally common where there are larch trees.
Edibility: Edible but mediocre, it is best to peel the cuticle before cooking.

Comments: This distinctive bolete forms a specific mycorrhizal association with larch trees. It occurs with native American larch, which reaches the southern limit of its natural range in Preston County, West Virginia, and also with introduced European larch. *Suillus luteus* (p.287) is similar but grows under pines. The specific epithet honors Scottish botanist Robert K. Greville (1794–1866).

Latin name: *Suillus intermedius* (A.H. Sm. & Thiers) A.H. Sm. & Thiers
Synonym: *Suillus acidus* (Peck) Singer var. *intermedius* A.H. Sm. & Thiers
Common name: Sour-cap Suillus
Order: Boletales
Family: Suillaceae
Width of cap: 1½–3½ in. (4–9 cm)
Length of stalk: 1½–3½ in. (4–9 cm)
Cap: Convex with an incurved margin at first, becoming nearly flat, sometimes with a low umbo; straw-yellow to tawny yellow or ochre; surface smooth, viscid to glutinous when wet, becoming blotchy or streaked as the gluten dries, margin at first ragged with fragments of a partial veil; **flesh** thick, cream to yellowish or pale ochraceous, odor not distinctive, cap cuticle taste sour.
Pore surface: Pale yellow to ochre, unchanging or slightly reddish brown when bruised; pores small, 2 per mm, often exuding moisture droplets when young.
Stalk: Equal or tapering upward from a swollen base, solid; more or less colored like the cap, with resinous dots and smears that darken to reddish brown in age; surface floccose to scurfy, partial veil leaving a persistent gluten-covered membranous ring near the apex.
Spore print: Dull cinnamon.
Spores: 8–11 x 3–5 μm, elliptical to spindle-shaped, smooth, pale brown.
Occurrence: Scattered or in small groups on the ground under pines, especially red pine (*Pinus resinosa*); mycorrhizal; summer-fall; occasional to locally common.
Edibility: Edible.

Comments: This small- to medium-sized bolete is recognized by the persistent ring on the stalk and the sour tasting cap cuticle. *Suillus salmonicolor* (not illustrated) is similar but has a darker cap, yellow-orange to orange flesh, and a thicker and wider "cigar band" ring on the stalk. *Intermedius* means "intermediate between other species" or "between extremes," alluding to the salient features of this mushroom.

Latin name: *Suillus luteus* (L.) Gray
Common name: Slippery Jack
Order: Boletales
Family: Suillaceae
Width of cap: 2–5 in. (5–12.5 cm)
Length of stalk: 1½–3 in. (4–7.5 cm)
Cap: Hemispheric at first, becoming convex to broadly convex, or nearly flat in age; yellow-brown, orange-brown, to reddish brown; surface smooth, viscid, often streaked, margin with remnants of partial veil when young; **flesh** white to pale yellow, odor and taste not distinctive.
Pore surface: Whitish at first, soon yellow to ochre, darkening in age, unchanging when bruised; pores small, 1–2 per mm; covered at first with a thick membranous partial veil.
Stalk: More or less equal, solid; white to yellowish, often with brownish stains near the base; surface with pinkish brown resinous dots and smears that darken with age, and are conspicuous above a flaring—or collar-like ring—on the middle or upper stalk; partial veil thick, membranous to somewhat glutinous, white, typically with purplish tones on the underside.
Spore print: Cinnamon-brown.
Spores: 7–9 x 2.5–3 μm, elliptical to spindle-shaped, smooth, pale brown, non-amyloid.
Occurrence: Gregarious on the ground under pines, often in caespitose clusters, also reported to occur under spruce; mycorrhizal; late summer-fall, rarely in spring; common.
Edibility: Edible for most people (see comments below).

Comments: This common edible bolete can often be found in large numbers in the fall. Although popular with mushroom hunters, consuming it can produce a strong laxative effect in some individuals. This problem can usually be avoided by discarding the peelable cap pellicle and spongy tube layer of older specimens before cooking. *Suillus brevipes* (not illustrated) is similar and often found in the same habitat at the same time, but it lacks a ring and colored resinous dots on the stalk. Compare with *Suillus grevillei* (p.285), which grows under larch trees. *Luteus* means "yellow."

Latin name: *Suillus pictus* (Peck) A.H. Sm. & Thiers
Synonym: *Suillus spraguei* (Berk. & M.A. Curtis) Kuntze
Common name: Painted Bolete
Order: Boletales
Family: Suillaceae
Width of cap: 1½–4½ in. (4–11.5 cm)
Length of stalk: 1½–3½ in. (4–9 cm)
Cap: Convex with an incurved margin at first, becoming nearly flat in age, with or without a low umbo, margin fringed with fragments or flaps of whitish partial veil when young; red to purplish red or rose red, fading in age to buff or ochraceous; surface dry, cottony-velvety, soon breaking into fibrils and scales and revealing yellowish flesh beneath; **flesh** yellow, sometimes staining slightly reddish, odor and taste not distinctive.
Pore surface: Yellow, becoming dingy yellow to ochraceous in age, staining reddish to brownish when bruised; pores large, 0.5–5 mm wide, angular to elongated, radially arranged, often slightly decurrent; covered at first with a whitish, cottony or web-like partial veil.
Stalk: More or less equal, sometimes with a swollen base, solid; colored like the cap or paler, surface floccose-scaly, the ornamentation sometimes arranged in concentric bands from the base up to a whitish or grayish cottony, flaring ring near the apex.
Spore print: Olive-brown.
Spores: 8–12 x 3.5–5 µm, elliptical, smooth, pale brown, non-amyloid.
Occurrence: Scattered and often gregarious on the ground under white pine (*Pinus strobus*); mycorrhizal; late spring-fall; common.
Edibility: Edible.

Comments: Although it has soft flesh that turns blackish when cooked, some mycophagists think the Painted Bolete is among the better edibles in the genus *Suillus. Pictus* means "brightly marked" or "painted."

Boletes with a Red to Brown Pore Surface

Four genera of boletes are represented in this section. They are placed together here because all have red or brown tube mouths. This is a prominent feature that is not shared by many boletes. The group includes *Boletus subvelutipes,* one of the few boletes known to be poisonous. Although the pore color itself has nothing to do with toxicity, the red-pored boletes have long been viewed with fear and suspicion.

Latin name: *Boletus frostii* J.L. Russell
Synonym: *Boletus alveolatus* Berk. & M.A. Curtis
Common name: Frost's Bolete
Order: Boletales
Family: Boletaceae
Width of cap: 1½–5 in. (4–12.5 cm)
Length of stalk: 2–5 in. (5–12.5 cm)
Cap: Hemispheric to convex at first, becoming broadly convex to flattened with age; deep blood red to candy apple red, often with a narrow yellow band at the margin, cap fading and sometimes developing yellowish areas with age; surface smooth and shiny, tacky when moist; **flesh** yellow, quickly staining blackish blue when exposed, odor not distinctive, taste sour, acidic.
Pore surface: Dark red, becoming paler red with age, often exuding golden moisture droplets when young, instantly staining blackish blue when bruised, pores small, 2–3 per mm.
Stalk: More or less equal or tapering upward from an enlarged base, solid; dark red to pinkish red with a white to yellow base; surface coarsely and deeply reticulate over the entire length.
Spore print: Olive-brown.
Spores: 11–17 x 4–5 μm, elliptic, smooth, pale brown.
Occurrence: Solitary or in groups under broad-leaved trees, especially oak; mycorrhizal; summer-fall; occasional to locally common.
Edibility: Edible.

Comments: This is one of the most beautiful and distinctive boletes in North America. It can hardly be mistaken for any other. The specific epithet honors Charles C. Frost (1805–1880), a Vermont shoe manufacturer who took up the study of fungi as a hobby and was the first to describe several North American boletes.

Latin name: *Boletus subluridellus* A.H. Sm. & Thiers
Common name: None
Order: Boletales
Family: Boletaceae
Width of cap: 2–5 in. (5–12.5 cm)
Length of stalk: 1½–4 in. (4–10 cm)
Cap: Convex becoming broadly convex to nearly flat; bright red to pinkish red, orange-red, or brick red, instantly bruising blackish blue; surface dry, more or less velvety, slightly resinous to the touch; cuticle extends beyond the tube layer to form a narrow band of sterile tissue, which is inrolled at first; **flesh** yellow, instantly staining blue when exposed, odor not distinctive or slightly pungent, taste not distinctive or metallic.
Pore surface: Rose red to pinkish red or dark red, fading to orange-red or brownish, instantly bruising blackish blue; pores small, 2–3 per mm; tubes depressed at the stalk with age.
Stalk: Nearly equal, solid; pale yellow, sometimes with reddish dots; surface dry, faintly scurfy.
Spore print: Olive-brown.
Spores: 11–15 x 4–6 µm, somewhat spindle-shaped with an apical germ pore, smooth, dull yellowish, non-amyloid.
Occurrence: Solitary or in small groups on the ground in broad-leaved or mixed woods, especially under oak; mycorrhizal; summer-fall; occasional to locally common in some years.
Edibility: Unknown.

Comments: The entire fruitbody of this bolete instantly bruises blackish blue. Compare with *Boletus subvelutipes* (p.292), which has a more orange to brownish cap, and usually reddish bristle-like hairs at the base of the stalk. *Subluridellus* means "somewhat like *Boletus luridellus,*" which is a southern species known from North Carolina southward. *Luridellus* means "drab yellow to dirty brown."

Latin name: *Boletus subvelutipes* Peck
Common name: Red-mouth Bolete
Order: Boletales
Family: Boletaceae
Width of cap: 2½–5 in. (6.5–12.5 cm)
Length of stalk: 2–4 in. (5–10 cm)
Cap: Convex, becoming broadly convex to flattened; color variable from yellow-brown, to brownish orange, or reddish brown, often dingy brown with age; surface dry, more or less velvety when young, sometimes cracking with age, instantly staining blackish blue when bruised or handled; **flesh** yellow, instantly changing to dark blackish blue when exposed, then fading to white, odor not distinctive, taste mild to somewhat acidic.
Pore surface: Various rich shades of red to orange, often with a yellow zone near the margin, instantly bruising blackish blue; pores small, round, 2–3 per mm.
Stalk: Nearly equal to enlarged toward the base, solid; color is a mixture of yellow and red, yellow at the apex, instantly bruising dark blackish blue; surface pruinose, sometimes with dotted lines that intersect in a net-like pattern but not truly reticulate, base often with short, reddish or yellow bristle-like hairs.
Spore print: Olive-brown.
Spores: 13–18 x 5–6.5 μm, elliptical with a minute germ pore, smooth, pale brown.
Occurrence: Solitary to gregarious on the ground in woods, especially with oak, beech, and hemlock; mycorrhizal; early summer-fall; common.
Edibility: Poisonous.

Comments: This is the most common bolete known to be poisonous. It is found in a variety of habitats and is quite variable in color, but the red tube mouths, red or yellow bristly hairs at the stalk base, and instant blackish blue bruising on all parts are helpful field features. It is one of the first boletes of the season to make an appearance. *Boletus luridiformis* (not illustrated) is similar but has a very dark brown, velvety cap. *Subvelutipes* means "somewhat velvety foot," referring to the surface texture of the stalk.

Latin name: *Boletus vermiculosoides* A.H. Sm. & Thiers
Common name: None
Order: Boletales
Family: Boletaceae
Width of cap: 1½–4½ in. (4–11.5 cm)
Length of stalk: 1½–4 in. (4–10 cm)
Cap: Convex to broadly convex, becoming nearly flat with age, margin incurved at first; dull yellow to yellow-brown or date brown; surface dry, dull, more or less velvety when young; **flesh** yellow, instantly changing to dark blue when cut or broken, odor slightly pungent, taste astringent or not distinctive.
Pore surface: Dark brown, coffee-brown, or orange-brown, instantly bruising blackish blue; pores minute, angular, 2–3 per mm.
Stalk: More or less equal, solid; brownish pruinose over a dull whitish to pale yellow ground color, sometimes with olive tones; surface staining blue then dark brown where bruised or handled.
Spore print: Olive-brown.
Spores: 9–12 x 3–4 μm, elliptical, smooth.
Occurrence: Solitary to scattered in groups on the ground under oaks; mycorrhizal; summer-fall; occasional to fairly common in some years.
Edibility: Unknown.

Comments: *Boletus vermiculosus* (not illustrated) is very similar but lacks yellow tones on the cap and has larger spores (11–15 x 4–6 μm). *Boletus fagicola* (not illustrated) is also similar but has reticulation on the upper stalk. *Vermiculosoides* means "resembling *Boletus vermiculosus.*" *Vermiculosus* means "infested with vermin" (insect larvae), which both of these species often are.

Latin name: *Chalciporus piperatoides* (A.H. Sm. & Thiers) T.J. Baroni & Both
Synonym: *Boletus piperatoides* A.H. Sm. & Thiers
Common name: Peppery Bolete
Order: Boletales
Family: Strobilomycetaceae
Width of cap: ¾–2 in. (2–5 cm)
Length of stalk: 1–2 in. (2.5–5 cm)
Cap: Convex, becoming nearly flat with age, sometimes slightly umbonate, margin with a narrow edge of sterile tissue, inrolled at first; color variable from orange to cinnamon, buff, yellow-brown, or rusty brown; surface slightly viscid when moist, shiny; **flesh** pinkish buff to brown, turning dark blue above the tubes, odor not distinctive, taste peppery (sometimes slowly).
Pore surface: Reddish brown to cinnamon-brown or yellowish brown, bruising blackish blue; pores medium, up to 1.5 mm wide, somewhat radially arranged.
Stalk: Nearly equal; colored like the cap or paler with bright yellow mycelium at the base; surface dry, slightly scurfy to dotted.
Spore print: Dark olive (cinnamon when dry).
Spores: 6–10 x 3–4 µm, somewhat spindle-shaped, smooth, yellowish to pale brown.
Occurrence: Solitary or in small groups on the ground in woods, grassy woodland margins, and parklands under broad-leaved and conifer trees, especially oak, beech, and hemlock; mycorrhizal; early summer-fall; common.
Edibility: Not edible.

Comments: This small bolete is easily recognized by the acrid, peppery taste of the flesh, the blackish blue staining of the bruised pore surface, and the bright yellow mycelium at the stalk base. *Chalciporus piperatus* (not illustrated) is nearly identical but its pore surface does not bruise blackish blue. *Chalciporus rubinellus* (not illustrated) is somewhat similar but has a broadly conical cap, a pinkish red pore surface that does not bruise blue, and mild tasting flesh. *Piperatoides* means "resembling *Chalciporus piperatus*." The genus name *Chalciporus* means "copper-colored pores." *Piperatus* means "peppery."

Latin name: *Suillus punctipes* (Peck) Singer
Synonym: *Boletus punctipes* Peck
Common name: Spicy Suillus
Order: Boletales
Family: Suillaceae
Width of cap: 1–2½ in. (2.5–6.5 cm)
Length of stalk: 1½–3 in. (4–7.5 cm)
Cap: Convex, becoming nearly flat and often with an upturned margin with age; dull yellow to ochraceous or tawny; surface smooth, often appearing somewhat velvety with a grayish bloom at first, viscid when wet; **flesh** pale yellow, with a spicy odor, taste not distinctive.
Pore surface: Grayish brown or cinnamon-brown to dull ochre, often exuding darker brown droplets when fresh; pores small, 2 per mm.
Stalk: Nearly equal to tapering upward from an enlarged base, frequently curved, solid; colored more or less like the cap; surface covered with glandular dots and clusters, somewhat sticky or waxy to the touch.
Spore print: Olive-brown.
Spores: 7.5–12 x 3–4 µm, elliptical to spindle-shaped, smooth, pale brown.
Occurrence: Usually gregarious under pines, especially white pine, also reported to occur under spruce and balsam fir; mycorrhizal; summer-fall; occasional to locally common.
Edibility: Edible but little appreciated.

Comments: The Spicy Suillus is one of the few boletes having a characteristic odor. *Suillus subaureus* (not illustrated) is similar but has an orangish yellow pore surface and lacks a distinctive odor. *Punctipes* means "dotted foot."

Latin name: *Tylopilus eximius* (Peck) Singer
Synonym: *Boletus eximius* Peck
Common name: Lilac-brown Bolete
Order: Boletales
Family: Boletaceae
Width of cap: 2–4½ in. (5–11.5 cm)
Length of stalk: 2–3½ in. (5–9 cm)
Cap: Convex to broadly convex, becoming nearly flat with age, margin incurved at first; purplish brown to grayish brown, often with a whitish bloom when young; surface glabrous to slightly velvety; **flesh** whitish, becoming grayish brown or tinged reddish when cut, odor not distinctive, taste not distinctive or slightly bitter.
Pore surface: Dark purple-brown to chocolate-brown; pores minute, up to 3 per mm; pore surface often pitted.
Stalk: Stout, more or less equal or enlarged downward, solid; pale purplish gray densely covered with darker purple-brown dots, flecks, or tiny scabers.
Spore print: Reddish brown to amber.
Spores: 11–17 x 3.5–5 μm, elliptical to narrow spindle-shaped, smooth, hyaline to pale brown.
Occurrence: Solitary or in groups on the ground under conifers, especially hemlock, also in mixed woods with oak; mycorrhizal; summer-fall; occasional.
Edibility: Not recommended (see comments below).

Comments: Although this bolete has been regarded as a safe edible in the past, there are recent reports of some collections from the northeast having caused poisonings. Those who would experiment with eating this species should do so with caution. Because of its speckled stalk, the Lilac-brown Bolete might be mistaken for a species of *Leccinum,* which differ by having brownish to yellow-brown spores. *Tylopilus plumbeoviolaceus* (not illustrated) has a smooth purplish stalk, and intensely bitter tasting flesh. *Eximius* means "distinguished" or "excellent in size or beauty."

Boletes with a White or Yellow Pore Surface and Net-like Ornamentation on the Stalk

The net-like pattern (reticulation) on the stalk of many boletes is usually an obvious feature, but it can vary in color, mesh size, and the extent of the stalk that is covered. With some boletes the reticulation is limited to the very apex of the stalk. Occasionally it may be so fine or delicate that a hand lens is helpful to determine its presence. Species that only rarely show a slight reticulation at the apex are distributed in other sections.

Latin name: *Austroboletus betula* (Schwein.) E. Horak
Synonym: *Boletellus betula* (Schwein.) Gilbert
Common name: Shaggy-stalked Bolete
Order: Boletales
Family: Strobilomycetaceae
Width of cap: 1¼–3 in. (3–7.5 cm)
Length of stalk: 3–7 in. (7.5–18 cm)
Cap: Hemispheric at first, becoming convex to flattened or depressed with age; color variable from bright yellow to yellow-brown, reddish orange, red, or reddish brown, often with a yellow rim on the margin; surface viscid, glabrous, shiny; **flesh** pale yellow or tinged with red, odor not distinctive, taste somewhat acidic.
Pore surface: Yellow, becoming greenish yellow to olivaceous with age; pores fairly large, 1 mm wide; tube layer depressed at the stalk, tubes deep (up to 1.5 cm).
Stalk: Slender, equal or slightly tapering upward, solid; surface with a coarse to almost shaggy yellow reticulation over a rose or pinkish ground color with copious white mycelium at the base; reticulation sometimes reddish with age, especially on the lower portion.
Spore print: Olive-brown.
Spores: 15–18 x 6–9 μm, elliptical with an apical germ pore, surface pitted, pale brown.
Occurrence: Solitary or gregarious on the ground in a variety of broad-leaved and mixed woods, especially low, wet woods along rivers and streams; mycorrhizal; summer-fall; occasional to common locally.
Edibility: Edible.

Comments: This spectacular bolete is not easily confused with any other. The cap is quite small in relation to the long stalk. The much rarer *Boletellus russellii* (not illustrated) is similar in stature, but has a dry cap that becomes cracked or scaly with age. *Betula* is the genus of birch trees, and the name is applied to this bolete because of its somewhat shaggy, birch-bark-like stalk.

Latin name: *Boletus auriflammeus* Berk. & M.A. Curtis
Synonym: *Pulveroboletus auriflammeus* (Berk. & M.A. Curtis) Singer
Common name: Flaming Gold Bolete
Order: Boletales
Family: Boletaceae
Width of cap: 1–3½ in. (2.5–9 cm)
Length of stalk: 2–3½ in. (5–9 cm)
Cap: Convex to becoming nearly flat with age; brownish orange to golden yellow, or chrome yellow; surface velvety to powdery at first, sometimes becoming finely cracked; **flesh** white to cream or pale yellow, odor not distinctive, taste mild or somewhat acidic.
Pore surface: Yellow to yellow-orange, sometimes pore mouths tinged with red; pores medium, typically more than 1 mm wide at maturity, radially elongated near the stalk to angular; tube layer often depressed at the stalk, or sometimes slightly decurrent.
Stalk: More or less equal or tapering in either direction, solid; colored like the cap or more intensely yellow-orange, white mycelium at the base; surface somewhat powdery at first, coarsely reticulated, at least on the upper portion, reticulation sometimes faint or, more rarely, lacking completely.
Spore print: Olive-brown to ochre-brown.
Spores: 8–12 x 3–5 μm, elliptical to somewhat spindle-shaped, smooth, nearly hyaline.
Occurrence: Solitary or in small groups or caespitose clusters on the ground in broad-leaved or mixed woods, especially with oak; summer-fall; uncommon.
Edibility: Edible.

Comments: This small to medium size bolete is striking for its overall deep, rich yellow-orange color, and coarse reticulation on the stalk. Handling fresh specimens will stain fingers yellow. Compare with *Boletus ornatipes* (p.304). *Auriflammeus* means "flaming gold."

Latin name: *Boletus auripes* Peck
Common name: Butter-foot Bolete
Order: Boletales
Family: Boletaceae
Width of cap: 1½–4½ in. (4–11.5 cm)
Length of stalk: 1½–3 in. (4–7.5 cm)
Cap: Convex becoming broadly convex to nearly flat with age; chestnut-brown to yellow-brown, paler with age; surface dry, velvety at first; **flesh** yellow, firm, odor and taste not distinctive.
Pore surface: Yellow, unchanging when bruised; pores minute, typically less than 1 mm wide.
Stalk: Nearly equal to enlarged at the base; bright yellow or yellow tinged with brown; surface fibrous-striate to somewhat scurfy, with fine yellow reticulation on the upper portion.
Spore print: Yellow-brown to olive-brown.
Spores: 10–14 x 3–5 μm, elliptical to somewhat spindle-shaped, smooth, yellowish.
Occurrence: Solitary or in groups on the ground in broad-leaved and mixed woods, especially with oak; mycorrhizal; early summer-fall; occasional.
Edibility: Edible.

Comments: The rich brown and yellow colors of this handsome bolete and the lack of staining anywhere when bruised are distinctive. *Boletus subglabripes* (not illustrated) and *Boletus hortonii* (p.326) have a similar color scheme but lack reticulation on the stalk. *Auripes* means "golden yellow foot."

Latin name: *Boletus edulis* Bull.
Common names: King Bolete; Cep; Porcini; Steinpilz
Order: Boletales
Family: Boletaceae
Width of cap: 1½–7½ in. (4–20 cm)
Length of stalk: 1½–4 in. (4–10 cm)
Cap: Convex becoming broadly convex to nearly flat with age; pale brown to reddish brown; surface viscid when moist, slightly tacky to the touch, smooth to somewhat lumpy or pitted; **flesh** white, odor not distinctive, taste mild.
Pore surface: White at first, becoming yellow then olivaceous with age, usually unchanging when bruised but sometimes staining dull orange-cinnamon to pale yellowish brown; pores small, 2–3 per mm, larger in older specimens; tube layer often depressed at the stalk.
Stalk: Club-shaped to bulbous, solid; white to pale brown; surface covered with a fine white reticulum, at least on the upper third.
Spore print: Olive-brown.
Spores: 13–19 x 4–6.5 μm, elliptical, smooth, pale yellowish brown.
Occurrence: Usually gregarious on the ground under conifers, especially with pine and Norway spruce; mycorrhizal; early summer-fall; occasional to locally common.
Edibility: Edible and choice.

Comments: This is one of the world's most popular edible mushrooms. There are several color forms and varieties, at least three of which occur in the central Appalachians. The features that they all have in common are a white pore surface that becomes yellow at maturity, white reticulation on the stalk, mild tasting flesh, and association with conifers. *Boletus edulis* var. *clavipes* (not illustrated) has a comparatively long, yellowish orange stalk. *Boletus subcaerulescens* (not illustrated) has a reddish brown, coarsely reticulated stalk and a pore surface that bruises slowly blue. *Tylopilus felleus* (p.310), which is sometimes mistaken for the King Bolete, has brown reticulation on the stalk, bitter tasting flesh, and pinkish brown spores. Compare also with *Boletus nobilis* (p.303) and *Boletus variipes* (p.307), both of which are most often associated with broad-leaved trees. *Edulis* means "edible," which in this case is an understatement.

Latin name: *Boletus griseus* Frost
Common name: Gray Bolete
Order: Boletales
Family: Boletaceae
Width of cap: 2–5 in. (5–12.5 cm)
Length of stalk: 1½–4 in. (4–10 cm)
Cap: Convex becoming broadly convex to nearly flat; gray to grayish brown, or charcoal, sometimes with yellow or ochraceous tints; surface dry, dull, somewhat velvety at first, often cracking into small scales or patches with age; **flesh** whitish, dark brown where insect larvae have tunneled, odor and taste not distinctive.
Pore surface: Whitish to pale gray, becoming darker with age, unchanging or slightly brownish when bruised; pores small, 1–2 per mm.
Stalk: Nearly equal but often pinched or tapered at the base, solid; whitish with a yellow base, or infrequently reddish at the base; surface covered with a prominent, coarse, yellowish reticulum that darkens to brown with age.
Spore print: Olive-brown.
Spores: 9–13 x 3–5 μm, oblong, smooth, pale brown.
Occurrence: Solitary or in small groups on the ground in broad-leaved woods, especially with oak and beech; mycorrhizal; early summer-fall; common in some years.
Edibility: Edible and good, but the flesh is often insect damaged.

Comments: This species is typical of the summer boletes that inhabit oak woodlands. It is easily recognized by the grayish cap, a coarsely netted white stalk with a yellow base, and mild tasting flesh. Gray forms of *Boletus ornatipes* (p.304) are similar but have entirely yellow stalks and bitter tasting flesh. *Tylopilus felleus* (p.310) has a brown cap, bitter tasting flesh, and pinkish brown spores. *Griseus* means "gray."

Latin name: *Boletus nobilis* Peck
Common name: Noble Bolete
Order: Boletales
Family: Boletaceae
Width of cap: 2–6 in. (5–15 cm)
Length of stalk: 2½–6 in. (6.5–15 cm)
Cap: Convex to broadly convex or nearly flat with age; yellow-brown to reddish brown, becoming ochraceous to reddish ochraceous with age; surface dry, smooth to somewhat pitted; **flesh** white or yellowish where the tubes meet, odor and taste not distinctive.
Pore surface: White when young, becoming yellow to ochraceous with age; unchanging when bruised; pores small, 1–3 per mm.
Stalk: More or less equal or enlarged downward, solid; white to pale ochre, sometimes with pale brown streaking, especially near the base; surface smooth, with a delicate white reticulation on the upper portion, or occasionally without reticulation.
Spore print: Ochre-brown to dull rusty brown.
Spores: 11–16 x 4–5 μm, somewhat spindle-shaped, smooth, pale olivaceous.
Occurrence: Solitary or in groups on the ground in broad-leaved woods, especially with oak and beech; mycorrhizal; summer-fall; occasional.
Edibility: Edible.

Comments: This bolete is easily and harmlessly mistaken as a color form of *Boletus edulis* (p.301), which has a moist to slightly sticky cap, more prominent reticulation, and is associated with conifers. *Boletus gertrudiae* (not illustrated) is very similar but has a broad yellow band at the apex of the stalk. Compare with *Boletus variipes* (p.307), which is also similar but has more conspicuous reticulation that usually covers most, if not all of the stalk. *Nobilis* means "noble" or "grand."

Latin name: *Boletus ornatipes* Peck
Common name: Ornate-stalked Bolete
Order: Boletales
Family: Boletaceae
Width of cap: 1½–8 in. (4–20 cm)
Length of stalk: 2–6 in. (5–15 cm)
Cap: Convex becoming broadly convex to nearly flat or slightly depressed in the center with age; color variable from yellow, mustard yellow, olive-yellow, to gray; surface dull to somewhat powdery or slightly velvety; **flesh** yellow, odor not distinctive, taste bitter.
Pore surface: Deep, rich yellow to lemon yellow, quickly staining yellow-orange to orange-brown when bruised; pores small, 1–2 per mm, or larger on fully expanded caps.
Stalk: Variable from nearly equal or swollen in the middle, to tapered toward the base, or less often with a swollen or club-shaped base, solid; yellow to somewhat brownish, staining darker where bruised or handled; surface prominently and coarsely reticulated, usually over the entire length.
Spore print: Olive-brown to dark yellow-brown.
Spores: 9–13 x 3–4 μm, elliptical, smooth, pale brown.
Occurrence: Solitary or more often in groups, sometimes in caespitose clusters on the ground in broad-leaved woods, especially with oak; mycorrhizal; early summer-fall; common.
Edibility: Not edible.

Comments: Handling this showy bolete will stain fingers yellow. Although sometimes reported to be edible, most collections in the central Appalachians have bitter, unpalatable flesh. Grayish forms bear a strong resemblance to *Boletus griseus* (p.302), which has mild tasting flesh and is yellow only at the stalk base. Both species often appear in the same habitat at the same time. Compare with *Boletus auriflammeus* (p.299), which is similar but generally smaller and has a bright orange-yellow cap and stalk. *Ornatipes* means "ornate foot."

Latin name: *Boletus projectellus* (Murrill) Murrill
Synonym: *Boletellus projectellus* (Murrill) Singer
Common name: None
Order: Boletales
Family: Boletaceae
Width of cap: 1½–8 in. (4–20 cm)
Length of stalk: 3–9 in. (7.5–23 cm)
Cap: Convex, becoming broadly convex to nearly flat; pinkish brown to grayish brown at first, becoming reddish brown; surface dry and somewhat velvety at first, often cracking with age, with a sterile margin that extends beyond the tube layer; **flesh** whitish to pale rose just beneath the cuticle, thick in the center, odor not distinctive, taste acidic, somewhat "lemony."
Pore surface: Pale yellow, becoming greenish yellow to olivaceous with age; pores large, 0.5–2 mm wide; tube layer depressed at the stalk.
Stalk: Equal to tapering upward from an enlarged base; solid; colored like the cap or paler, with copious white mycelium at the base; surface somewhat granular, viscid when wet, covered at least on the upper portion with a coarse, wide-meshed reticulum.
Spore print: Olive-brown.
Spores: 18–33 x 7.5–12 µm, oval to spindle-shaped, smooth, pale brown.
Occurrence: Usually in groups on the ground under pines, especially planted Scots pine (*Pinus sylvestris*); mycorrhizal; late summer-fall; uncommon.
Edibility: Edible.

Comments: This bolete is distinctive for its overlapping cap cuticle, reddish brown cap and stalk, association with pines, and late summer or fall appearance. *Boletus atkinsonianus* (not illustrated) is similar but lacks the prominent stalk reticulation and grows in broad-leaved woods, or mixed woods with hemlock. *Boletus badius* (p.315) is smaller, has a smooth stalk, and has a pale yellow pore surface that bruises blue. The name *projectellus* refers to the cap cuticle that projects beyond the tube layer.

Latin name: *Boletus speciosus* Frost var. *brunneus* Peck
Synonym: *Boletus pseudopeckii* A.H. Sm. & Thiers
Common name: None
Order: Boletales
Family: Boletaceae
Width of cap: 1½–5 in. (4–12.5 cm)
Length of stalk: 2–4 in. (5–10 cm)
Cap: Convex, becoming broadly convex to nearly flat; reddish brown or yellow-brown to olive-brown; surface dry, dull, smooth; **flesh** yellow, quickly staining blue when exposed, odor and taste not distinctive.
Pore surface: Bright yellow to medium yellow, instantly bruising dark blue; pores small, 2–3 per mm.
Stalk: More or less equal or enlarged at the base, solid; yellow at least on the upper portion, often merging with pinkish red to purplish red areas toward the base; surface covered with a fine reticulum on the upper stalk, and extending downward for at least half the length.
Spore print: Olive-brown.
Spores: 10–15 x 3.5–5 μm, oblong to spindle-shaped, smooth, ochraceous.
Occurrence: Usually gregarious, but sometimes solitary on the ground under broad-leaved and conifer trees, especially hemlock; mycorrhizal; early summer-fall; occasional to locally common.
Edibility: Edible.

Comments: This is one of the first boletes of the season to appear, often in early June. Although edible, its tender flesh is eagerly sought by numerous insect larvae, snails, slugs, and small mammals. Unblemished specimens are seldom found. *Speciosus* means "showy." *Brunneus* means "russet-brown."

Latin name: *Boletus variipes* Peck
Common name: None
Order: Boletales
Family: Boletaceae
Width of cap: 2–6 in. (5–15 cm)
Length of stalk: 2½–5 in. (6.5–12.5 cm)
Cap: Convex, becoming broadly convex to nearly flat or depressed in the center with age; gray-brown to yellow-brown, ochraceous, or tan; surface somewhat velvety when young, often cracking, sometimes with rather large fissures in dry weather; **flesh** white, odor and taste not distinctive.
Pore surface: White, becoming yellow to greenish yellow with age, unchanging when bruised; pores small, 1–2 per mm.
Stalk: Equal to club-shaped, solid; whitish to variously brownish; surface dry, covered with a more or less distinct white or brown reticulum.
Spore print: Olive-brown.
Spores: 12–18 x 4–6 μm, elliptical to spindle-shaped, smooth, yellow.
Occurrence: Usually gregarious on the ground, typically under broad-leaved trees, especially oak, but sometimes under pine, hemlock, or Norway spruce; mycorrhizal; late spring-fall; common.
Edibility: Edible.

Comments: *Tylopilus felleus* (p.310) is similar but has bitter tasting flesh and a pore surface that becomes dull pinkish as the spores mature. *Xanthoconium affine* (p.342) is similarly colored but typically has a smooth stalk. Compare with *Boletus edulis* (p.301), which has a reddish brown, somewhat tacky cap cuticle, and grows only with conifers. Compare also with *Boletus nobilis* (p.303), which has a taller stature and a delicate reticulation that is normally limited to the upper stalk. *Variipes* means "various foot" referring to the shape, color, and degree of reticulation on the stalk.

Latin name: *Strobilomyces floccopus* (Vahl.) P. Karst.
Common names: Old Man of the Woods; Pine Cone Bolete
Order: Boletales
Family: Boletaceae
Width of cap: 1½–6 in. (4–15 cm)
Length of stalk: 2–5 in. (5–12.5 cm)
Cap: Hemispherical at first, soon convex, becoming flattened with age; surface dry, densely covered with soft gray to black appressed scales on a whitish to grayish ground color; margin shaggy with remnants of partial veil material; **flesh** whitish, staining orangish red, then turning black when exposed, odor and taste not distinctive.
Pore surface: White to grayish white, becoming darker to nearly black with age; staining red then black when bruised; pores medium to large, 1–2 mm wide, angular.
Stalk: Equal or enlarged at the base, solid; colored like the cap, surface coarsely scaly-shaggy; partial veil leaving a shaggy ring zone on the upper stalk, reticulate above the ring zone.
Spore print: Black.
Spores: 9.5–15 x 8.5–12 μm, broadly elliptical with an apical germ pore, surface reticulate, grayish.
Occurrence: Solitary or loosely scattered on the ground in broad-leaved and mixed woods, especially with oak, infrequently with conifers; mycorrhizal; early summer-fall; common.
Edibility: Edible, but mediocre.

Comments: *Strobilomyces confusus* (not illustrated) is nearly identical but has more pointed, erect scales on the cap, and an incomplete reticulation on the spores. *Floccopus* means "woolly leg."

Latin name: *Suillus castanellus* (Peck) A.H. Sm. & Thiers
Synonyms: *Boletinus castanellus* Peck; *Boletinus squarrosoides* Snell & Dick
Common name: None
Order: Boletales
Family: Suillaceae
Width of cap: 1–3 in. (2.5–7.5 cm)
Length of stalk: ¾–3 in. (2–7.5 cm)
Cap: Convex, becoming nearly flat or shallowly depressed; reddish brown to burgundy or yellow-brown; surface dry, densely tomentose, becoming nearly glabrous with age; **flesh** white, odor and taste not distinctive.
Pore surface: Pinkish brown to golden brown, tan, or buff, bruising ochraceous tawny; pores angular to elongated, up to 2 mm wide, radially arranged; tube layer subdecurrent.
Stalk: Nearly equal to tapering downward, central or eccentric; colored like the cap but usually paler; surface distinctly reticulate, at least on the upper portion.
Spore print: Yellow-brown.
Spores: 8–11 x 4.5–5.5 µm, ovate to narrowly elliptical, smooth, yellowish.
Occurrence: Solitary or in groups, sometimes in caespitose clusters, on the ground under oak in woods and parklands; mycorrhizal; summer-fall; occasional to locally common.
Edibility: Unknown.

Comments: This small- to medium-sized bolete is fairly easy to recognize. The reticulated stalk is unusual for boletes in the genus *Suillus,* and some mycologists think it should be placed in a separate genus. *Castanellus* means "small chestnut," referring to the cap color.

Latin name: *Tylopilus felleus* (Bull.) P. Karst.
Common name: Bitter Bolete
Order: Boletales
Family: Boletaceae
Width of cap: 2–8 in. (5–20 cm)
Length of stalk: 1½–6½ in. (4–16.5 cm)
Cap: Convex, becoming nearly flat; sometimes pinkish to pale reddish purple at first, soon turning brown to yellow-brown or ochraceous; surface dry, smooth, somewhat suede-like; **flesh** white, sometimes staining slightly reddish when cut or broken, odor not distinctive, taste intensely bitter.
Pore surface: White at first, becoming dull pinkish with age, often bruising brownish; pores small, 1–2 per mm, rounded to angular.
Stalk: Nearly equal to bulbous, solid; colored like the cap or paler, often developing olive-brown stains when bruised; surface covered with a coarse, brown reticulation, usually over most of its length.
Spore print: Pinkish brown to reddish brown.
Spores: 11–17 x 3–5 μm, elliptical, smooth, hyaline to pale brown.
Occurrence: Solitary or in groups on the ground or on decaying stumps and logs in broad-leaved and conifer woods; mycorrhizal; summer-fall; common.
Edibility: Non-poisonous.

Comments: The Bitter Bolete is sometimes confused with the mild tasting *Boletus edulis* (p.301), which has fine white reticulation on the stalk. *Tylopilus indecisus* (not illustrated) is similar but has mild tasting flesh and finer, less prominent reticulation on the stalk. *Tylopilus rubrobrunneus* (p.340) has purplish or violaceous tones on the mature cap, and reticulation only near the apex of the stalk. Compare also with *Tylopilus variobrunneus* (p.312). The specific epithet is derived from the Latin word *fel* (gall bladder), relating here to the bitter flesh. Apparently some individuals who cannot detect bitterness enjoy eating this mushroom. I am reminded of a remark overheard at a mushroom foray social gathering, "if you like the bitter bolete, you're gonna love this wine."

Latin name: *Tylopilus intermedius* A.H. Sm. & Thiers
Common name: Bitter Parchment Bolete
Order: Boletales
Family: Boletaceae
Width of cap: 2–6 in. (5–15 cm)
Length of stalk: 2–5 in. (5–12.5 cm)
Cap: Convex with an incurved margin at first, becoming broadly convex to nearly flat with age; whitish to buff or pale tan, developing yellow-brown to olive-brown stains; surface dry, dull, uneven, wrinkled to somewhat lumpy; **flesh** firm, white, slowly discoloring brownish when exposed, odor not distinctive, taste intensely bitter.
Pore surface: White to grayish buff, becoming pinkish brown, slowly bruising brownish; pores 1–2 per mm; tube layer depressed at the stalk.
Stalk: Club-shaped to bulbous, or sometimes nearly equal, solid; dull white to dingy brown, developing yellow-brown to olive-brown stains when bruised or handled, white mycelium at the base; surface more or less smooth with a fine white reticulum near the apex, or infrequently without reticulation.
Spore print: Pinkish brown.
Spores: 10–15 x 3–5 μm, nearly oblong, smooth, hyaline to pale brown.
Occurrence: Usually in groups in broad-leaved woods, especially with oak and hickory; mycorrhizal; summer-early fall; uncommon.
Edibility: Not edible.

Comments: Unless found in young, perfect condition, this bolete often has a "soiled" appearance. The specific epithet means "intermediate between two species," referring here to *Tylopilus peralbidus* (not illustrated) and *Tylopilus rhoadsiae* (not illustrated), both of which occur in the southeastern United States.

Latin name: *Tylopilus variobrunneus* W.C. Roody, A.R. Bessette, & A.E. Bessette
Common name: Brown-net Bolete
Order: Boletales
Family: Boletaceae
Width of cap: 1½–4½ in. (4–11.5 cm)
Length of stalk: 1½–4 in. (4–10 cm)
Cap: Convex, becoming broadly convex to nearly flat; dark reddish brown to chestnut-brown, often with olive tones when young, paler with age; surface dry, somewhat velvety when young; **flesh** white, staining brownish pink when exposed, odor not distinctive, taste mild to somewhat bitter.
Pore surface: Dull white, becoming brownish pink at maturity, staining brownish rose to brown when bruised; pores small, 2–3 per mm.
Stalk: Nearly equal to somewhat club-shaped, solid; dull white on the upper portion becoming brown toward the base, mycelium at the base white; surface dry, covered with prominent white to brown reticulation.
Spore print: Pinkish brown to brown.
Spores: 9–13 x 3–4.5 μm, somewhat elliptical to spindle-shaped, smooth, pale ochraceous.
Occurrence: Solitary or more often in groups on the ground in oak or mixed oak-pine woods; mycorrhizal; summer-early fall; occasional to locally common.
Edibility: Unknown.

Comments: This handsome bolete has a compact stature and firm flesh when young. It might be confused with *Tylopilus felleus* (p.310), which has very bitter tasting flesh. *Tylopilus badiceps* (not illustrated) has a reddish brown cap and stalk, and more delicate reticulation on the stalk. *Tylopilus indecisus* (not illustrated) is more ochraceous brown and taller, typically having less prominent reticulation on the stalk. *Variobrunneus* means "various shades of brown."

Latin name: *Xanthoconium separans* (Peck) Halling & Both
Synonym: *Boletus separans Peck*
Common name: Lilac Bolete
Order: Boletales
Family: Boletaceae
Width of cap: 2–6 in. (5–15 cm)
Length of stalk: 2–5 in. (5–12.5 cm)
Cap: Convex to broadly convex; pinkish brown to lilac-brown, reddish brown or sometimes dark purple, often paler near the margin and becoming yellowish brown with age; surface somewhat velvety to glabrous, often pitted, wrinkled, or lumpy; **flesh** white, odor and taste not distinctive.
Pore surface: White at first, becoming yellowish to ochraceous with age, unchanging when bruised; pores small, 1–2 per mm.
Stalk: More or less equal to tapering upward from an enlarged base, solid; colored like the cap or paler, often with lilac tones in the mid-portion, or sometimes purplish overall; surface with white reticulation over most of the length, or at least covering the upper half of the stalk.
Spore print: Brownish ochre to pale reddish brown.
Spores: 12–16 x 3.5–5 µm, narrowly spindle-shaped, smooth, pale brown.
Occurrence: Solitary or in groups on the ground in mixed woods, oak woods, or occasionally with pine; mycorrhizal; early summer-fall; common.
Edibility: Edible and good.

Comments: The medium- to large-sized Lilac Bolete is a beautiful mushroom with a delicately netted stalk. The color of the cap and stalk is variable, but lilac tones are usually present. The epithet *separans* refers to the tube layer that sometimes pulls away from the stalk as the cap expands.

Boletes with a White or Yellow Pore Surface That Stains Blue or Greenish Blue When Bruised

A bluish or greenish blue bruising of the pore surface is not uncommon in boletes. Although the intensity of the staining varies, it is readily apparent when the pore surface is rubbed or scratched. With most species the reaction is immediate and pronounced. With others, especially on older and dryer specimens, the staining may be slow and weak. It is best to check this feature using fresh boletes in good condition.

Latin name: *Boletus badius* Fr.
Synonym: *Xerocomus badius* (Fr.) Kühner
Common name: Bay Bolete
Order: Boletales
Family: Boletaceae
Width of cap: 1½–4 in. (4–10 cm)
Length of stalk: 1½–3½ in. (4–9 cm)
Cap: Convex to broadly convex, becoming nearly flat to depressed in the center, margin upturned with age; chestnut-brown, reddish brown, or yellow-brown, sometimes with olive tones; surface smooth to somewhat velvety, dry, or slightly viscid when wet; **flesh** soft, white, often staining bluish near the tubes when cut, odor and taste not distinctive.
Pore surface: Pale yellow to greenish yellow, bruising greenish blue; pores angular, 1–2 per mm.
Stalk: Equal or enlarged toward the base, solid; colored like the cap with white mycelium at the base; surface smooth to somewhat pruinose.
Spore print: Olive-brown.
Spores: 10–14 x 4–5 μm, elliptical to spindle-shaped, smooth, yellow.
Occurrence: Solitary or gregarious on the ground, or sometimes on decaying stumps in conifer woods, especially with pine and spruce, infrequently reported in broad-leaved woods; mycorrhizal; early summer-fall; common.
Edibility: Edible and good. The flesh is not often insect damaged.

Comments: *Boletus projectellus* (p.305) has similar colors but is typically more robust and has a coarsely reticulated stalk. *Austroboletus gracilis* (p.324) has a white pore surface at first that becomes pinkish in age and does not bruise blue. *Badius* means "bay-brown," which is reddish brown or chestnut brown.

Latin name: *Boletus bicolor* var. *bicolor* Peck
Common name: Red and Yellow Bolete
Order: Boletales
Family: Boletaceae
Width of cap: 2–5 in. (5–12.5 cm)
Length of stalk: 1½–4 in. (4–10 cm)
Cap: Convex to broadly convex, becoming nearly flat in age, margin incurved at first; dark red to rose red, fading to ochraceous or tan; surface dry, dull, somewhat velvety when young, often cracking in dry weather; **flesh** yellow, unchanging or slowly and weakly staining blue when exposed, odor and taste not distinctive.
Pore surface: Bright yellow, becoming olive-yellow in age, quickly bruising blue; pores angular, minute, 1–2 per mm; tubes layer shallow.
Stalk: Nearly equal to club-shaped, solid; purplish red to rusty red or rosy red, yellow at the apex or sometimes nearly yellow overall; surface dry, smooth, or slightly reticulate at the apex in *Boletus bicolor* var. *subreticulatus* (not illustrated).
Spore print: Olive-brown.
Spores: 8–12 x 3.5–5 μm, oblong to spindle-shaped, smooth, pale brown, non-amyloid.
Occurrence: Solitary or in groups, sometimes clustered on the ground in broadleaved or mixed woods, especially with oak; mycorrhizal; early summer-fall; common.
Edibility: Edible with caution. Some individuals have reported having gastrointestinal disturbance from eating *Boletus bicolor* or a closely related species.

Comments: The Red and Yellow Bolete is generally considered to be a good edible, but it is easily confused with a number of similar species of uncertain edibility. Important distinguishing features are the yellow flesh that doesn't change when exposed, or at most turns blue very slowly and weakly, and the shallow tube layer with minute pores. The similarly colored *Boletus sensibilis* (not illustrated), which is reputed to be poisonous, quickly bruises blue on all parts and often has a sweet odor of fenugreek or curry. *Bicolor* means "two colors."

Latin name: *Boletus chrysenteron* Bull.
Synonym: *Xerocomus chrysenteron* (Bull.) Quél.
Common name: Red-cracked Bolete
Order: Boletales
Family: Boletaceae
Width of cap: 1¼–3 in. (3–7.5 cm)
Length of stalk: 1½–2½ in. (4–6.5 cm)
Cap: Convex, becoming broadly convex to nearly flat; dark olive to olive-brown or grayish brown, often with a reddish zone at the margin; surface velvety at first, soon becoming cracked with red to pinkish color in the cracks; **flesh** soft, white to yellow, or sometimes pinkish beneath the cuticle, slowly turning blue when exposed, odor and taste not distinctive.
Pore surface: Yellow, becoming olivaceous with age, bruising greenish blue, sometimes slowly; pores 1 mm wide, irregular to angular; tube layer depressed at the stalk.
Stalk: Slender, more or less equal to slightly enlarged downward, solid; yellow on the upper portion, usually with reddish streaks and red near the base, white mycelium at the base; surface dry, granular to somewhat scurfy.
Spore print: Olive-brown.
Spores: 9–13 x 3.5–4.5 μm, oblong, smooth, pale brown.
Occurrence: Solitary to thinly scattered on the ground in broad-leaved and conifer woods, also on roadside banks; mycorrhizal; early summer-fall; common but rarely found in large numbers.
Edibility: Edible but poor.

Comments: This bolete is commonly infected with the parasitic mold *Hypomyces chrysospermus* (p.494). The similar *Boletellus chrysenteroides* (not illustrated), which often grows on decaying logs, stumps, or at the base of standing trunks, has a darker cap that also cracks when mature, but does not show reddish tints in the fissures. *Boletus subtomentosus* (not illustrated) is similar but rarely has a cracked cap surface, and has a yellow pore surface that bruises slightly greenish blue if at all. *Chrysenteron* means "gold intestine," perhaps referring to the meandering pattern of the cracks on the cap?

Latin name: *Boletus pallidus* Frost
Common name: Pale Bolete
Order: Boletales
Family: Boletaceae
Width of cap: 1½–5 in. (4–12.5 cm)
Length of stalk: 2–5 in. (5–12.5 cm)
Cap: Convex, becoming broadly convex to nearly flat, margin incurved at first; whitish to buff or pale grayish brown when young, dingy brown in age, infrequently with a pale reddish tinge; surface dry, slightly velvety, smooth or sometimes becoming cracked with age; **flesh** white to pale yellow, odor and taste not distinctive.
Pore surface: Whitish to pale greenish yellow, bruising greenish blue then grayish brown; pores 1–2 per mm.
Stalk: Nearly equal or enlarged downward, solid; whitish, often with brownish streaks, especially near the base; surface smooth or slightly reticulate at the apex.
Spore print: Olive-brown.
Spores: 9–15 x 3–5 μm, narrowly oval to somewhat spindle-shaped, smooth, pale brown.
Occurrence: Solitary or gregarious, sometimes in caespitose clusters on the ground in broad-leaved woods, especially with oak; mycorrhizal; summer-early fall; fairly common in most years.
Edibility: Edible and good.

Comments: This bolete usually appears in mid to late summer and is sometimes abundant in oak woods. The mild tasting flesh and greenish blue bruising of the pore surface separate it from other whitish boletes. The somewhat similar *Tylopilus intermedius* (p.311) has a grayish buff pore surface and intensely bitter tasting flesh. *Pallidus* means "pallid," and refers to the overall pale color of the fruitbody.

Latin name: *Boletus pseudosensibilis* A.H. Sm. & Thiers
Common name: None
Order: Boletales
Family: Boletaceae
Width of cap: 2–5 in. (5–12.5 cm)
Length of stalk: 2½–5 in. (6.5–12.5 cm)
Cap: Convex to broadly convex; dull rusty brown to dull reddish brown, cinnamon, or yellow-brown; surface dry, dull, smooth or sometimes cracking in dry weather or in age; **flesh** bright yellow, quickly turning blue when cut or broken, odor not distinctive, taste mild.
Pore surface: Bright yellow, becoming olivaceous ochre in age, instantly bruising blue; pores minute, 1–3 per mm; tube layer shallow, often somewhat decurrent.
Stalk: Equal to spindle-shaped or enlarged toward the base, solid; pale yellow with reddish or reddish brown tones near the base; surface smooth.
Spore print: Olive-brown.
Spores: 9–12 x 3–4 μm, oblong to somewhat spindle-shaped, smooth, pale brown.
Occurrence: Solitary or in groups on the ground in broad-leaved woods, especially with oak, infrequently with pine; mycorrhizal; early summer-fall; occasional.
Edibility: Not recommended (see comments below).

Comments: Although reported to be edible, this bolete can be confused with similar species of unknown edibility. A drop of household ammonia on the cap surface will turn greenish blue, eliminating confusion with the poisonous *Boletus sensibilis* (not illustrated), which does not exhibit this reaction. Compare with *Boletus bicolor* (p.316), which has a more reddish cap, and flesh that stains slowly and weakly blue, if at all. *Pseudosensibilis* means "false *Boletus sensibilis.*" *Sensibilis* means "sensitive," referring to the readily bruising tissue.

Latin name: *Boletus pulverulentus* Opatowski
Common name: None
Order: Boletales
Family: Boletaceae
Width of cap: 1½–3 in. (4–7.5 cm)
Length of stalk: 1½–3 in. (4–7.5 cm)
Cap: Convex, becoming nearly flat in age; dark yellow-brown to dark brown or reddish brown, instantly bruising blackish blue; surface somewhat velvety or powdery at first, becoming smooth and slightly shiny in age; **flesh** yellow, instantly changing to deep blue when cut or broken, odor and taste not distinctive.
Pore surface: Yellow to golden yellow, instantly bruising blue; pores 1–2 per mm.
Stalk: Nearly equal, solid; yellow on the upper portion, reddish orange to reddish brown below with white mycelium at the base; surface somewhat granular to powdery, sometimes with raised vertical lines, instantly bruising blackish blue then brown when handled.
Spore print: Olive-brown.
Spores: 11–15 x 4–6 μm, elliptical to spindle-shaped, yellowish.
Occurrence: Solitary or in groups in broad-leaved and conifer woods, especially with oak, beech, and hemlock; mycorrhizal; summer-fall; occasional.
Edibility: Edible.

Comments: The entire fruitbody of this bolete instantly bruises dark blue to blackish blue when handled. *Pulverulentus* means "covered with powder," referring to the somewhat powdery appearance of the young cap and stalk. A form of this species with a reticulated stalk has been reported from Tennessee.

Latin name: *Gyrodon merulioides* (Schwein.) Singer
Synonym: *Boletinellus merulioides* (Schwein.) Murrill
Common name: Ash Tree Bolete
Order: Boletales
Family: Paxillaceae
Width of cap: 1½–5 in. (4–12.5 cm)
Length of stalk: ¾–2 in. (2–5 cm)
Cap: Often kidney-shaped in outline, convex with an incurved margin at first, becoming depressed in the center to nearly funnel-shaped in age; yellow-brown to olive-brown or reddish brown; surface slightly velvety at first, smooth, viscid and shiny when moist; **flesh** thick in the center, thin at the margin, yellow; odor and taste not distinctive.
Pore surface: Yellow to dull gold or olive, usually slowly bruising greenish blue; pore surface made up of prominent, almost gill-like ridges and crosswalls forming irregular to elongated pores that are radially arranged; pores 1 mm or more wide, becoming larger as the cap expands; pore layer thin, decurrent, not readily separating from the cap flesh.
Stalk: Short, solid, eccentric to nearly lateral; colored like the pore surface above, brownish to blackish at the base.
Spore print: Olive-brown.
Spores: 7–10 x 6–7.5 μm, broadly elliptical to nearly globose, pale yellow.
Occurrence: Solitary or more often in groups on the ground beneath ash trees in woods, parklands, urban areas, and lawns, rarely on stumps or on the base of tree trunks; saprobic; early summer-fall; common.
Edibility: Edible, but mediocre.

Comments: This unique and fascinating bolete is easily recognized by the dark reddish brown cap, off center stalk, radiating yellow pore surface, and association with ash trees. Although almost invariably found beneath ash trees, it is not mycorrhizal with ash. The connection is with aphids that feed on the roots of ash. The fungus feeds on a sugary exudation produced by the aphids. The specific epithet means "resembling *Merulius*," which is a genus of resupinate crust fungi having a wrinkled and ridged fertile surface.

Latin name: *Gyroporus cyanescens* (Bull.) Quél.
Common name: Bluing Bolete
Order: Boletales
Family: Gyroporaceae
Width of cap: 1½–4 in. (4–10 cm)
Length of stalk: 1½–4 in. (4–10 cm)
Cap: Convex to broadly convex, becoming nearly flat or depressed in the center with age; usually whitish to cream, but sometimes buff, pale yellowish, or pale tan, quickly bruising greenish blue to dark blue; surface dry, more or less roughened to coarsely tomentose or floccose-scaly; **flesh** whitish to pale yellow, quickly changing to blue when cut or broken, odor and taste not distinctive.
Pore surface: White to yellowish, becoming tan, quickly bruising greenish blue to blue; pores 1–2 per mm; tubes depressed at the stalk or sometimes free from the stalk.
Stalk: More of less equal or often swollen in the mid-portion or below; chambered within, stuffed with cottony pith, soon becoming hollow; colored like the cap; surface somewhat floccose-scaly at first, becoming smoother in age, quickly bruising blue; flesh hard, brittle.
Spore print: Pale yellow.
Spores: 8–10 x 5–6 μm, elliptical, smooth, hyaline.
Occurrence: Solitary or in groups in the ground, often on sandy soil in broad-leaved and mixed woods, especially with birch and hemlock, also along wooded road banks and mossy woodland margins; mycorrhizal; summer-early fall; occasional to fairly common in some years.
Edibility: Edible.

Comments: This bolete is easily recognized by its pale color, rapid deep blue bruising of all parts, and hard, brittle flesh, especially noticeable in the stalk. A relatively rare non-bluing variety occurs in the mountains of North Carolina. *Cyanescens* means "becoming deep blue."

Boletes with a White to Yellow Pore Surface That Does Not Stain When Bruised, or Bruises Colors Other Than Blue

These boletes have a pore surface that is white to yellow when young, but with age may develop other colors in the brown to gray, vinaceous, or olivaceous range. This is a fairly large section that contains species from several genera. They are placed here as a catch-all group that lack features used to define boletes in other sections.

Latin name: *Austroboletus gracilis* (Peck) Wolfe
Synonym: *Tylopilus gracilis* (Peck) Hennings
Common name: Graceful Bolete
Order: Boletales
Family: Boletaceae
Width of cap: 1¼–4 in. (3–10 cm)
Length of stalk: 2–6 in. (5–15 cm)
Cap: Convex to broadly convex; usually maroon to reddish brown or cinnamon, but sometimes more tawny or yellow-brown; surface dry, velvety to granulose, often cracking with age; **flesh** soft, white, odor not distinctive, taste mild or slightly tart.
Pore surface: White at first, becoming pinkish brown at maturity; pores 1–2 per mm; tube layer thick, depressed at the stalk.
Stalk: Slender, often curved, tapering upward or nearly equal, solid; colored like the cap or paler with white mycelium at the base; surface more or less smooth to roughened, often with raised lines that sometimes merge to form a somewhat net-like pattern, but it is not truly reticulate.
Spore print: Reddish brown.
Spores: 10–17 x 5–8 μm, elliptical, minutely pitted, pale brown.
Occurrence: Solitary to scattered on the ground, or less often on well-decayed wood in conifer and mixed woods, especially with hemlock and spruce; mycorrhizal; early summer-fall; common.
Edibility: Edible.

Comments: The genus *Austroboletus* was segregated from *Tylopilus* to accommodate species having ornamented spores. *Gyroporus castaneus* (p.331) is similarly colored but has firm, brittle flesh, a hollow stalk, and produces a yellow spore print. Compare also with *Boletus badius* (p.315), which has a pale to dark yellow pore surface that bruises instantly blue. *Boletus projectellus* (p.305) is similar but has a yellow pore surface and a prominently reticulated stalk. *Gracilis* means "slender" or "graceful."

Latin name: *Boletus auriporus* Peck
Synonym: *Boletus viridiflavus* Coker & Beers
Common name: None
Order: Boletales
Family: Boletaceae
Width of cap: 1–3 in. (2.5–7.5 cm)
Length of stalk: 1½–4 in. (4–10 cm)
Cap: Convex to broadly convex, becoming flat or depressed in the center with age, margin with a narrow band of sterile tissue; pinkish cinnamon to dark reddish brown, often with a whitish margin when young; surface smooth, viscid when wet, the peelable cuticle has an acidic taste; **flesh** white to pale yellow, vinaceous beneath the cuticle, odor somewhat pungent, taste not distinctive.
Pore surface: Brilliant yellow, becoming golden yellow to greenish yellow with age, typically bruising slowly dull brick red; pores 1–2 per mm, irregular to angular.
Stalk: Enlarged toward the base to nearly equal, solid; pale yellow near the apex, pinkish brown to reddish brown below, often streaked, white mycelium at the base; surface viscid and slippery when fresh.
Spore print: Olive-brown.
Spores: 11–16 x 4–6 μm, elliptical to spindle-shaped, smooth, pale brown.
Occurrence: Solitary to scattered in small groups on the ground under oak; mycorrhizal; summer-fall; occasional.
Edibility: Edible.

Comments: This species has often been confused with *Boletus innixus* (p.327), which has a brown, dry, non-peelable cap that often cracks with age. *Auriporus* means "golden pores."

Latin name: *Boletus hortonii* A.H. Sm. & Thiers
Synonym: *Boletus subglabripes* var. *corrugis* Peck
Common names: Corrugated Bolete; Horton's Bolete
Order: Boletales
Family: Boletaceae
Width of cap: 1½–4 in. (4–10 cm)
Length of stalk: 1½–3½ in. (4–9 cm)
Cap: Convex to broadly convex; tan to ochre-brown or reddish brown to reddish tan; surface dry to somewhat viscid when wet, deeply pitted to corrugated; **flesh** whitish to pale yellow, odor and taste not distinctive.
Pore surface: Yellow at first, becoming olive-yellow with age, rarely bruising slowly and weakly bluish; pores 2–3 per mm, round to angular.
Stalk: Club-shaped to more or less equal, solid; pale yellow to tan, sometimes reddish at the base; surface smooth to lightly pruinose.
Spore print: Olive-brown.
Spores: 12–15 x 3.5–4.5 μm, somewhat boat-shaped, smooth, yellowish.
Occurrence: Solitary or in scattered groups on the ground in broad-leaved and mixed woods, especially with oak, beech, and hemlock; mycorrhizal; early summer-early fall; occasional to fairly common in some years.
Edibility: Edible.

Comments: The deeply corrugated tan to brownish cap of this bolete is distinctive. *Boletus subglabripes* var. *subglabripes* (not illustrated) is nearly identical except that it has a smooth cap. *Leccinum rugosiceps* (not illustrated) has a pitted to cracked cap, flesh that stains dull reddish when cut, and brownish, scurfy points (scabers) on the stalk. The specific epithet honors American mycologist Charles Horton Peck (1833–1917).

Latin name: *Boletus innixus* Frost
Synonym: *Boletus caespitosus* Peck
Common name: Clustered Brown Bolete
Order: Boletales
Family: Boletaceae
Width of cap: 1¼–3 in. (3–7.5 cm)
Length of stalk: 1½–3½ in. (4–9 cm)
Cap: Convex to nearly flat; dull reddish brown to dull cinnamon, or yellow-brown, often purplish or reddish near the margin; surface dull to somewhat velvety when dry, slightly viscid when wet, often cracking with age; **flesh** white to pale yellow, or tinged vinaceous beneath the cuticle, odor pungent, somewhat like witch hazel, taste not distinctive.
Pore surface: Bright yellow when young, becoming dull yellow, unchanging when bruised; pores round to angular, small to medium, 1–3 per mm when young, up to 2 mm wide in older specimens.
Stalk: Stout, club-shaped to swollen in the mid-portion above a tapered base, solid; yellowish, streaked with brown, basal mycelium yellow (when present); surface more or less smooth to fibrillose roughened, somewhat viscid near the base, becoming more so when rubbed.
Spore print: Olive-brown.
Spores: 8–11 x 3–5 μm, elliptical, smooth, pale brown.
Occurrence: Solitary to scattered, often in caespitose clusters of 2 or more, on the ground in broad-leaved woods, especially with oak; mycorrhizal; early summer-fall; common.
Edibility: Edible.

Comments: Compare with *Boletus auriporus* (p.325), which has a viscid cap with a peelable cuticle, and copious white mycelium at the stalk base. *Innixus* means "reclining," alluding to the fruitbodies that are often leaning, especially when tightly clustered.

Latin name: *Boletus longicurvipes* Snell & A.H. Sm.
Common name: None
Order: Boletales
Family: Boletaceae
Width of cap: 1–2 in. (2.5–5 cm)
Length of stalk: 2–3½ in. (5–9 cm)
Cap: Convex to broadly convex; yellow-orange to brownish orange or ochraceous, sometimes with olive-green tones in wet weather; surface viscid and shiny, smooth to somewhat pitted or wrinkled; **flesh** white to pale yellow, odor and taste not distinctive.
Pore surface: Pale yellow, becoming dull greenish yellow to greenish gray, sometimes bruising brownish; pores small, 2 per mm.
Stalk: Slender, nearly equal to tapering upward, often curved, solid; whitish, often with pinkish brown to reddish brown tones, especially toward the base; surface scurfy, ornamented with scabrous yellow dots that darken to reddish brown, white mycelium at the base.
Spore print: Olive-brown.
Spores: 13–17 x 4–5 μm, oblong to somewhat spindle-shaped, smooth, pale brown.
Occurrence: Solitary or in scattered groups on the ground in oak-pine woods or other mixed woods, especially with white pine; mycorrhizal; summer-early fall; fairly common.
Edibility: Edible.

Comments: This bolete is recognized by its shiny, orangish cap, slender, curved stalk, and frequent association with pine. Because of the dotted stalk, it is sometimes mistaken for a species of *Leccinum*. *Longicurvipes* means "long, curved foot."

Latin name: *Boletus morrisii* Peck
Common name: Red-speckled Bolete
Order: Boletales
Family: Boletaceae
Width of cap: 1¼–3½ in. (3–9 cm)
Length of stalk: 1½–3 in. (4–7.5 cm)
Cap: Convex to broadly convex with a narrow, overlapping band of sterile tissue at the margin; olive-brown to reddish brown, often yellow to olive-gold on the margin; surface dry, somewhat velvety at first, becoming glabrous with age; **flesh** pale yellow, slowly staining reddish when exposed, odor and taste not distinctive.
Pore surface: Yellow to yellow-orange, or tinged reddish near the stalk, becoming orange to brownish orange with age; pores small, 2–3 per mm; tube layer deeply depressed at the stalk.
Stalk: Nearly equal to enlarged downward, solid; yellow beneath a surface covering of reddish or reddish brown dots, scales, or patches; basal mycelium yellow.
Spore print: Olive-brown.
Spores: 10–16 x 4–6 µm, elliptical to spindle-shaped, smooth, yellowish.
Occurrence: Solitary or sometimes in small groups on the ground, especially in sandy soil, in broad-leaved woods with oak and beech, or in mixed woods with oak and pine; mycorrhizal; summer-fall; uncommon, possibly rare.
Edibility: Reported to be edible but too rare to collect for food.

Comments: The scaber-like stalk ornamentation on this unusual and distinctive bolete is similar to those in the genus *Leccinum,* which have brown to blackish stalk scabers. The species was named for the collector of the type specimen, George Edward Morris (1853–1916).

Latin name: *Boletus parasiticus* Bull.
Synonym: *Xerocomus parasiticus* (Bull.) Quél.
Common names: Parasitic Bolete; Earthball Bolete
Order: Boletales
Family: Boletaceae
Width of cap: 1–2½ in. (2.5–6.5 cm)
Length of stalk: 1–2½ in. (2.5–6.5 cm)
Cap: Hemispherical to convex, margin incurved when young; ochre-brown to tawny olive, or like tarnished brass; surface dry, minutely tomentose, becoming glabrous and shiny with age, viscid when wet; **flesh** pale yellow, odor and taste not distinctive.
Pore surface: Yellow, becoming olivaceous with age, sometimes with reddish or rust-colored stains, unchanging or very rarely slightly bluish when bruised; pores angular, 1–2 mm wide on mature specimens; tube layer often somewhat decurrent.
Stalk: Nearly equal, usually curved, solid; yellow to brown (like the cap); surface glabrous to fibrillose.
Spore print: Olive-brown.
Spores: 12–18.5 x 3.5–5 μm, elliptical, smooth, pale brown.
Occurrence: Solitary or in small groups or clusters of 2 or 3 attached at the base of the earthball, *Scleroderma citrinum;* parasitic; summer-fall; occasional to common in most years.
Edibility: Non-poisonous.

Comments: This small bolete is unmistakable because it is always found attached to *Scleroderma citrinum* (p.448). Although the association with *Scleroderma* is not fully understood, it is thought to be parasitic, and the specific epithet refers to this relationship.

Latin name: *Gyroporus castaneus* (Bull.) Quél.
Common name: Chestnut Bolete
Order: Boletales
Family: Gyroporaceae
Width of cap: 1–3 in. (2.5–7.5 cm)
Length of stalk: 1–3 in. (2.5–7.5 cm)
Cap: Convex to broadly convex or nearly flat, sometimes slightly depressed in the center, margin often splitting; yellow-brown to orange-brown or reddish brown; surface dry, more or less tomentose to nearly glabrous; **flesh** brittle, white, odor not distinctive or faintly pungent, taste not distinctive.
Pore surface: White, becoming buff or yellowish, unchanging or bruising brownish; pores small, 1–3 per mm, round; tube layer depressed at the stalk to nearly free.
Stalk: More or less equal to swollen in the mid-portion or below, at times compressed, soon becoming hollow; surface glabrous, often uneven, colored like the cap or paler; flesh firm and brittle.
Spore print: Pale yellow.
Spores: 8–13 x 5–6 μm, elliptical, smooth, hyaline.
Occurrence: Solitary or scattered in groups on the ground, usually in broad-leaved woods, especially with oak, but also in conifer woods; mycorrhizal; early summer-fall; fairly common.
Edibility: Edible.

Comments: The hard, brittle flesh, and the hollow stalk are good field characters. The specific epithet relates to *Castanea,* the genus of chestnut trees, referring to the chestnut-brown color of the fruitbodies.

Latin name: *Leccinum albellum* (Peck) Singer
Common name: None
Order: Boletales
Family: Boletaceae
Width of cap: 1–2½ in. (2.5–6.5 cm)
Length of stalk: 2–4 in. (5–10 cm)
Cap: Convex, becoming broadly convex or nearly flat; color variable from whitish to grayish buff, pale tan, or brownish gray; surface dry, smooth to pitted, often cracked with age; **flesh** white, unchanging when cut, odor and taste not distinctive.
Pore surface: White when young, becoming yellowish gray to brownish at maturity, unchanging when bruised; pores small, less than 1 mm wide, angular; tube layer depressed at the stalk.
Stalk: Slender, equal or enlarged downward, solid; whitish or more or less colored like the cap; surface roughened with whitish to brown scabers, often arranged in lines.
Spore print: Brown to olive-brown.
Spores: 14–22 x 4–6 μm, cylindric to spindle-shaped, smooth, pale-yellow.
Occurrence: Solitary or in groups on the ground in broad-leaved woods, especially with oak; mycorrhizal; summer-fall; fairly common.
Edibility: Edible but uninteresting.

Comments: The similar but more brownish *Leccinum scabrum* (not illustrated) is associated with birch trees, is overall more robust, and usually shows bluish green staining at the base of the stalk. Compare with *Leccinum snellii* (p.334), which occurs under birch and has a much darker, nearly black cap and stalk. *Albellum* means "whitish."

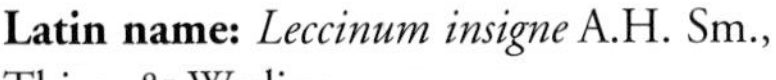

Latin name: *Leccinum insigne* A.H. Sm., Thiers & Watling
Common name: Aspen Bolete
Order: Boletales
Family: Boletaceae
Width of cap: 1½–5 in. (4–12.5 cm)
Length of stalk: 3–6 in. (7.5–15 cm)
Cap: Convex to broadly convex, with a sterile band of tissue that extends beyond the tube layer and remains as a hanging flap on the rim of mature caps; orange to rusty cinnamon or brick red, becoming brownish orange; surface dry, somewhat granular to fibrous; **flesh** white, staining purplish gray to purplish black when cut or broken, odor and taste not distinctive.
Pore surface: Whitish at first, becoming yellowish to olive-brown with age, unchanging or brownish when bruised; pores small, 1–3 per mm.
Stalk: Equal to swollen in the mid-portion, or enlarged downward, solid; whitish beneath a covering of brownish to blackish scabers; surface dry, roughened with projecting scabers.
Spore print: Yellow-brown to olive-brown.
Spores: 11–16 x 4–5 μm, elliptical, smooth, yellowish.
Occurrence: Scattered or in groups on the ground in woods and parklands beneath aspen and birch; mycorrhizal; summer-fall; occasional to locally common.
Edibility: Edible with caution.

Comments: Although the flesh of this sturdy bolete turns black when cooked, it is generally considered to be a good edible. However, there are reports of individuals having experienced gastrointestinal distress from eating closely related species. *Leccinum aurantiacum* (not illustrated), which grows with aspen or pine, is very similar but has flesh that stains reddish to vinaceous pink before turning blackish when exposed to the air. *Leccinum piceinum* (not illustrated) is nearly identical but has a duller orange cap and grows in association with Norway spruce. *Insigne* means "remarkable" or "striking."

Latin name: *Leccinum snellii* A.H. Sm., Thiers & Watling
Common name: Snell's Leccinum
Order: Boletales
Family: Boletaceae
Width of cap: 1¼–3 in. (3–7.5 cm)
Length of stalk: 1½–4½ in. (4–11.5 cm)
Cap: Convex to broadly convex, becoming nearly flat with age; dark smoky brown to nearly black, fading to dark yellowish brown, often with cream-colored spots; surface dry, dull, fibrillose, becoming more glabrous with age; **flesh** white, slowly staining orange-red, especially near the apex of the stalk, then turning black, odor and taste not distinctive.
Pore surface: White at first, becoming grayish to grayish brown, bruising yellowish to brownish; pores small, 2–3 per mm.
Stalk: Nearly equal or tapering slightly upward, solid; surface densely covered with dark grayish brown to blackish scabers, whitish beneath; cut flesh slowly staining bright orange-red on the upper portion and blue-green at the base.
Spore print: Brown.
Spores: 16–22 x 5–7.5 μm, somewhat spindle-shaped, smooth, pale brown.
Occurrence: Solitary or in small groups on the ground in mixed woods with birch, especially wet woods with yellow birch (*Betula alleghaniensis*); mycorrhizal; late spring-fall; common.
Edibility: Edible.

Comments: Snell's Leccinum is one of the first boletes to appear in the spring. It is recognized by the dark, almost black cap and stalk, and the orangish red staining of the cut flesh at the junction of the cap and stalk that slowly becomes more intense before eventually turning black. *Leccinum scabrum* (not illustrated) has a paler brown cap, and its flesh typically does not stain reddish when exposed. This species was named in honor of the American mycologist Walter H. Snell (1889–1980), who specialized in the study of boletes.

Latin name: *Suillus americanus* (Peck) Snell
Common names: Chicken-fat Suillus; American Slippery Jack
Order: Boletales
Family: Suillaceae
Width of cap: 1¼–3 in. (3–7.5 cm)
Length of stalk: 1¼–3 in. (3–7.5 cm)
Cap: Broadly conical to convex, becoming nearly flat, sometimes with a broad umbo, margin often adorned with fragments of cottony veil tissue; yellow, typically with cinnamon to reddish streaks and patches; surface viscid to glutinous, especially when wet, margin sometimes with clusters of fibrils that aggregate into scales; **flesh** yellow, staining purplish brown when cut, odor and taste not distinctive.
Pore surface: Yellow, becoming dull ochraceous, slowly bruising reddish brown; pores small, 2–3 per mm, larger on medium to large specimens, angular to sometimes radially elongated, tube layer more or less decurrent.
Stalk: Slender, nearly equal, becoming somewhat hollow; colored like the cap with reddish brown resinous dots and smears on the surface.
Spore print: Brown.
Spores: 8–11 x 3–4 μm, spindle-shaped, smooth, pale brown.
Occurrence: Gregarious, sometimes in clusters of 2 or 3 on the ground under eastern white pine; mycorrhizal; mid-summer-fall; common.
Edibility: Edible but not popular.

Comments: The material that hangs from the cap margin of young specimens is not a remnant of a true partial veil since it is never connected to the stalk and leaves no ring. It has been reported that handling this bolete may cause an allergic contact dermatitis in some individuals. *Suillus americanus* var. *reticulipes* (not illustrated) has been reported from North Carolina. It is nearly identical except that it has a coarse reticulum on the stalk and a strongly decurrent tube layer. *Suillus subaureus* (not illustrated) is similar but has a paler yellow to buff-yellow cap and can occur in broad-leaved woods, apparently forming mycorrhiza with oak as well as pine. *Americanus* means "of America."

Latin name: *Suillus granulatus* (L.) Snell
Synonym: *Boletus granulatus* Bull.
Common names: Butterball; Granulated Slippery Jack
Order: Boletales
Family: Suillaceae
Width of cap: 1½–3½ in. (4–9 cm)
Length of stalk: 1½–3 in. (4–7.5 cm)
Cap: Convex to broadly convex, becoming nearly flat, margin incurved at first; pale cinnamon to vinaceous brown, orange-brown, pale yellow, or tan; surface viscid to glutinous when wet, often with streaks or blotches or dried gluten; **flesh** white to pale yellow, odor and taste not distinctive.
Pore surface: Whitish to pinkish buff, becoming yellowish to yellowish brown with age, often exuding milky droplets when young, unchanging or staining dull cinnamon when bruised; pores small, 1 mm wide, angular.
Stalk: Equal, solid; whitish at first, becoming pale yellow especially near the apex; surface covered with brownish to pinkish tan resinous dots and smears.
Spore print: Brown.
Spores: 7–10 x 2.5–3.5 μm, oblong to spindle-shaped, smooth.
Occurrence: Usually gregarious on the ground under white pine (*Pinus strobus*), also reported to occur under other pines, spruce, and hemlock; mycorrhizal; early summer-late fall; common, often abundant.
Edibility: Edible. Discarding the cap cuticle and tube layer of older specimens is recommended.

Comments: This bolete forms mycorrhiza with white pine and possibly with other conifers. *Suillus neoalbidipes* (not illustrated) is similar except that it has a roll of white, cottony veil material on the cap margin of young specimens. *Suillus brevipes* (not illustrated) has a dark brown to vinaceous brown or cinnamon-brown cap, and a short, stubby stalk that lacks colored resinous dots. *Suillus hirtellus* (not illustrated) is more slender in stature, has fibrils or scales integrated with a viscid cap cuticle, and has pores that are more or less radially arranged. *Suillus placidus* (not illustrated) has a viscid white to yellowish white cap and occurs only with white pine. *Granulatus* means "covered with granules," referring to the dotted stalk.

Latin name: *Tylopilus alboater* (Schwein.) Murrill
Synonym: *Boletus alboater* Schwein.
Common name: Black Velvet Bolete
Order: Boletales
Family: Boletaceae
Width of cap: 2–6 in. (5–15 cm)
Length of stalk: 1½–4½ in. (4–11.5 cm)
Cap: Convex to broadly convex, becoming nearly flat to shallowly depressed; dark grayish brown to black, paler with age, often with a whitish or purplish bloom when very young; surface dry, velvety at first, often cracking with age; **flesh** white, firm when young, soft with age, staining reddish gray then black when cut or broken, odor not distinctive, taste mild.
Pore surface: Whitish, becoming dull pinkish as the spores mature, bruising reddish then black; pores small, 1–2 per mm.
Stalk: Variable, from nearly equal to swollen in the mid-portion or somewhat bulbous, solid; colored like the cap or paler, especially at the apex; surface smooth or rarely with a slight reticulation at the apex.
Spore print: Dull pink.
Spores: 7–11 x 3.5–5 µm, narrowly oval, smooth, hyaline.
Occurrence: Solitary or scattered in groups, sometimes in caespitose clusters on the ground under oak in broad-leaved or mixed woods; mycorrhizal; early summer-fall; common.
Edibility: Edible and good.

Comments: Most members of the genus *Tylopilus* have bitter tasting flesh, but the Black Velvet Bolete is pleasantly mild. Although fairly common in most years, it is easily overlooked due to its dark color. When young the compact flesh is notably dense and heavy, and has a firm texture when cooked. Compare with *Tylopilus atronicotianus* (p.338), which is quite similar but has an olive-brown cap. *Alboater* means "white and black."

Latin name: *Tylopilus atronicotianus* Both
Common name: False Black Velvet Bolete
Order: Boletales
Family: Boletaceae
Width of cap: 1½–6 in. (4–15 cm)
Length of stalk: 1½–6 in. (4–15 cm)
Cap: Hemispheric with an inrolled margin at first, becoming convex to flattened, margin expanding and forming an overlapping sterile flap; olive-brown, bronze-brown, to pale brown, sometimes with very pale brown to nearly whitish areas, bruising blackish brown; surface glabrous, slightly shiny to minutely velvety; **flesh** white, slowly staining pinkish red then gray to black when exposed, odor not distinctive, taste mild.
Pore surface: White, becoming vinaceous to vinaceous brown with age, bruising rusty brown to black; pores small, up to 1.5 mm wide, angular to irregular; tube layer depressed around the stalk.
Stalk: Cylindrical to somewhat compressed, more or less equal or rarely bulbous; grayish to brownish, or almost black near the base; surface minutely velvety.
Spore print: Vinaceous russet to cocoa-brown.
Spores: 7.7–10.5 x 3.8–5.3 μm, ovate, smooth, pale yellow-brown in Melzer's regent.
Occurrence: Solitary to gregarious, sometimes clustered on the ground in broad-leaved and mixed woods, especially with red oak, beech, and hemlock; mycorrhizal; summer; uncommon.
Edibility: Unknown.

Comments: This recently described species is sometimes present in very dry weather when few other boletes are found. *Tylopilus alboater* (p.337) is similar but has a blackish cap that lacks brown tones. *Atronicotianus* means "dark, like nicotine."

Latin name: *Tylopilus chromapes* (Frost) A.H. Sm. & Thiers
Synonym: *Leccinum chromapes* (Frost) Singer
Common name: Chrome-footed Bolete
Order: Boletales
Family: Boletaceae
Width of cap: 2–5 in. (5–12.5 cm)
Length of stalk: 2–5 in. (5–12.5 cm)
Cap: Hemispheric at first, becoming convex to broadly convex, or nearly flat; pink to rose-colored, fading to tan with age; surface dry to slightly viscid when moist; **flesh** white, odor not distinctive, taste mild to slightly acidic (lemony).
Pore surface: White becoming pale to dingy pink as the spores mature, unchanging or at times slightly rose-colored when bruised; pores small, 2–3 per mm.
Stalk: Nearly equal or tapering in either direction, often crooked, solid; white to pinkish, covered with pink to reddish scabrous dots, bright chrome yellow at the base.
Spore print: Pinkish to pinkish brown.
Spores: 11–17 x 4–4.5 μm, oblong to oval, smooth, hyaline to pale brown.
Occurrence: Solitary or in small groups on the ground in broad-leaved and conifer woods; mycorrhizal; early summer-fall; occasional to fairly common in some areas.
Edibility: Edible.

Comments: This pretty bolete is easily recognized by its pinkish cap, pink to reddish dotted stalk, and the brilliant yellow stalk base. *Chromapes* means "chrome yellow foot."

Latin name: *Tylopilus rubrobrunneus* Mazzer & A.H. Sm.
Common name: Reddish brown Bitter Bolete
Order: Boletales
Family: Boletaceae
Width of cap: 3–10 in. (7.5–25 cm)
Length of stalk: 2½–9 in. (6.5–23 cm)
Cap: Convex to broadly convex, becoming nearly flat; dark purple at first, becoming purple-brown to vinaceous brown, fading to pinkish cinnamon with age; surface dry, somewhat velvety when young, becoming more glabrous and sometimes cracking with age; **flesh** white, odor not distinctive, taste intensely bitter.
Pore surface: White to pale yellow or pinkish gray at first, becoming dull pinkish to vinaceous or pinkish brown as the spores mature, bruising brown; pores small, 1–2 per mm.
Stalk: Equal to club-shaped, solid; vinaceous brown to chestnut-brown, aging dingy olive-brown, or developing olive-brown stains from handling, especially near the base; surface glabrous, smooth to sometimes slightly reticulate at the apex.
Spore print: Reddish brown to pinkish brown.
Spores: 10–14 x 3–4.5 μm, oblong to somewhat spindle-shaped, smooth, pale brown.
Occurrence: Solitary or in groups, sometimes in caespitose clusters on the ground in broad-leaved and mixed woods, often under oak; mycorrhizal; summer-early fall; common.
Edibility: Unknown.

Comments: This bolete is probably not poisonous, but the flesh is intensely bitter tasting. It is one of the largest boletes in the genus, at times having a cap as wide as a dinner plate. The equally bitter tasting *Tylopilus violatinctus* (p.341) is similar but has a pale purplish to grayish violet, or pale brown cap, and it does not develop olive-brown stains on the stalk. *Tylopilus felleus* (p.310) has a prominent brown reticulation on the stalk. *Rubrobrunneus* means "reddish brown."

Latin name: *Tylopilus violatinctus* T.J. Baroni & Both
Common name: Pale Violet Bitter Bolete
Order: Boletales
Family: Boletaceae
Width of cap: 3–5½ in. (7.5–14 cm)
Length of stalk: 3–6 in. (7.5–15 cm)
Cap: Hemispheric at first, becoming convex to broadly convex, or nearly flat, margin with a narrow band of sterile tissue, incurved at first; grayish violet to bluish violet when young, becoming pale purplish, purplish pink, or pale brownish with age, bruising violet to dark violet; surface dry, more or less velvety when young, becoming more glabrous with age; **flesh** white, unchanging or becoming very pale slate-colored when exposed, odor not distinctive, taste intensely bitter.
Pore surface: White at first, becoming dull pink to brownish, unchanging when bruised; pores small, 1–2 per mm, nearly round; tube layer often depressed at the stalk but may have a slightly decurrent extension.
Stalk: Nearly equal to club-shaped or bulbous, solid; colored like the cap or paler, becoming brownish with age but apex and base typically whitish; surface smooth, sometimes with an inconspicuous reticulum at the apex.
Spore print: Reddish brown.
Spores: 7–10 x 3–4 μm, somewhat spindle-shaped, smooth, pale yellow.
Occurrence: Solitary or in groups on the ground in mixed woods often containing red oak and beech, spruce, or hemlock; mycorrhizal; summer-fall; occasional.
Edibility: Unknown, and too bitter to be of interest.

Comments: This is one of a small group of boletes with bitter tasting flesh and lilac to purple tones on the cap or stalk. *Tylopilus plumbeoviolaceus* (not illustrated) is similar but has a dark purple stalk. *Tylopilus rubrobrunneus* (p.340) is also similar but is typically more robust and develops olive-brown stains on the stalk, especially when handled. *Violatinctus* means "tinged violet."

Latin name: *Xanthoconium affine* var. *affine* (Peck) Singer
Synonym: *Boletus affinis* Peck
Common name: None
Order: Boletales
Family: Boletaceae
Width of cap: 1½–4½ in. (4–11.5 cm)
Length of stalk: 2½–4½ in. (6.5–11.5 cm)
Cap: Convex to broadly convex, becoming nearly flat; dark brown to chestnut-brown or ochre-brown; surface dry, somewhat velvety, smooth or often wrinkled; **flesh** white, unchanging when exposed, odor and taste not distinctive.
Pore surface: White at first, becoming yellowish to dingy yellow-brown, bruising dull yellow to brownish; pores small, 1–2 per mm, round.
Stalk: Nearly equal or tapering in either direction, solid; whitish, often with brownish streaks on the middle portion or base; surface smooth or infrequently with a slight reticulation at the apex.
Spore print: Yellow-brown.
Spores: 12–16 x 3–4 μm, cylindric to spindle-shaped, smooth, yellowish.
Occurrence: Usually in groups, sometimes in caespitose clusters on the ground in broad-leaved woods, especially with oak and beech, also under pine; mycorrhizal; early summer-early fall; common.
Edibility: Edible.

Comments: *Xanthoconium affine* var. *maculosus* (not illustrated) is nearly identical but has yellowish to cream-colored spots on the cap. *Xanthoconium purpureum* (not illustrated) is also similar except it has a purplish red to maroon cap. Compare with *Boletus variipes* (p.307), which has a similar color scheme, but has a prominently reticulated stalk. *Tylopilus badiceps* (not illustrated) has a reddish brown cap and stalk, and produces a purplish brown spore deposit. *Affine* means "related," referring to a similarity with *Boletus edulis.*

Stalked Polypores with Simple or Compound Fruitbodies

Stalked polypores differ from boletes in that their tube or pore layer is proportionately shallow and does not easily separate from the flesh of the cap. Most have leathery or fibrous, persistent fruitbodies, whereas boletes are always relatively soft and fleshy. Also, boletes are typically terrestrial and only infrequently found growing on wood. In contrast, most polypores grow on or from woody substrates. Those that appear to be growing on the ground are mostly saprobes that are attached to buried wood. A few terrestrial polypores form mycorrhiza with trees. Stalked polypores may have independent caps and stalks, or they can be compound structures with multiple caps and stalks that originate from a common base. Some polypores in this section can grow shelf-like when attached to vertical surfaces. The bracket and shelf fungi (Section 3) are *stalkless* polypores that are laterally attached to wood. However, the distinction between these two groups is sometimes blurred.

Latin name: *Albatrellus caeruleoporus* (Peck) Pouzar
Synonym: *Polyporus caeruleoporus* Peck
Common name: Blue Albatrellus
Order: Polyporales
Family: Albatrellaceae
Width of cap: 1–4 in. (2.5–10 cm)
Length of stalk: 1–3 in. (2.5–7.5 cm)
Cap: Convex to nearly flat or depressed in the center, circular to irregular in outline, margin inrolled, caps sometimes fused together; indigo to bluish gray at first, becoming grayish brown to tan in age, sometimes with orangish tones; surface smooth to slightly scaly, dry; **flesh** whitish to pale buff, unchanging when exposed, odor and taste not distinctive.
Pore surface: Colored like the cap or paler; pores small, 2–3 per mm, angular; tube layer decurrent.
Stalk: More or less equal or tapering in either direction, central or eccentric, solid; colored like the cap or paler; surface smooth to pitted on the upper portion.
Spore print: White.
Spores: 4–6 x 3–5 μm, subglobose to oval, smooth, hyaline, non-amyloid.
Occurrence: Solitary, or more often in groups or clusters on the ground in conifer and mixed woods, often with hemlock; mycorrhizal; late summer-fall; uncommon.
Edibility: Edible.

Comments: This distinctive mushroom is not likely to be confused with any other. *Boletopsis subsquamosa* (p.347) is somewhat similar but lacks bluish tones, has a whitish pore surface, and has bitter tasting flesh. *Caeruleoporus* means "having blue pores."

Latin name: *Albatrellus cristatus* (Schaeff.) Kotl. & Pouzar
Synonym: *Polyporus cristatus* (Pers.) Fr.
Common name: Crested Polypore
Order: Polyporales
Family: Albatrellaceae
Width of cap: 2–8 in. (5–20 cm), compound fruitbodies are often broader
Length of stalk: 1–1½ in. (2.5–4 cm)
Cap: Convex to flat or somewhat funnel-shaped, more of less circular to irregular in outline, often compound with multiple caps fused together, margin wavy; yellow-brown to yellowish olive, usually zonate; surface dry, glabrous to slightly velvety, sometimes breaking into coarse scales or cracks; **flesh** white, staining yellowish green when cut, odor not distinctive, taste bitter or not distinctive.
Pore surface: White, often with yellowish or greenish yellow stains; pores small; tube layer often decurrent.
Stalk: Short, stout, irregular, solid; whitish to more or less colored like the cap; surface smooth.
Spore print: White.
Spores: 5–7 x 4–5 µm, broadly elliptical to subglobose, smooth, hyaline, weakly amyloid.
Occurrence: Solitary, or more often in fused clusters on the ground in broad-leaved and mixed woods, especially with oak, also reported to occur infrequently in conifer woods; mycorrhizal; summer-fall; common.
Edibility: Not edible.

Comments: This polypore often grows with multiple overlapping, confluent caps and branched stalks. The stalk and pore layer usually exhibit some yellowish green staining. Compare with *Phaeolus schweinitzii* (p.357). *Cristatus* means "crested."

Latin name: *Albatrellus ovinus* (Schaeff.) Kotl. & Pouzar
Synonym: *Polyporus ovinus* (Schaeff.) Fr.
Common name: Sheep Polypore
Order: Polyporales
Family: Albatrellaceae
Width of cap: 2–6 in. (5–15 cm)
Length of stalk: 1–2½ in. (2.5–6.5 cm)
Cap: Convex with an incurved margin at first, sometimes becoming funnel-shaped in age, circular to irregular in outline, occasionally having multiple fused caps; chalk white to cream when young, becoming buff to yellow or tan in age; surface dry, dull, often cracking; **flesh** white to pale yellow, brittle, taste mild to bitter.
Pore surface: White to pale yellowish; pores minute; tube layer very shallow, decurrent.
Stalk: More or less equal or sometimes bulbous at the base, central or eccentric; colored more or less like the cap, sometimes with pink or vinaceous stains, especially near the base; surface smooth.
Spore print: White.
Spores: 3–5 x 3–3.5 μm, oval to subglobose, smooth, hyaline, non-amyloid.
Occurrence: Usually gregarious, sometimes in caespitose clusters, on the ground in conifer woods, especially with Norway spruce; mycorrhizal; summer-early fall; occasional to locally common.
Edibility: Edible when thoroughly cooked.

Comments: Pale forms of *Hydnum repandum* (p.398) are very similar in appearance as seen from above, but they have spines rather than pores beneath the cap. The name *ovinus* pertains to sheep. Its application here is vague, perhaps relating to the color.

Latin name: *Boletopsis subsquamosa* (Fr.) Pouzar
Synonym: *Boletopsis leucomelaena* (Pers.) Fayod
Common name: Kurotake
Order: Thelephorales
Family: Bankeraceae
Width of cap: 2–5 in. (5–12.5 cm)
Length of stalk: 1–2½ in. (2.5–6.5 cm)
Cap: Convex with an incurved margin at first, becoming broadly convex or flat with a wavy margin in age; color variable from dingy brown to grayish brown or grayish black, sometimes with olive tones; surface smooth to somewhat fibrillose-scaly on the disc in age, cracking in dry weather; **flesh** white to gray, odor not distinctive, taste bitter.
Pore surface: Whitish to pale gray; pores small (1–3 per mm); tube layer somewhat decurrent.
Stalk: Stout, more or less equal or swollen in the mid-portion, central or eccentric; colored like the cap; surface smooth or finely scaly.
Spore print: Whitish to pale brown.
Spores: 5–7 x 4–5 μm, angular, warted, hyaline, non-amyloid.
Occurrence: Solitary or in small scattered groups on the ground in broad-leaved and conifer woods; mycorrhizal; summer-fall; occasional.
Edibility: Non-poisonous.

Comments: The bitter tasting flesh of this terrestrial polypore is reported to be edible after being processed in brine. Experiments by the author in this respect have not been encouraging. *Albatrellus caeruleoporus* (p.344) has a bluish to violet cap and pore surface. *Subsquamosa* means "somewhat scaly."

Latin name: *Bondarzewia berkeleyi* (Fr.) Bondartsev & Singer
Synonym: *Polyporus berkeleyi* Fr.
Common names: Stump Blossoms; Berkeley's Polypore
Order: Russuales
Family: Bondarzewiaceae
Fruitbody: Massive, up to 3 feet or more across; typically a compound rosette with multiple shelf-like overlapping caps with thick, wavy margins, occasionally with a single cap; **upper surface** roughened to densely hairy or wooly, often wrinkled; tan to yellow-brown, or grayish brown, more or less zoned, often with a yellowish margin; **flesh** whitish, thick, taste mild when young, bitter in older specimens.
Pore Surface: Decurrent; whitish, when fresh exuding a white latex when cut or broken; pores 0.5–2 mm wide, circular to angular, or maze-like, often torn, especially near the stalk.
Stalk: Above ground portion 2–4 in. (5–10 cm), short, stout, irregularly shaped, rooting and usually arising from an underground sclerotium; yellow to yellowish brown.
Spore print: White.
Spores: 7–9 x 6–8 µm, globose, surface with warts or ridges, hyaline, amyloid.
Occurrence: Solitary or clustered at the base of living broad-leaved trees and around decaying stumps, especially oak, in broad-leaved and mixed woods; weakly parasitic causing a butt rot in living trees, or saprobic; early summer-early fall; occasional.
Edibility: Edible when young, later the flesh becomes bitter and fibrous.

Comments: This spectacular fungus first emerges from the ground in the form of irregular-shaped lumps or thick, knobby fingers that later expand into compound shelf-like rosettes. Compare with *Phaeolus schweinitzii* (p.357), which has reddish brown caps with a yellow-brown to greenish brown pore surface, and grows on or near conifers. *Grifola frondosa* (p.355) is more grayish brown to umber-brown, and has thinner and smaller individual caps. *Meripilus sumstinei* (not illustrated) is similar but its cap and pore surface stain blackish when bruised or handled. The specific epithet honors the British mycologist M.J. Berkeley (1803–1889), who in collaboration with American mycologist M.A. Curtis (1808–1872) first described many North American mushrooms and fungi.

Latin name: *Coltricia montagnei* (Fr.) Murrill
Synonym: *Polyporus montagnei* Quél.
Common name: Montagne's Polypore
Order: Hymenochaetales
Family: Hymenochaetaceae
Width of cap: 1½–5 in. (4–12.5 cm)
Length of stalk: 1–3 in. (2.5–7.5 cm)
Cap: Convex to flat or depressed in the center, sometimes with concrescent caps; yellow-brown to red-brown, or blackish brown; surface uneven, velvety to matted-hairy, more or less zonate; **flesh** tough, leathery to woody, brown, odor not distinctive, taste bitter.
Pore surface: Variable from poroid to concentric gill-like plates; whitish to grayish brown, becoming darker in age.
Stalk: Roughly cylindrical, often tapered toward the base, central or eccentric; colored like the cap; surface more or less velvety.
Spore print: Pale brown.
Spores: 9–15 x 5–7.5 µm, oblong to elliptical, smooth, pale brown, dextrinoid.
Occurrence: Solitary or in groups on the ground in broad-leaved and mixed woods; saprobic; summer-fall; occasional.
Edibility: Not edible.

Comments: Specimens having a concentrically arranged gill-like pore surface were formerly considered to be a separate variety (*Coltricia montagnei* var. *greenei*). The specific epithet honors the French mycologist Jean Pierre Francois Caruille Montagne (1784–1866).

Latin name: *Coltricia perennis* (L.) Murrill
Common name: Brown Funnel Polypore
Order: Hymenochaetales
Family: Hymenochaetaceae
Width of cap: 1–3½ in. (2.5–9 cm)
Length of stalk: ¾–2 in. (2–5 cm)
Cap: Flat to funnel-shaped, sometimes in clusters fused at the edges or in tiers, margin often wavy; color variable with concentric bands of yellow, brown, reddish brown, cinnamon, or orange-brown, margin typically whitish to yellowish; **flesh** thin, tough, leathery, brown.
Pore surface: Grayish to beige or brown, paler near the margin, bruising brown; margin with a narrow band of sterile tissue lacking pores; pores small (2–4 per mm), angular; tube layer shallow, somewhat decurrent.
Stalk: Slender, cylindrical to compressed, more or less equal, solid, tough; colored like the cap; surface uneven, velvety.
Spore print: Pale yellow-brown.
Spores: 6–9 x 3–5 μm, elliptical, smooth, pale yellow-brown, dextrinoid.
Occurrence: Solitary, or more often in small groups on the ground in conifer and mixed woods, along trails, on burnsites, on mossy ground, and on sandy soil along streams, also on well-decayed woody debris; saprobic; spring-fall, fruitbodies can persist through the winter; fairly common.
Edibility: Not edible.

Comments: *Coltricia cinnamomeus* (not illustrated) is similar but generally smaller and thinner with a shiny, satiny cap. *Onnia tomentosa* (p.356), which grows from spreading roots of conifers, has a larger, thicker, less distinctly zoned cap, and white spores. *Perennis* means "perennial."

Latin name: *Fistulina hepatica* (Schaeff.) Fr.
Common names: Beefsteak Polypore; Ox-tongue
Order: Agaricales
Family: Fistulinaceae
Fruitbody: Annual; semicircular to fan-shaped, convex to flattened, at times with an upturned margin with age; **upper surface** finely roughened, sometimes with radial furrows, viscid when moist, sticky; orange-red to blood red when young, becoming wine red to brownish red with age.
Pore surface: Composed of individual closely packed tubes with small round pores (1–3 per mm); pores whitish at first, becoming yellowish to pinkish tan, bruising and aging reddish brown, exuding clear or pinkish droplets when fresh; **flesh** up to 2½ in. (6.5 cm) thick, soft, somewhat rubbery, juicy; whitish to yellowish when young, soon becoming wine red with paler marbling, odor not distinctive, taste sour, acidic.
Stalk: Sometimes lacking, lateral when present, often twisted, 1–3 in. (2.5–7.5 cm) long, fleshy, colored like the upper surface of the fruitbody.
Spore print: Pinkish salmon.
Spores: 4–6 x 3–4 μm, broadly elliptical, smooth, hyaline, non-amyloid.
Occurrence: Solitary or in small overlapping clusters on stumps and logs of broad-leaved trees, especially oaks. Also at the base of living or dead standing tree trunks; weakly parasitic and saprobic; summer-fall; fairly common.
Edibility: Edible, with a unique citrus-like flavor.

Comments: This tongue-like fleshy polypore is unlikely to be mistaken for any other mushroom. It causes a heartwood rot in oak trees, but the decay process is very slow. The infected wood becomes richly stained before it is structurally weakened and is known to cabinet-makers as "brown oak." *Fistulina* means "hollow, pipe-like." *Hepatica* refers to the likeness of the fruitbody to a liver.

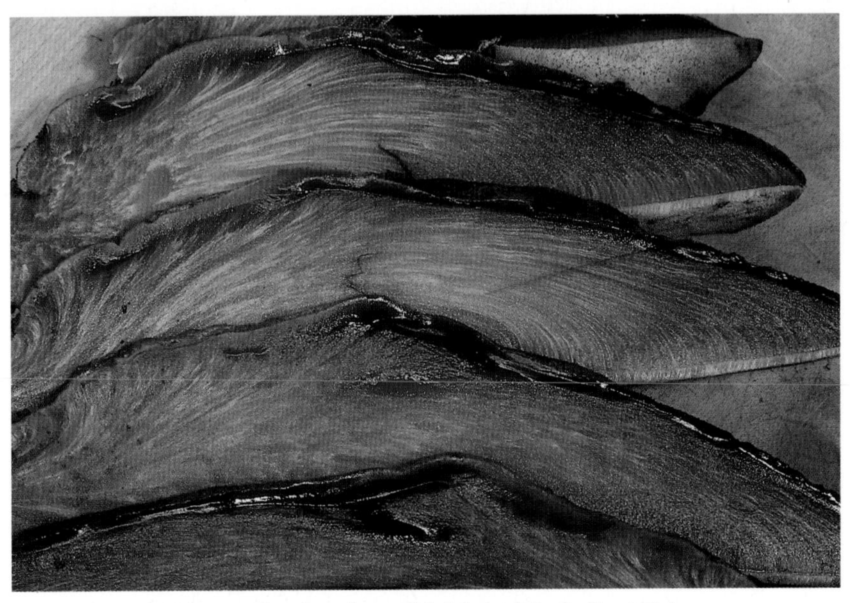

Cross sections detailing the marbled flesh of the Beefsteak Polypore.

Latin name: *Ganoderma tsugae* Murrill
Common names: Hemlock Polypore; Hemlock Varnish Shelf
Order: Polyporales
Family: Ganodermataceae
Width of cap: 3–12 in. (7.5–30 cm)
Length of stalk: 1½–6 in. (4–15 cm)
Cap: Nearly globose at first, soon expanding to spathulate, fan- or kidney-shaped; brownish orange to reddish brown with a white to yellowish margin when young; surface glabrous, shiny "varnished," often wrinkled, more or less concentrically zoned and furrowed; **flesh** white, soft, spongy at first, becoming firm and corky.
Pore surface: Whitish, bruising and becoming dingy yellow to brownish with age; pores round to angular, minute (4–6 per mm).
Stalk: More of less cylindrical to somewhat flattened, usually lateral but eccentric and erect when growing from horizontal surfaces; colored like the cap, often darker toward the point of attachment.
Spore print: Rusty brown.
Spores: 9–11 x 6–8 μm, elliptical with a truncated end, double-walled; outer surface roughened, hyaline.
Occurrence: Solitary or in groups on standing dead or dying trunks, logs, and stumps of conifers, especially hemlock; saprobic; spring-fall, fruitbodies persist throughout the year; common.
Edibility: Inedible.

Comments: This beautiful fungus will retain its color and varnished appearance for many months after being collected. The very similar *Ganoderma lucidum* (not illustrated) has a more distinctly zoned, chestnut-colored cap and grows on wood of broad-leaved trees. Although the flesh is fibrous and inedible, a "tea" that is reputed to be a health tonic can be brewed from the fruitbodies of both *Ganoderma tsugae* and *Ganoderma lucidum*. *Ganoderma lucidum* is especially valued in Chinese folk medicine and can be found in Asian markets and health food stores under the name Ling Chih. *Tsugae* refers to the genus of Hemlock trees.

Latin name: *Grifola frondosa* (Dicks.) Gray
Synonym: *Polyporus frondosus* (Dicks.) Fr.
Common names: Hen of the Woods; Sheep's Head
Order: Polyporales
Family: Meripilaceae
Fruitbody: Up to 2 feet or more wide; compound clusters of overlapping, spathulate- to fan-shaped caps attached laterally to a branched stalk; **upper surface** of the caps smooth to fibrous and radially streaked, ochre-brown to gray-brown or dark brown, more or less zoned; **flesh** white, firm, becoming fibrous in age, odor and taste not distinctive.
Pore surface: Decurrent; white to ochraceous; pores small (2–3 per mm), round to angular.
Stalk: ¾–2 in. (2–5 cm) long, up to 4 in. (10 cm) thick; stout, repeatedly branched; white.
Spore print: White.
Spores: 5–7 x 3.5–5 μm, broadly elliptical, smooth, hyaline, non-amyloid.
Occurrence: In rosette-like clusters, often massive, at the base of oak trees, or around decaying oak stumps, also reported to occur with other broad-leaved trees; parasitic; late summer-fall; occasional to locally common.
Edibility: Edible and good when fresh.

Comments: This fungus is a weak parasite of oaks, causing a white rot of the root system and trunk base. Its fruitbodies can often be found year after year at the base of the same tree. *Meripilus sumstinei* (not illustrated), which is similar except that all parts stain blackish when bruised, is also edible when young, but inferior. *Polyporus umbellatus* (p.366) has numerous small, rounded caps with individual stalks that branch from a rooting primary stalk. *Frondosa* means "leafy."

Latin name: *Onnia tomentosa* (Fr.) P. Karst.
Synonym: *Inonotus tomentosus* (Fr.) S. Teng
Common name: Wooly Velvet Polypore
Order: Hymenochaetales
Family: Hymenochaetaceae
Width of cap: 1½–6 in. (4–15 cm)
Length of stalk: ¾–1½ in. (2–4 cm)
Cap: Convex to flat or depressed, circular to semicircular or irregularly lobed in outline, often concrescent; yellow-brown to rust-brown, margin paler; surface velvety to hairy, uneven, at times more or less zonate; **flesh** tough, leathery to woody, yellow-brown to rust-brown.
Pore surface: Decurrent; buff to tan, becoming darker in age; pores angular, minute (2–4 per mm).
Stalk: Tapered toward the base, central or eccentric, sometimes poorly developed or merely a point of attachment; colored more or less like the cap; surface velvety, gnarled.
Spore print: Pale yellow to pale brown.
Spores: 5–6 x 3–4 μm, elliptical, smooth, hyaline.
Occurrence: Usually in groups on the ground attached to the spreading roots of conifers, especially Norway spruce, less often on conifer stumps; parasitic; summer-fall; common.
Edibility: Not edible.

Comments: This parasitic fungus attacks sapwood and may eventually cause the death of infected trees. Compare with *Phaeolus schweinitzii* (p.357), which is overall thicker and more substantial, has a greenish yellow to brown pore surface, and often grows in overlapping rosettes. *Coltricia perennis* (p.351) has a thinner, decidedly zonate cap. *Tomentosa* means "hairy" or "velvety."

Latin name: *Phaeolus schweinitzii* (Fr.) Pat.
Common names: Dyer's Polypore; Velvet Top Fungus
Order: Polyporales
Family: Polyporaceae
Width of cap: Individual caps 3–10 in. (7.5–25 cm), compound clusters may reach 2 feet or more across.
Length of stalk: ½–3 in. (1–7.5 cm) when present.
Fruitbody: Usually in compound overlapping clusters of circular, semicircular, or fan-shaped flattened caps, sometimes growing singly and more or less top-shaped; **upper surface** rough and uneven, densely hairy when young, less so with age; ochre to orange or reddish brown with a yellow margin when young, becoming dark brown to dark reddish brown in age, more or less zonate; **flesh** yellow to brown, tough and spongy, but brittle when dry, odor not distinctive, taste sour.
Pore surface: Decurrent; greenish yellow to mustard yellow, or grayish brown, bruising dark brown, becoming dark brown to rusty brown with age, often exuding clear droplets when fresh; pores angular, 1–3 per mm.
Stalk: Short, thick, solid, irregular to somewhat cylindrical, or bulbous, sometimes branched, central or eccentric, sometimes rooting; colored like the cap; surface velvety.
Spore print: White.
Spores: 5–9 x 3–5 μm, elliptical, smooth, hyaline.
Occurrence: Solitary or in clustered rosettes on the ground arising from buried roots, or at the base of living conifers, especially pine, often encompassing nearby plant stems, twigs or branches, occasionally growing directly on decaying stumps; also reported to occur rarely on the wood of broad-leaved trees (cherry); parasitic and saprobic; early summer-fall; common.
Edibility: Not edible.

Continued on next page

Comments: This aggressive pathogen causes a root and butt rot of conifers and is capable of killing living trees. It is sometimes found in large clusters of multiple rosettes. Compare with *Onnia tomentosa* (p.356), which has smaller and thinner caps, and a buff to tan pore surface. *Phaeolus schweinitzii* contains pigments that can be extracted for use as fabric dye. The specific epithet honors the American mycologist Lewis D. von Schweinitz (1780–1834).

Latin name: *Polyporus arcularius* (Batsch) Fr.
Common name: Spring Polypore
Order: Polyporales
Family: Polyporaceae
Width of cap: ½–3 in. (1–7.5 cm)
Length of stalk: ¾–2½ in. (2–6.5 cm)
Cap: Broadly convex to somewhat funnel-shaped or umbilicate, circular or nearly so in outline, margin thin and fringed with bristles; yellow-brown to ochre-brown or dark brown, paler beneath the scales; surface dry, covered with small radiating scales, becoming nearly glabrous with age; **flesh** cream-colored, tough, leathery, odor and taste not distinctive.
Pore surface: White to yellowish cream or darker in age; pores medium (1–2 per mm), angular to polygonal (honeycombed), radially arranged; pore layer shallow, more or less decurrent.
Stalk: Cylindrical, more or less equal or with a swollen base, central or nearly so, solid; colored like the cap; surface scurfy to scaly, sometimes almost smooth.
Spore print: White.
Spores: 7–11 x 2–3 μm, cylindrical, smooth, hyaline, non-amyloid.
Occurrence: Solitary or more often in small groups and clusters on dead wood of broad-leaved trees, especially on fallen branches and partly buried wood; saprobic; spring-early summer; common.
Edibility: Non-poisonous but too tough for eating.

Comments: *Polyporus brumalis* (p.361) has a much darker cap and smaller, randomly arranged pores. *Polyporus mori* (p.363) has an orange-brown cap and a short, squatty stalk. *Arcularius* pertains to "vaults or boxes," referring to the chambered pore surface.

Latin name: *Polyporus badius* (Pers.) Schwein.
Synonym: *Polyporus picipes* Fr.
Common name: Bay-colored Polypore
Order: Polyporales
Family: Polyporaceae
Width of cap: 2–8 in. (5–20 cm)
Length of stalk: ½–2 in. (1.3–5 cm)
Cap: Convex with an incurved margin at first, becoming flattened to somewhat funnel-shaped, irregular in outline with a thin, wavy margin; reddish brown to grayish brown or tan, usually paler toward the margin and dark reddish brown in the center; surface glabrous, smooth to slightly wrinkled, dry, more or less shiny; **flesh** thin, tough, leathery, white, odor and taste not distinctive.
Pore surface: White to cream or pale buff; pores minute (6–8 per mm); pore layer very shallow, decurrent.
Stalk: Short, usually eccentric, sometimes merely a point of attachment, equal to tapered toward the base, sometimes slightly bulbous; upper portion colored like the pore surface, brown to blackish at the base; surface smooth to finely velvety.
Spore print: White.
Spores: 6–10 x 3–5 μm, cylindrical to elliptical, smooth, hyaline.
Occurrence: Solitary or in groups, sometimes concrescent, on logs, stumps, and fallen branches of broad-leaved trees, or from buried wood; saprobic; spring-fall; common.
Edibility: Inedible.

Comments: *Polyporus varius* (not illustrated) is smaller and has a yellow to ochre cap that fades to whitish when old, and a blackish stalk base. *Badius* means "reddish brown."

Latin name: *Polyporus brumalis* Rostk.
Common name: Winter Polypore
Order: Polyporales
Family: Polyporaceae
Width of cap: 1–3 in. (2.5–7.5 cm)
Length of stalk: ½–1 in. (1.3–2.5 cm), or longer when attached to buried wood.
Cap: Convex to flat or umbilicate with an incurved, often crenulate margin; yellow-brown to gray-brown, or nearly black; surface smooth, finely hairy or minutely scaly, at times obscurely zonate; **flesh** tough, leathery, white.
Pore surface: White to cream-colored; pores small (2–3 per mm), oblong to somewhat angular; tube layer very shallow, more of less decurrent.
Stalk: Cylindrical, equal or swollen at the point of attachment, central or eccentric, firmly attached to the substrate; colored like the cap or paler; surface smooth to scurfy or finely hairy.
Spore print: White.
Spores: 6–8 x 2–2.5 μm, cylindrical, sometimes slightly curved, smooth, hyaline.
Occurrence: Solitary or in groups or clusters on decaying wood of broad-leaved trees, especially fallen trunks and branches; saprobic; late winter-spring; fairly common.
Edibility: Not edible.

Comments: This cold weather polypore is often present in early spring when few other mushrooms are fruiting. Compare with *Polyporus arcularius* (p.359), which has a paler, scaly cap with a fringed margin. *Brumalis* means "of the winter."

Latin name: *Polyporus craterellus* Berk. & M.A. Curtis
Synonym: *Polyporus fagicola* Murrill
Common name: None
Order: Polyporales
Family: Polyporaceae
Width of cap: 2–4 in. (5–10 cm)
Length of stalk: ¾–2 in. (2–5 cm)
Cap: Flat to depressed in the center or funnel-shaped; ochre to tan; upper surface smooth to somewhat radially wrinkled, glabrous or with sparsely scattered tufts of hairs or small reddish brown scales, margin fringed; **flesh** white, odor and taste not distinctive or somewhat fruity like melon rind.
Pore surface: Whitish to yellowish or yellowish brown; pores large (about 1 mm wide), angular; tube layer decurrent.
Stalk: Cylindrical, nearly equal or tapering toward the base, central or eccentric; colored more or less like the cap; surface scurfy to tomentose, smoother in age.
Spore print: White.
Spores: 10–14 x 4–6 μm, cylindrical to oblong-elliptical, smooth, hyaline.
Occurrence: Solitary or in small groups of 2 or 3 on dead wood of broad-leaved trees, especially on medium-sized fallen branches; saprobic; late spring-fall, uncommon.
Edibility: Edible.

Comments: The general aspect of this uncommon mushroom is similar to *Polyporus squamosus* (p.365) and *Polyporus mori* (p.363), but the cap has fewer scales than either, and it is intermediate in size. It has a more prominent stalk than *Polyporus mori* and lacks the reddish orange tones of that species. *Craterellus* means "goblet-shaped."

Latin name: *Polyporus mori* Pollini
Synonym: *Favolus alveolaris* Quél.
Common name: Hexagonal-pored Polypore
Order: Polyporales
Family: Polyporaceae
Width of cap: ¾–2½ in. (2–6.5 cm)
Length of stalk: ⅜–1 in. (1–2.5 cm)
Cap: Convex to depressed in the center, or somewhat funnel-shaped, circular to kidney- or fan-shaped in outline, margin inrolled at first; orange-yellow to reddish orange, fading to tan or pale yellow with age or when pigments are leached out; surface appressed-scaly to radially fibrillose, margin more or less fringed; **flesh** cream-colored, tough, odor and taste not distinctive.
Pore surface: Cream to pale yellow; pores medium to large (1–2 mm wide), angular to polygonal or hexagonal and radially arranged; tube layer more or less decurrent, at times extending down the entire stalk.
Stalk: Short, eccentric to nearly lateral, occasionally central, cylindrical, solid; whitish.
Spore print: White.
Spores: 9–11 x 3–3.5 μm, cylindrical, smooth, hyaline, non-amyloid.
Occurrence: Solitary or in groups on small- to medium-sized fallen branches of broad-leaved trees, especially oak, beech, and poplar; saprobic; spring-early summer; common.
Edibility: Edible but the flesh is tough.

Comments: *Pycnoporellus fulgens* (p.382) lacks a stalk, is larger and thicker, and has pores that become torn and tooth-like with age. *Polyporus arcularius* (p.359) has a yellow-brown to dark brown cap and a more developed stalk. Compare also with *Polyporus craterellus* (p.362). The name *mori* refers to *Morus alba* (white mulberry), on which this species was first collected.

Latin name: *Polyporus radicatus* (Schwein.) Fr.
Common name: Rooting Polypore
Order: Polyporales
Family: Polyporaceae
Width of cap: 2–8 in. (5–20 cm)
Length of stalk: Aboveground portion 2–8 in. (5–20 cm)
Cap: Convex to flattened or shallowly depressed with a broad central umbo; yellow-brown to reddish brown; surface dry, more or less velvety to fibrillose or finely scaly, smoother with age; **flesh** tough, leathery, white, odor not distinctive, taste mild to bitter.
Pore surface: White to pale creamy yellow; pores small (2–3 per mm), angular; pore layer decurrent.
Stalk: Rooting, more or less cylindrical above ground, irregular and often twisted and knobby below ground; colored like the pore surface above, blackish below ground; surface scurfy.
Spore print: White.
Spores: 12–15 x 6–8 μm, oblong to elliptical, smooth, hyaline, non-amyloid.
Occurrence: Solitary or in small groups on the ground, arising from buried wood and roots of broad-leaved trees, often in the vicinity of decaying stumps; saprobic; summer-fall; fairly common.
Edibility: Not edible.

Comments: At first glance this polypore might be mistaken for a bolete. However boletes have soft flesh and a tube layer that is easily removable. *Polyporus brumalis* (p.361) will occasionally have a rooting stalk but is smaller overall, fruits in the spring, and has a much darker brown cap. *Radicatus* means "rooting."

Latin name: *Polyporus squamosus* (Huds.) Fr.
Common names: Dryad's Saddle; Pheasant-back Polypore
Order: Polyporales
Family: Polyporaceae
Width of cap: 4–18 in. (10–45 cm), or sometimes larger.
Length of stalk: 1–3½ in. (2.5–9 cm)
Cap: Convex to depressed, kidney- or fan-shaped, or sometimes nearly circular in outline; ground color pale yellow to ochraceous overlaid with concentrically arranged zones of blackish brown to reddish brown appressed scales; **flesh** thick, thin at the margin, compact, white, soft at first, soon becoming tough and leathery, odor sweetish or fruity, like melon rind.
Pore surface: Whitish to cream, yellowish in age; decurrent; pores large (0.5–1 per mm), angular, radially elongated, sometimes exuding clear moisture droplets in wet weather.
Stalk: Lateral to eccentric, more or less cylindrical, thick, solid, tenaciously attached to the substrate; colored like the cap above, becoming darker to nearly black toward the point of attachment.
Spore print: White.
Spores: 12–16 x 4–6 μm, cylindrical to sausage-shaped, smooth, hyaline, non-amyloid.
Occurrence: Sometimes solitary but more often growing in groups or overlapping clusters on living or dead trunks (often high up), logs, and stumps of broad-leaved trees; parasitic and saprobic; spring-summer, fruitbodies persistent; common.
Edibility: Edible when young.

Comments: This large and conspicuous polypore appears fairly early in the spring and can be a consolation prize for the hapless morel hunter. Only the young fruitbodies and tender cap margins are suitable for eating. The somewhat similar *Polyporus craterellus* (p.362) is smaller, has a central stalk, and grows on fallen branches of broad-leaved trees. *Squamosus* means "full of scales."

Latin name: *Polyporus umbellatus* (Pers.) Fr.
Synonyms: *Grifola umbellata* (Pers.) Pilát; *Dendropolyporus umbellatus* (Pers.) Jülich.
Common name: Umbrella Polypore
Order: Polyporales
Family: Polyporaceae
Width of cap: Individual caps ¾–1½ in. (2–4 cm), compound fruitbodies may reach 16 in. (40 cm) or more across.
Length of stalk: 1–3 in. (2.5–7.5 cm)
Fruitbody: A compound cluster, hemispherical to nearly globose in outline, composed of numerous small, circular caps with individual stalks that branch from a central stalk; **caps** grayish brown to yellow-brown or at times dull white, depressed in the center; surface radially fibrillose to finely scaly; **pore surface** decurrent; white to cream-colored; pores small (1–3 per mm); **stalk** central or eccentric, somewhat cylindrical to irregularly shaped with extensive branching, colored more or less like the pore surface or more brownish, attached at the base to a buried sclerotium; **flesh** soft but becoming more fibrous with age, whitish, odor and taste not distinctive.
Spore print: White.
Spores: 7–9.5 x 3–4 μm, cylindrical, smooth, hyaline, non-amyloid.
Occurrence: Solitary or in groups on the ground in mixed woods with beech; parasitic and saprobic on the roots of beech, also reported to occur on oak; spring-fall; uncommon.
Edibility: Edible and very good when young.

Comments: This distinctive polypore grows from an underground, blackish sclerotium that is easily broken off when the fruitbody is picked. It can exist as a weak parasite causing a white rot of infected roots, but doesn't seem to be seriously detrimental to the host. Fruitbodies may appear repeatedly over many years in the same location, especially following heavy rains. *Umbellatus* means "with umbrellas."

Stereum ostrea, p.384

Lignicolous Bracket and Shelf Polypores with Lateral Attachment

The polypores in this section have simple or compound fruitbodies that grow on wood. Some are perennials that continue to grow and produce spores over a period of several years by adding a new fertile portion each season. Others are annuals and are functional for one season only, though their tough fruitbodies often persist for much longer. Normally these polypores are laterally attached to the substrate and are stalkless, however at times, such as when growing on the upper side of a fallen log or other horizontal surfaces, they may have a more or less central point of attachment, or an abbreviated stalk. The fruitbodies are mostly fibrous and woody or leathery, but a few are fleshy when young. Many are important decomposers and recyclers of dead wood. Others are wound parasites that cause heartrot in living trees. All are essential to the dynamics of a healthy forest.

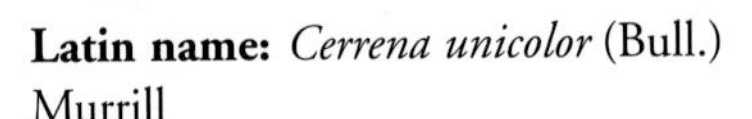

Latin name: *Cerrena unicolor* (Bull.) Murrill
Synonyms: *Trametes unicolor* (Bull.) Pilát; *Daedalea cinerea* Fr.
Common name: Mossy Maze Polypore
Order: Polyporales
Family: Polyporaceae
Fruitbody: Annual; up to 4 in. (10 cm) wide; semicircular, stalkless, wavy bracket, sometimes merely a projecting margin from a resupinate form; **upper surface** densely hairy, whitish to gray or grayish brown, concentrically zoned, often covered with green algae; **pore surface** whitish to ashy gray, typically maze-like (labyrinthine) to irregularly poroid, sometimes becoming almost tooth-like in age; **flesh** white, separated from the hairy surface by a thin dark zone, odor and taste not distinctive.
Spore print: White.
Spores: 5.5–7 x 2.5–3.5 μm, elliptical, smooth, hyaline.
Occurrence: In overlapping clusters on broad-leaved trees, stumps, and logs, rarely on conifer wood; parasitic on living trees or saprobic; year-round; occasional.
Edibility: Not edible.

Comments: The underside of this fungus is distinctive for the ashy gray, maze-like pore surface. *Daedaleopsis confragosa* (p.371) is thicker, more woody, and has a whitish to buff pore surface that bruises pinkish. On horizontal substrates the Mossy Maze Polypore can be almost entirely resupinate. *Unicolor* means "one color."

Latin name: *Daedaleopsis confragosa* (Bolton) J. Schröt
Synonyms: *Daedalea confragosa* Pers.; *Trametes rubescens* (Alb. & Schwein.) Fr.
Common name: Thin-walled Maze Polypore
Order: Polyporales
Family: Polyporaceae
Fruitbody: Annual; 1½–5 in. (4–12.5 cm) wide and up to ¾ in. (2 cm) thick; semicircular to fan-shaped in outline, or nearly circular when growing on horizontal surfaces, slightly convex to flat with a thin, often wavy margin, sometimes concrescent; **upper surface** somewhat radially wrinkled, glabrous to finely hairy; concentrically zoned with grayish to brown or yellow-brown bands, becoming reddish brown with age, margin whitish when actively growing; **pore surface** whitish to buff or gray, becoming grayish brown or darker brown with age, bruising pinkish brown when fresh; pores variable from round, angular, elongated, or labyrinthine, to gill-like when old; **flesh** tough, leathery to corky, whitish to brownish, odor not distinctive.
Spore print: White.
Spores: 7–10 x 2–3 μm, cylindrical to sausage-shaped, smooth, hyaline, non-amyloid.
Occurrence: Solitary or in groups on decaying wood of broad-leaved trees, especially birch, occasionally growing from wounds on living trees; saprobic; actively growing in summer and fall but can be found throughout the year; common.
Edibility: Not edible.

Comments: When fresh this bracket fungus is easy to recognize by the pinkish brown to reddish stains that develop on the bruised pore surface. *Daedalea quercina* (not illustrated), which usually grows on oak stumps, forms much thicker brackets with thicker pore walls, and has a non-staining pore surface. *Trametes elegans* (not illustrated) is nearly white overall and does not bruise on the pore surface. *Gloeophyllum saepiarium* (not illustrated), which grows on conifer wood, especially spruce, has a dark reddish brown, hairy upper surface with a yellow to orange margin, and thick, woody gill-like pores beneath. *Cerrena unicolor* (p.370) has a densely hairy upper surface and a white to gray maze-like pore surface. *Confragosa* means "rough."

Latin name: *Fomes fomentarius* (L.) J.J. Kickx
Common names: Tinder Fungus; Amadou
Order: Polyporales
Family: Polyporaceae
Fruitbody: Perennial; hard, woody hoof-shaped brackets with a blunt margin up to 8 in. (20 cm) or more broad, often longer vertically than wide; broadly attached to the substrate; **upper surface** crust-like; gray to blackish gray or brown; prominently zoned and furrowed, glabrous; **pore surface** cream-colored at first, becoming ochraceous to brown or reddish brown; pore surface concave; pores round, small (2–4 per mm). **flesh** woody, fibrous, yellow-brown, zonate, odor not distinctive.
Spore print: White.
Spores: 12–20 x 4.5–7 μm, cylindrical, smooth, hyaline.
Occurrence: Solitary or in groups on standing or fallen trunks and stumps of broad-leaved trees, especially beech, birch, and maple; parasitic and saprobic; year-round; common.
Edibility: Not edible.

Comments: The tinder fungus is a weak parasite that infects trees that are themselves weakened from other causes. It produces a heartrot that eventually reduces the structural strength of living trees. It can also function for many years as a saprobe on fallen trunks and stumps. *Phellinus igniarius* (not illustrated) is somewhat similar but has a blackened, fissured upper surface, darker rusty brown flesh, and smaller subglobose spores (5.5–7 x 4.5–6 μm). *Fometopsis pinicola* (p.373) has a reddish, resinous crust on the upper surface and a cream-colored pore surface. *Ganoderma applanatum* (p.374) is a relatively flat bracket with a whitish pore surface that bruises dark brown when scratched. *Fomentarius* means "material to feed a fire," referring to the past use of this fungus as tinder.

Latin name: *Fomitopsis pinicola* (Swartz) P. Karst.
Common name: Red-belted Polypore
Order: Polyporales
Family: Fomitopsidaceae
Fruitbody: Perennial; large semicircular woody bracket up to 18 in. (45 cm) wide, with a thick, rounded margin; broadly attached to the substrate; **upper surface** glabrous, shiny at first with a somewhat sticky resinous crust, later dull, concentrically grooved; at first orange-red to rusty red with a yellowish to ochraceous margin, later becoming gray to dull dark reddish brown or nearly black with a reddish marginal band; **pore surface** white to cream-colored, becoming brownish in age; pores small (4–5 per mm), round, often exuding clear droplets when young; **flesh** cream-colored, woody, odor and taste bitter.
Spore print: White to pale yellow.
Spores: 6–9 x 3.5–4.5 μm, cylindrical to elliptical, smooth, hyaline.
Occurrence: Solitary or in groups on living or dead trunks, stumps and logs of conifer trees, especially spruce, or less often on wood of broad-leaved trees; parasitic and saprobic; year-round; common.
Edibility: Not edible.

Comments: The resinous crust of the Red-belted Polypore will melt if a match is held to it. It is a brown rot fungus that infects damaged or dying trees and is an important decomposer of cellulose. *Fomitopsis cajanderi* (not illustrated) is a much smaller and thinner pinkish brown to grayish brown bracket fungus that grows on conifers and has a distinctive pinkish red pore surface. Compare with *Fomes fomentarius* (p.372). *Pinicola* means "pine dwelling."

Latin name: *Ganoderma applanatum* (Pers.) Pat.
Synonym: *Ganoderma lipsiense* (Batsch) G.F. Atk.
Common name: Artist's Conk
Order: Polyporales
Family: Ganodermataceae
Fruitbody: Perennial; semicircular to fan-shaped woody bracket up to 30 in. (75 cm) broad, proportionately flat with a thin margin; **upper surface** crust-like with lumps and wrinkles, often cracking, concentrically zoned and furrowed; dull grayish brown to pale brown or cinnamon-brown with a white margin when actively growing; **pore surface** white, smooth, bruising dark brown; pores minute (5–6 per mm); **flesh** brown, often with white flecks, darker above the pore layer, corky.
Spore print: Rusty brown.
Spores: 7–11 x 5–7.5 μm, broadly elliptical with a truncated end, double-walled, inner wall faintly warted, pale brown.
Occurrence: Solitary or in groups, sometimes compound or in overlapping clusters, on standing or fallen trunks and stumps of broad-leaved trees, reported to rarely occur on conifer wood; saprobic or weakly parasitic; year-round; common.
Edibility: Not edible.

Comments: This large bracket fungus can be up to 2 feet or more across. It produces astronomical numbers of rusty brown spores that often accumulate on the upper surface of the bracket and on surrounding vegetation. Because the pores are so small, the undersurface appears smooth to the naked eye. This white surface permanently stains dark brown when bruised or scratched. The brackets are popular with artists who etch drawings on the pore surface. Compare with *Fomitopsis pinicola* (p.373), which is similar but grows mainly on conifers. Compare also with *Fomes fomentarius* (p.372). *Applanatum* means "flattened."

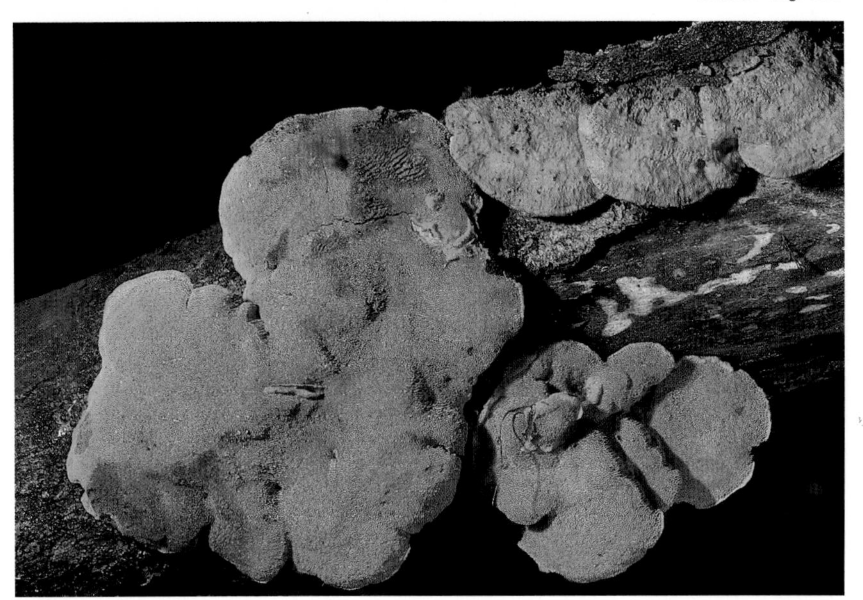

Latin name: *Hapalopilus rutilans* (Pers.) P. Karst
Synonym: *Hapalopilus nidulans* (Fr.) P. Karst.
Common name: Tender Nesting Polypore
Order: Polyporales
Family: Hapalopilaceae
Fruitbody: Annual; small, semicircular to kidney-shaped bracket up to 4 in. (10 cm) wide and 1 in. (2.5 cm) thick, broadly attached to the substrate; **upper surface** convex to flat, more or less matted hairy, uneven, sometimes obscurely concentrically furrowed; pinkish ochre to cinnamon-brown; **pore surface** ochraceous to cinnamon-brown, more or less bruising reddish brown when fresh; pores small (2–4 per mm), angular; **flesh** colored like the upper surface, soft and spongy at first, becoming hard and brittle when dry, odor and taste sweetish or not distinctive.
Spore print: White.
Spores: 3.5–5 x 2–2.5 µm, subglobose to elliptical, smooth, hyaline.
Occurrence: Solitary or in clusters, often fused and overlapping, on fallen branches, logs, and trunks of broad-leaved trees, especially oak, beech, and birch, also reported to rarely occur on conifer wood; saprobic; early summer-fall, fruitbodies persistent; fairly common.
Edibility: Not edible.

Comments: These plump little brackets are rather dull-colored and easily overlooked. All parts instantly turn deep reddish purple in contact with KOH. The Tender Nesting Polypore contains concentrated pigments and is a favorite species for those who use fungi for natural fabric dyes. *Hapalopilus croceus* (not illustrated) is yellow-orange, more robust, and grows on oak (also reported on hemlock). It also turns reddish purple with the application of KOH. *Rutilans* means orange-red.

Latin name: *Inonotus hispidus* (Bolton) P. Karst.
Common name: Shaggy Polypore
Order: Hymenochaetales
Family: Hymenochaetaceae
Fruitbody: Annual; semicircular bracket up to 12 in. (30 cm) wide and 4 in. (10 cm) thick broadly attached to the substrate; **upper surface** somewhat hairy to densely shaggy; yellowish red to reddish orange with a yellowish margin when young, becoming rusty brown, eventually blackish; **pore surface** pale yellow to ochraceous or sienna brown, at times with olivaceous tones, bruising brown when fresh; pores round to angular, small (2–3 per mm), often exuding watery droplets when young and fresh; **flesh** soft and watery at first, becoming fibrous and hard when old, ochraceous, staining brown when cut, odor and taste mild or acidic.
Spore print: Ochre to brown.
Spores: 7.5–11.5 x 6.5–9 μm, subglobose, smooth, brownish.
Occurrence: Solitary or in clusters, often located high on the trunks of living broad-leaved trees, especially oaks, also on fallen trunks and large branches; parasitic and saprobic; summer-fall; occasional.
Edibility: Not edible.

Comments: This attractive bracket fungus is a wound parasite on living trees, but also functions as a saprobe on fallen deadwood. It is often encountered in urban areas, cemeteries, and parklands, as well as in woodlands. The fruitbodies are rich in pigment and can be used for dying textiles. *Inonotus radiatus* (not illustrated), which is usually less than 3 in. (7.5 cm) wide, has a velvety reddish brown upper surface with a yellowish margin and grows in overlapping tiers on broad-leaved and conifer wood. *Hispidus* means "bristly."

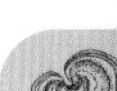

Latin name: *Ischnoderma resinosum* (Schrad.) P. Karst.
Common name: Resinous Polypore
Order: Polyporales
Family: Hapalopilaceae
Fruitbody: Annual; medium to large semicircular or fan-shaped brackets up to 10 in. (25 cm) wide and 1 in. (2.5 cm) thick; margin rounded and thick; **upper surface** velvety when young, later almost smooth, often radially wrinkled; rusty brown to dark brown or nearly black, concentrically zoned; **pore surface** creamy white to ochraceous, quickly bruising brown; pores round to angular, minute (4–6 per mm), often exuding amber droplets when young and fresh, especially near the margin; **flesh** whitish to pale brown, soft and sappy at first, becoming tough and hard on drying, odor and taste not distinctive.
Spore print: Whitish.
Spores: 5–7 x 1.5–2 μm, cylindrical to sausage-shaped, smooth, hyaline, non-amyloid.
Occurrence: Solitary or in groups, often in overlapping tiers on logs, stumps, or standing dead trunks of broad-leaved trees, also reported to occur on conifer wood; saprobic; late summer-late fall, occasionally in spring; fairly common.
Edibility: Not edible.

Comments: *Ganoderma tsugae* (p.354) has a shiny, "varnished" upper surface and grows on hemlock. *Ganoderma applanatum* (p.374) has a smooth, hard crust on the upper surface, and produces rusty brown spores. *Resinosum* means "resinous" in reference to the sappy flesh.

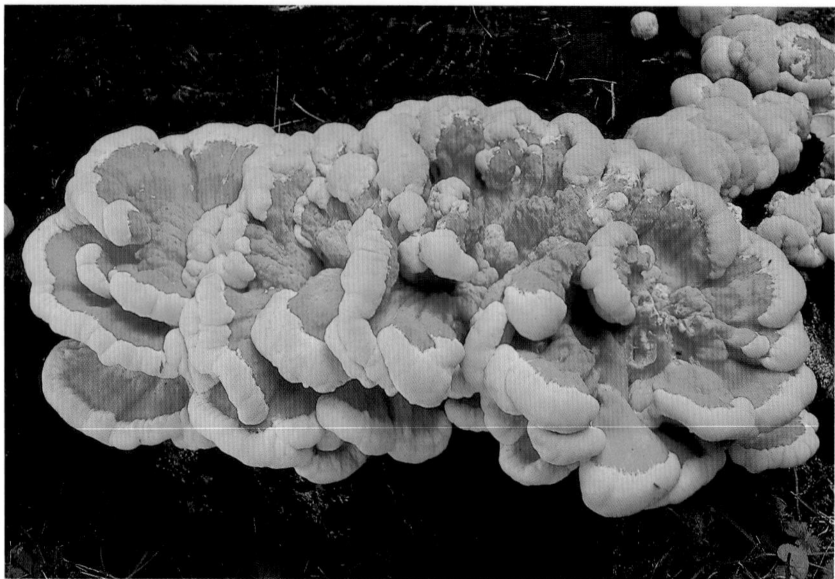

Latin name: *Laetiporus sulphureus* var. *sulphureus* (Bull.) Murrill
Synonym: *Polyporus sulphureus* (Bull.) Fr.
Common names: Sulphur Shelf; Chicken of the Woods; Rooster Comb
Order: Polyporales
Family: Polyporaceae
Fruitbody: Annual; semicircular to fan-shaped brackets with blunt, wavy margins, usually in compound overlapping clusters; individual caps up to 12 in. (30 cm) wide; **upper surface** velvety, uneven, radially wrinkled, more or less zonate; brilliant orange-yellow when fresh, fading to pale orange or whitish in age; **pore surface** sulphur yellow; pores round to oblong, small (3–5 per mm); **flesh** soft and yellow when young, later becoming fibrous, eventually whitish and crumbly, odor and taste not distinctive.
Spore print: White.
Spores: 5–7 x 3.5–5 μm, subglobose, smooth, hyaline, non-amyloid.
Occurrence: Sometimes solitary but usually in compound clusters on living or dead tree trunks, logs, stumps, and larger fallen branches of broad-leaved trees; parasitic and saprobic; spring-fall; common.
Edibility: The entire fruitbody is very good when young, but later, as it matures and becomes fibrous, only the tender margin is edible. (Note: some individuals cannot tolerate this species when consumed with alcohol.)

Comments: The Sulphur Shelf is one of the most colorful bracket fungi in our woods. When fruiting in massive clusters it can be spotted from a long distance away. It first appears as bright yellow irregular lumps or finger-like projections, but soon expands into flattened brackets. *Laetiporus huronensis* (not illustrated) is nearly identical but grows on conifer wood. The less common *Laetiporus cincinnatus,* also known as *Laetiporus sulphureus* var. *semialbinus* (see photo), is similar but has a whitish pore surface and often grows in compound rosettes. It too is edible and popular. *Sulphureus* means "pertaining to sulphur," referring to the color of the pore surface.

Laetiporus cincinnatus

Latin name: *Phellinus gilvus* (Schwein.) Pat.
Synonym: *Polyporus gilvus* (Schwein.) Fr.
Common name: Mustard-yellow Polypore
Order: Hymenochaetales
Family: Hymenochaetaceae
Fruitbody: Annual or perennial; small to medium semicircular or fan-shaped brackets up to 6 in. (15 cm) wide and 1 in. (2.5 cm) thick; **upper surface** velvety to hairy and rough, becoming smoother in age; rusty yellow to ochre or dark reddish brown with a yellow margin, blackish in age, more or less zonate; **pore surface** grayish brown to reddish brown; pores round, minute (5–8 per mm); **flesh** leathery to corky, bright ochre to yellow-brown, odor not distinctive.
Spore print: White.
Spores: 4–5 x 2.5–3.5 μm, elliptical, smooth, hyaline.
Occurrence: Solitary or in groups, often in rows or overlapping clusters on dead wood of broad-leaved trees, occasionally on conifer wood; saprobic; year-round; common.
Edibility: Not edible.

Comments: The similar *Inonotus radiatus* (not illustrated) is more conspicuously zoned and has rusty brown flesh, larger pores (2–4 per mm) that are whitish when young, and yellow spores. *Phaeolus schweinitzii* (p.357) is much larger overall and typically grows in compound rosettes at the base of conifers. *Gilvus* means "dull pale yellow."

Latin name: *Piptoporus betulinus* (Bull.) P. Karst.
Common names: Birch Polypore; Razor Strop Fungus
Order: Polyporales
Family: Fomitopsidaceae
Fruitbody: Annual; 4–10 in. (10–25 cm) wide, 1–3 in. (2.5–7.5 cm) thick; shell-shaped or convex kidney-shaped to hoof-shaped bracket with an inrolled margin that overlaps the tube layer; sessile or narrowly attached to the substrate by a rudimentary stalk; **upper surface** smooth and glabrous when young, later cracking and crust-like; dingy white to pale gray-brown; **pore surface** white to cream; pores small (3–4 per mm), round; **flesh** white, soft at first but soon becoming corky and tough, odor not distinctive, taste increasingly bitter with age.
Spore print: White.
Spores: 5–7 x 1.5–2 μm, sausage-shaped, smooth, hyaline, non-amyloid.
Occurrence: Solitary, or more often in groups on living or dead birch trees, often appearing high on the trunk, also on fallen logs and stumps; parasitic and saprobic; actively grows in the fall but fruitbodies persist throughout the year; common.
Edibility: Edible when young and tender but not often collected due to the bitter flesh.

Comments: This bracket fungus is easily recognized by its distinctive cushion shape and specific association with birch trees. It is a weak parasite that infects old and otherwise weakened birch trees, usually gaining entrance through broken branch stubs. When the fruitbodies first appear they are nearly globose, looking almost like puffballs. In times past the light corky flesh was cut into strips for use as razor strops. It has also been used as a mounting material for insect collections and for polishing metal. The name *betulinus* pertains to "*Betula*," the genus of birch trees.

Latin name: *Pycnoporellus fulgens* (Fr.) Donk
Synonym: *Polyporus fibrillosus* P. Karst.
Common name: None
Order: Polyporales
Family: Polyporaceae
Fruitbody: Annual; small to medium semicircular or fan-shaped stalkless brackets from 1½–4 in. (4–10 cm) wide; **upper surface** radially fibrillose to tomentose, often zonate; pale to dark orange-red or orange-brown; **pore surface** cream to pale orange or brownish orange; pores round to angular, sometimes becoming torn and tooth-like, small (1–3 per mm); tubes 2–6 mm long; **flesh** pale orange to rusty red, staining red then quickly black in contact with KOH, spongy at first, becoming ridged and brittle when dry, odor not distinctive.
Spore print: Whitish.
Spores: 6–10 x 3–4 μm, oblong-elliptical, smooth, hyaline, non-amyloid.
Occurrence: Solitary or in small, overlapping clusters on decaying wood of conifers and less often on broad-leaved trees; saprobic; spring-fall; occasional.
Edibility: Not edible.

Comments: This bracket often fruits on the ends of cut or broken branches or logs. Because the pore surface can be ragged, at times it may appear more like a tooth fungus than a polypore. *Fulgens* means "shining" or "glistening."

Latin name: *Pycnoporus cinnabarinus* (Jacq.) Fr.
Synonym: *Polyporus cinnabarinus* (Jacq.) Fr.
Common name: Cinnabar-red Polypore
Order: Polyporales
Family: Polyporaceae
Fruitbody: Annual; small to medium semicircular to kidney-shaped, flat, stalkless bracket up to 4 in. (10 cm) wide, laterally attached, or centrally attached when growing on upper surface of a horizontal substrate; **upper surface** glabrous to slightly tomentose, smooth to uneven or warted, vaguely zoned; brilliant cinnabar-red to orange-red or pale orange, fading to whitish with age; **pore surface** cinnabar-red, fading less with age than the upper surface; pores small (2–4 per mm), round to angular, occasionally labyrinthine; **flesh** cinnabar-red, tough, leathery to corky, becoming hard when dry, odor not distinctive.
Spore print: Whitish.
Spores: 5–6 x 2–2.5 μm, cylindrical to sausage-shaped, smooth, hyaline.
Occurrence: Solitary or in groups, often fused and overlapping, on deadwood of broad-leaved trees, rarely on conifer wood; saprobic; year-round; common.
Edibility: Not edible.

Comments: This colorful bracket fungus is especially common in open, sunny areas where logging slash and brush are piled. *Cinnabarinus* means "cinnabar-red."

Latin name: *Stereum ostrea* Nees
Common name: False Turkey Tail
Order: Russulales
Family: Stereaceae
Fruitbody: Annual; shell-shaped to petal-shaped or semicircular bracket from ¾–3 in. (2–7.5 cm) wide; **upper surface** with fine, silky hairs, zonate; variously colored with reddish brown, orange, gray, or yellow bands, margin often whitish; **fertile undersurface** smooth or with small bumps; grayish yellow to buff or reddish brown; **flesh** thin, leathery, hard when dry, odor and taste not distinctive; **stalk** lacking.
Spore print: White
Spores: 5–7.5 x 2–3 µm, cylindrical, smooth, hyaline.
Occurrence: Usually numerous in overlapping clusters on decaying logs, stumps, and fallen branches of broad-leaved trees; saprobic; early summer-early winter, fruitbodies persistent; common.
Edibility: Not edible.

Comments: *Stereum striatum* (not illustrated), which usually grows on small twigs and branches of blue beech (*Carpinus caroliniana*), is similar but smaller and has a silky whitish to pale gray upper surface. *Trametes versicolor* (p.385) has a white undersurface with minute pores. *Ostrea* means "an oyster," referring to the shape and manner of growth.

Latin name: *Trametes versicolor* (L.) C.G. Lloyd
Synonym: *Coriolus versicolor* (L.) Quél.
Common names: Turkey Tail; Many-zoned Polypore
Order: Polyporales
Family: Polyporaceae
Fruitbody: 1–3 in. (2.5–7.5 cm) wide; annual; small, thin semicircular to fan-shaped bracket, or more or less circular when growing on a flat horizontal surface; **upper surface** varies from silky and smooth to matt and velvety; multicolored and variable with reddish brown, yellowish, gray, greenish, blue, or blackish concentric bands, the outermost zone is usually pale yellow or tan; **pore surface** white at first, becoming yellowish to pale orange in age; pores round to angular or elongated, minute (1–5 per mm); **flesh** thin, cream-colored, leathery, tough, odor not distinctive.
Spore print: Cream.
Spores: 5–6 x 1.5–2 μm, cylindrical to sausage-shaped, smooth, hyaline, non-amyloid.
Occurrence: Gregarious in overlapping tiers, or in fused rosettes on fallen branches, logs, and stumps of broad-leaved trees, or on the standing trunks of dying trees, also reported to occur rarely on conifer wood; saprobic or weakly parasitic; year-round; common.
Edibility: Not edible.

Continued on next page

Comments: This colorful bracket fungus is a vigorous and important decomposer of wood. The fruitbodies are primarily active (producing spores) during summer and fall but are found throughout the year and can persist for years. The pore surface appears to be nearly smooth but a hand lens will reveal the minute pores. There are many similar concentrically banded "Turkey Tails" including *Trametes hirsuta* (not-illustrated), which has stiff hairs on a grayish to yellowish upper surface. *Trametes suaveolens* (not illustrated) has a paler whitish to gray upper surface and a pronounced anise-like odor when fresh. *Lenzites betulina* (not illustrated) has gills on the underside. *Bjerkandera adusta* (not illustrated) has a silver-gray pore surface that bruises black. Compare with *Trichaptum biforme* (p.387), which has violaceous tones on the fruitbodies and a pore surface that often becomes tooth-like in age. *Stereum* species have a completely smooth undersurface. *Versicolor* means "multicolored."

Latin name: *Trichaptum biforme* (Fr.) Ryvarden
Synonyms: *Hirschioporus pergamenus* (Fr.) Bondartsev & Singer; *Trichaptum pargamenum* (Fr.) G. Cunn.
Common name: Violet Toothed Polypore
Order: Polyporales
Family: Polyporaceae
Fruitbody: Annual; semicircular to fan-shaped flattened bracket up to 3 in. (7.5 cm) wide; **upper surface** velvety to hairy at first, becoming smoother in age; whitish to gray with a purplish margin, concentrically zoned, often partly covered with greenish algae; **pore surface** deep violet at first, fading to buff or brownish in age but usually retaining violaceous tones until very old, especially near the margin; pores round to angular (3–5 per mm), soon irregular and splitting into teeth-like projections; **flesh** white to ochraceous, thin, tough, leathery, odor not distinctive.
Spore print: White.
Spores: 6–8 x 2–2.5 μm, cylindrical to sausage-shaped, smooth, hyaline.
Occurrence: Usually clustered in overlapping concrescent tiers, on dead branches, trunks, and stumps of numerous broad-leaved trees; saprobic; year-round; common.
Edibility: Not edible.

Comments: *Trichaptum abietinum* (not illustrated) is very similar but grows on conifer wood, especially spruce. Compare with *Trametes versicolor* (p.385). The Violet Toothed Polypore is often inhabited by *Phaeocalicium polyporaeum* (not illustrated), a saprobic ascomycete that produces tiny black clubs on the upper surface of the polypore, especially near the margin (use a hand lens). *Biforme* means "having two forms," referring to the pore surface that can be either poroid or tooth-like.

Latin name: *Tyromyces chioneus* (Fr.) P. Karst.
Synonym: *Tyromyces albellus* (Peck) Bondartsev & Singer
Common name: White Cheese Polypore
Order: Polyporales
Family: Polyporaceae
Fruitbody: Annual; semicircular or fan-shaped, convex to flattened bracket up to 4 in. (10 cm) wide; **upper surface** tomentose or glabrous, whitish at first, becoming yellowish to grayish in age, at times obscurely zoned; **pore surface** whitish; pores round to angular, small (3–5 per mm); **flesh** white, soft and juicy when fresh, becoming somewhat hard and brittle when dry, odor pleasant, taste not distinctive.
Spore print: White
Spores: 4–5 x 1.5–2 μm, cylindric to sausage-shaped, smooth, hyaline, non-amyloid.
Occurrence: Solitary or in small groups in tiers or rows, often fused, on fallen branches, logs, and stumps of broad-leaved trees, especially birch; saprobic; early summer-early winter; common.
Edibility: Not edible.

Comments: *Tyromyces caesius* (not illustrated) is similar but grows on conifer wood and has a distinct bluish tinge, especially near the margin. The name *chioneus* pertains to snow, alluding to the color.

Latin name: *Tyromyces fragilis* (Fr.) Donk
Synonyms: *Postia fragilis* (Fr.) Jülich; *Oligoporus fragilis* (Fr.) Gilbert & Ryvarden
Common name: Brown-staining Cheese Polypore
Order: Polyporales
Family: Polyporaceae
Fruitbody: Annual; semicircular or fan-shaped, flattened to convex bracket up to 3½ in. (9 cm) wide; **upper surface** radially fibrillose or hairy to glabrous; yellowish ochre to orange-brown with a whitish margin, bruising reddish brown; **pore surface** white to buff, quickly bruising yellowish brown or reddish brown; tube layer often decurrent onto the substrate; pores angular to tooth-like or sometimes labyrinthine, small (2–4 per mm); **flesh** white, soft and elastic at first, becoming firm and brittle when dry, odor not distinctive or sweet on drying, taste not distinctive.
Spore print: Whitish.
Spores: 4–5 x 1.5–2 μm, cylindrical to broadly sausage-shaped, smooth, hyaline.
Occurrence: Solitary or in groups, sometimes fused in overlapping tiers or rows, on conifer wood; saprobic; summer-fall; occasional.
Edibility: Not edible.

Comments: Compare with *Pycnoporellus fulgens* (p.382), which is superficially similar. *Fragilis* means "fragile," perhaps referring to the instant bruising of the fruitbody when handled.

Pseudohydnum gelatinosum, p. 402

Tooth Fungi

The mushrooms in this section exhibit a variety of forms and occur on various substrates, but what they all have in common is a fertile surface composed of hanging or projecting spines or "teeth" on which the reproductive spores are produced. One bracket polypore, *Irpex lacteus,* is also included here because its poroid fertile surface soon becomes lacerated and tooth-like.

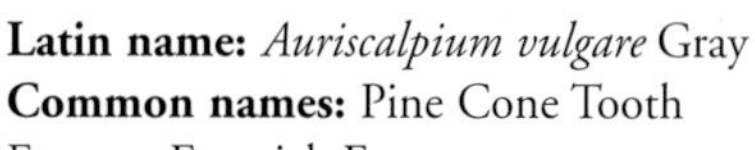

Latin name: *Auriscalpium vulgare* Gray
Common names: Pine Cone Tooth Fungus; Ear-pick Fungus
Order: Russulales
Family: Auriscalpiaceae
Fruitbody: With a distinct cap and stalk; **cap** ⅜–1 in. (1–2.5 cm) wide, kidney-shaped to nearly circular in outline, convex to flat, light brown to reddish brown, paler near the margin; surface densely covered with short erect hairs, becoming smoother with age; **flesh** whitish, tough, leathery, odor and taste not distinctive; **fertile surface** composed of short (2–3 mm), pointed spines; grayish white to pinkish brown or grayish brown; **stalk** up to 3 in. (7.5 cm) long, ⅛ in. (3 mm) thick; slender, solid, tough, attached to the cap near the margin; surface densely hairy, colored like the cap or darker blackish brown.
Spore print: White.
Spores: 4.5–5.5 x 3.5–4.5 μm, broadly elliptical, minutely spiny, hyaline, amyloid.
Occurrence: Solitary or clustered on fallen pine cones, especially Scots pine (*Pinus silvestris*), also reported to occur rarely on spruce cones; saprobic; spring-fall; fairly common.
Edibility: Not edible.

Comments: This unique fungus is easily overlooked and is probably more common than reported. The pine cones on which it grows are often at least partially buried. As many as 4 or 5 fruitbodies may grow from a single cone. *Vulgare* means "common."

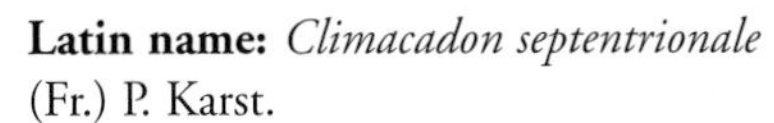

Latin name: *Climacadon septentrionale* (Fr.) P. Karst.
Synonym: *Steccherinum septentrionale* Fr.
Common name: Northern Tooth
Order: Polyporales
Family: Meruliaceae
Fruitbody: Annual; consists of multiple overlapping, semicircular to fan-shaped stalkless brackets; individual brackets are up to 12 in. (30 cm) wide and from 1–2 in. (2.5–5 cm) thick; white to yellowish, or ochraceous in age; surface roughened to somewhat hairy, especially toward the margin; **flesh** thick, tough and elastic when fresh, hard when dry, whitish, odor and taste not distinctive when fresh, becoming somewhat unpleasant (like old ham) on drying; **fertile surface** composed of slender, crowded spines, up to ¾ in. (2 cm) long, shorter near the margin, white to yellowish or ochraceous in age, tips lacerated.
Spore print: White.
Spores: 3.5–5.5 x 2.5–3 μm, elliptical, smooth, hyaline.
Occurrence: In overlapping clusters (sometimes massive) on the trunks of living broad-leaved trees, especially maple and beech, often 6–8 feet or more from the ground; parasitic; summer-fall; occasional.
Edibility: Not edible.
Continued on next page

Comments: This spectacular and unmistakable fungus is frequently encountered in urban areas and parklands as well as in woods. *Septentrionalis* means "northern."

Latin name: *Hericium erinaceum* (Bull.) Pers.
Common names: Satyr's Beard; Lion's Mane
Order: Russulales
Family: Hericiaceae
Fruitbody: A cushion-shaped to somewhat globose mass of hanging spines up to 10 inches (25 cm) or more wide; white at first, becoming yellowish to pale brownish in age, especially on the spine tips; **flesh** white, firm, elastic, sponge-like, odor and taste not distinctive when young, sour and unpleasant when old; **fertile surface** composed of crowded, pendulous spines from ¾–2 inches (2–5 cm) long (or longer); **stalk** short to rudimentary, tough, firmly attached laterally to the substrate.
Spore print: White.
Spores: 6–6.5 x 4–5.5 µm, subglobose, smooth to minutely warted, hyaline, amyloid.
Occurrence: Usually solitary, growing from wounds on the trunks of broad-leaved trees, especially beech, maple, and oak, also on cut stumps and log ends; parasitic and saprobic; summer-fall; occasional.
Edibility: Edible and good when young.

Comments: This beautiful fungus is always a treat to encounter. Cultivated versions, sometimes called "Pom Poms," are becoming increasingly available for growers and consumers. *Hericium ramosum* (p.396) is strongly branched and has shorter, more delicate spines. *Erinaceus* means "prickly" or "hedgehog-like."

Latin name: *Hericium ramosum* (Bull.) Letellier
Common name: Comb Tooth
Order: Russulales
Family: Hericiaceae
Fruitbody: Irregular-shaped mass of toothed branches up to 10 in. (25 cm) or more wide, intricately and repeatedly branched; white at first, becoming yellowish to tan in age; **fertile surface** small, delicate, "feathery" spines up to ½ in. (1.3 cm) or longer distributed more or less evenly along the branches; glistening white when young, dingy in age; **stalk** short or rudimentary, tough, laterally attached to the substrate; **flesh** white, soft, odor and taste not distinctive.
Spore print: White.
Spores: 3–5 x 3–4 μm, subglobose, smooth to minutely warted, hyaline, amyloid.
Occurrence: Solitary or in groups on fallen branches, logs, and stumps of broad-leaved trees, especially beech, oak, and maple; saprobic; summer-fall; occasional to fairly common in some years.
Edibility: Edible.

Comments: This is a good edible mushroom when young. It can sometimes be found in multiple clusters on one log or stump. *Hericium americanum* (not illustrated), which is also a good edible, has longer spines that are concentrated in tufts at the ends of its branches. Compare with *Hericium erinaceum* (p.395), which has longer spines and lacks extensive branching. *Ramosum* means "branched."

Latin name: *Hydnellum scrobiculatum* (Fr.) P. Karst.
Common name: Rough Hydnellum
Order: Thelephorales
Family: Bankeraceae
Fruitbody: Up to 3½ in. (9 cm) wide, turbinate (top-shaped) and convex at first, becoming flattened or depressed in the center; margin wavy and lobed, bruising dark brown; whitish to pinkish brown or reddish brown, at times exuding a red juice when fresh; surface irregular with pits, trenches, and projections, or smaller auxiliary caps, center somewhat scaly; **flesh** tough, spongy, zonate, reddish brown often with rows of white dots, odor not distinctive; **fertile surface** with short spines up to ¼ in. (6 mm) long, more or less decurrent; whitish at first, becoming buff-brown to purplish brown or darker in age; **stalk** up to 2 in. (5 cm) long and ½ in. (1.3 cm) thick, cylindrical, usually with an enlarged spongy base beneath the soil; light brown to rusty brown, velvety.
Spore print: Brown.
Spores: 4.5–5.5 x 5.5–7 μm, subglobose to somewhat angular, warted, light brown.
Occurrence: Solitary to gregarious, sometimes in fused clusters, on the ground in conifer and mixed woods; mycorrhizal; summer-fall; occasional.
Edibility: Not edible.

Comments: The exudations of red juice that are sometimes present on the upper surface of fresh specimens is a good field character. Similar species, such as *Hydnellum ferrugineum* (not illustrated) and *Hydnellum spongiosipes* (not illustrated), are macroscopically difficult to distinguish. *Scrobiculatum* means "having a pitted or grooved surface."

Latin name: *Hydnum repandum* var. *album* (Quél.) Rea
Common names: White Wood Hedgehog
Order: Cantharellales
Family: Hydnaceae
Fruitbody: Cap ¾–3 in. (2–7.5 cm) wide, convex to nearly flat or somewhat depressed in the center, margin even or lobed, often wavy; chalk white to creamy white, bruising orange; surface dull, felted or glabrous; **flesh** firm, brittle, white, bruising orange, taste mild or bitter; **fertile surface** with vertical spines ⅛–¼ in. (3–6 mm) long; white to pale buff, fragile; **stalk** 1–3 in. (2.5–7.5 cm) long; more or less equal or enlarged toward the base, often crooked, solid; colored like the cap, bruising orange.
Spore print: White.
Spores: 7–8.5 x 5.5–7 μm, subglobose, smooth, hyaline.
Occurrence: Solitary or in small groups on the ground in conifer and broad-leaved woods; mycorrhizal; summer-fall; occasional.
Edibility: Edible but sometimes bitter.

Comments: *Hydnum repandum* var. *repandum* (not illustrated) is more robust and has a creamy white to pale orange or pale ochre cap. *Repandum* means "bent back," in reference to the wavy cap margin. *Album* means "dead white" or "the white of an egg."

Latin name: *Hydnum umbilicatum* Peck
Synonym: *Dentinum umbillicatum* (Peck) Pouzar
Common name: Navel Hedgehog
Order: Cantharellales
Family: Hydnaceae
Fruitbody: Cap 1–2 in. (2.5–5 cm) wide, convex to flat, becoming depressed in the center forming a navel-like cavity (umbilicate); orange to brownish orange or reddish buff; surface often irregular, wavy or pitted, dry, matt, sometimes cracking; **flesh** thick, brittle, colored like the cap or paler, to nearly white, staining orange when cut, odor and taste not distinctive; **fertile surface** with vertical spines up to ⅜ in. (1 cm) long; creamy buff to pale orange; **stalk** 1–2 in. (2.5–5 cm) long, nearly equal or enlarged downward, often crooked; pale buff to pale orange or nearly white, bruising orange.
Spore print: White.
Spores: 6–7.5 x 8.5–9.5 μm, subglobose, smooth, hyaline.
Occurrence: Usually in groups or clusters on the ground in wet woods, often with sphagnum moss; mycorrhizal; summer-fall; fairly common.
Edibility: Edible.

Comments: The cap of *Hydnum repandum* var. *repandum* (not illustrated) can be slightly depressed in the center but is never umbilicate. *Hydnum rufescens* (not illustrated) is nearly identical but lacks a navel-like central depression. *Umbilicatum* means "having a navel."

Latin name: *Irpex lacteus* (Fr.) Fr.
Common name: Milk-white Toothed Polypore
Order: Polyporales
Family: Steccherinaceae
Fruitbody: Fan-shaped to shell-shaped, convex, stalkless bracket up to 1½ in. (4 cm) wide, typically projecting at the margin of a resupinate spreading crust; **upper surface** velvety-hairy, dry, radially furrowed, white to cream or pale grayish, becoming dingy yellowish brown with age, obscurely zoned; **flesh** thin, leathery, whitish; **fertile surface** irregularly poroid to labyrinthine at first and remaining so near the margin, elsewhere the tubes soon splitting into flattened tooth-like projections up to ¼ in. (6 mm) long; white to dingy yellow or ochraceous in age.
Spore print: White.
Spores: 5–6.5 x 2–3 µm, cylindrical to narrowly elliptical, smooth, hyaline, non-amyloid.
Occurrence: On fallen branches and dead wood of broad-leaved trees; saprobic; year-round; occasional.
Edibility: Not edible.

Comments: The brackets are often laterally fused in rows. Sometimes distinct caps are not present, or may be limited to slightly projecting flaps on the margin of a resupinate spreading mass of "teeth." *Lacteus* means "milk white."

Latin name: *Mycorrhaphium adustum* (Schwein.) Maas Geest.
Synonym: *Steccherinum adustum* (Schwein.) C.S. Bi & Zheng
Common name: Kidney-shaped Tooth
Order: Polyporales
Family: Steccherinaceae
Fruitbody: Kidney-shaped to nearly circular, flat to depressed bracket up to 3 in. (7.5 cm) wide, sometimes concrescent; white to yellowish white or tan, blackish on the margin; surface minutely velvety to glabrous, more or less concentrically zoned; **flesh** white, leathery, tough, elastic; **fertile surface** with short, densely packed, somewhat flattened spines, up to ⅛ in. (3 mm) long, often fused and appearing to be forked at the tip; white at first becoming rose to purplish, then dark brown in age, bruising blackish; **stalk** sometimes absent, up to 1 in. (2.5 cm) long when present, tapering upward, eccentric to central, velvety, tough, solid.
Spore print: White.
Spores: 1–1.2 x 3–5.8 μm, cylindrical, smooth, hyaline.
Occurrence: Solitary or in small groups on fallen logs and branches of broad-leaved trees, especially oak; saprobic; summer-fall; common.
Edibility: Not edible.

Comments: The caps of this distinctive fungus are often fused together or stacked in shallow tiers. *Adustum* means "scorched" or "burned."

Latin name: *Pseudohydnum gelatinosum* (Scop.) P. Karst.
Synonym: *Tremellodon gelatinosum* (Scop.) Pers.
Common names: Jelly Tooth; Jelly Tongue; Jelly Hedgehog
Order: Tremellales
Family: Exidiaceae
Fruitbody: Up to 3 in. (7.5 cm) or more wide; spoon-shaped to tongue-shaped, thick gelatinous bracket with an inrolled margin; **upper surface** downy to nearly smooth; whitish to grayish brown, darker in age; **flesh** soft, jelly-like, translucent, odor and taste not distinctive or slightly resinous; **fertile surface** composed of short, soft, pendulous spines; translucent white to watery gray, sometimes with a bluish tinge; **stalk** up to 2 in. (5 cm) long, usually a lateral extension of the cap tissue, or more or less central when growing from top of horizontal surfaces, sometimes absent; colored like the cap.
Spore print: White.
Spores: 4.5–5.5 x 5–6 µm, nearly globose, smooth, hyaline, non-amyloid.
Occurrence: Usually in small groups, or sometimes growing in extensive overlapping, fused clusters, on well-decayed conifer logs and stumps; saprobic; summer-fall; common.
Edibility: Edible, even raw, but consumed more as a novelty since it is essentially tasteless.

Comments: This is the only jelly-like fungus with spines and it is not likely to be confused with any other. *Gelatinosum* means "jelly-like."

Latin name: *Sarcodon atroviridis* (Morgan) Banker
Synonym: *Hydnum atroviride* Morgan
Common name: None
Order: Thelephorales
Family: Bankeraceae
Fruitbody: Cap 1½–5 in. (4–12.5 cm) wide, convex to nearly flat; dingy white to yellowish tan or grayish tan, becoming blackish with olive green tones in age, paler on the margin, bruising black; surface dull, felted, smooth to somewhat pitted; **flesh** thick, firm, brittle, grayish white at first, soon brown to blackish, taste bitter; **fertile surface** with fragile spines up to ⅝ in. (1.5 cm) long, free from the stalk, easily separating from the cap; grayish buff at first, becoming dark brown with buff tips; **stalk** 2–3½ in. (5–9 cm) long, cylindrical, nearly equal or tapering upward, solid at first, becoming hollow in age, surface smooth, colored like the cap, bruising black.
Spore print: Pale brown.
Spores: 6–8.5 μm, globose to subglobose, warted, pale brown, non-amyloid.
Occurrence: Solitary or in groups on the ground in broad-leaved and mixed woods; mycorrhizal; summer-fall; uncommon.
Edibility: Reported to be edible but of poor quality.

Comments: This mushroom is fairly distinctive and easily recognized by the overall brittle nature of the fruitbody and the olivaceous tones on the mature caps. *Sarcodon scabrosus* (p.404) has a chestnut-brown to blackish brown scaly cap and bluish green stains at the stalk base. *Bankera fuligineo-alba* (not illustrated), which grows only under conifers (especially pine), has a yellowish brown to reddish brown cap with a white, felted margin, and has white spores. *Atroviridis* means "blackish green."

Latin name: *Sarcodon scabrosus* (Fr.) Quél.
Synonym: *Hydnum scabrosum* Fr.
Common name: Bitter Hedgehog
Order: Thelephorales
Family: Bankeraceae
Fruitbody: Cap 2–6 in. (5–15 cm) wide, convex to flat or depressed, margin incurved, wavy; pinkish brown or reddish brown to chestnut-brown or blackish brown; surface soon becoming cracked and scaly showing yellowish brown to dingy white beneath, scales appressed to somewhat erect at the center; **flesh** whitish (grayish to blackish in the stalk base), firm, brittle, staining bluish green in contact with KOH, odor not distinctive, taste bitter; **fertile surface** with spines up to ⅜ in. (1 cm) long, light grayish to tan, becoming purplish brown in age; **stalk** up to 4 in. (10 cm) long, cylindrical, tapered at the base, central or eccentric, solid, firm; surface fibrillose to minutely scaly, colored like the cap on the upper portion, bluish green to blackish near the base.
Spore print: Brown.
Spores: 7–9 x 5.5–7.5 µm, subglobose to somewhat saddle-shaped, coarsely warted, brownish, non-amyloid.
Occurrence: Solitary or in groups, sometimes in caespitose clusters on the ground in conifer and broad-leaved woods; mycorrhizal; summer-fall; fairly common.
Edibility: Not edible.

Comments: *Sarcodon imbricatus* (not illustrated) has larger, coarser, more erect scales, and lacks bluish green at the base of the stalk. *Sarcodon underwoodii* (not illustrated) has a smoother more vinaceous cap, shorter spines, and lacks bluish green color at stalk base. *Scabrosus* means "rough."

Microglossum rufum, p. 413

Club Fungi

The mushrooms in this diverse group are more or less club-shaped or finger-like with simple stalks (rarely branched or lobed). Although superficially similar, they represent several different taxonomic groups. They can grow singly, clustered, or scattered on a variety of substrates. Tubular or narrowly club-shaped fungi that do not have clearly differentiated fertile portions are placed with the Coral Fungi (Section 6).

Latin name: *Clavariadelphus truncatus* Donk
Common name: Flat-topped Fairy Club
Order: Phallales
Family: Gomphaceae
Fruitbody: 2½–6 in. (6.5–15 cm) tall and up to 2½ in. (6.5 cm) wide; club-shaped to vase-shaped with a slightly convex to flattened or depressed top; **fertile surface** is on the vertical exterior of the fruitbody, which normally has folds and wrinkles, but at times may be nearly smooth; pale yellow to pinkish buff or ochraceous orange, sometimes with lilac tints; **flesh** whitish to ochre, firm at first, becoming spongy, odor not distinctive, taste mild to sweetish or sometimes bitter; **stalk** not differentiated from the fertile surface.
Spore print: Pale yellow to ochre.
Spores: 9–13 x 5–8 μm, broadly elliptical, smooth, hyaline.
Occurrence: Solitary or more often gregarious on the ground in woods, typically under conifers; mycorrhizal; summer-fall; occasional.
Edibility: Edible but sometimes bitter.

Comments: Specimens with broadly flaring tops might be mistaken for the smooth chanterelle, *Cantharellus lateritius* (not illustrated), which is yellow-orange overall, has much thinner flesh with a fruity odor, and a thin, wavy margin. *Clavariadelphus pistillaris* (not illustrated) is very similar but has a rounded top and bitter tasting flesh. *Gomphus floccosus* (p.136) is funnel-shaped with a scaly orange upper surface, and a strongly ribbed exterior. *Truncatus* means "cut off," referring to the flattened top of the fruitbodies.

Latin name: *Cordyceps militaris* (L.) Link
Common name: Orange Club Cordyceps
Order: Hypocreales
Family: Clavicipitaceae
Fruitbody: ¾–3 in. (2–7.5 cm) tall; club-like with a clearly differentiated fertile head and sterile stalk; **head** cylindrical to spindle-shaped or compressed; bright orange to orange-red or pale orange; surface roughened with raised dots (embedded flasks in which spores are formed); **flesh** pale orange-yellow, brittle; **stalk** cylindrical, often curved; paler orange than the fertile head.
Spores: 300–500 x 1–1.5 μm, thread-like, breaking into multiple segments after discharge, smooth, hyaline.
Occurrence: Solitary or in small groups arising from the buried pupae of moths (*Lepidoptera*); parasitic; summer-fall; fairly common.
Edibility: Not edible.

Comments: These bright orange clubs appear to be growing on the soil, but careful excavation will reveal the insect pupae below ground to which they are always attached. As many as 4 or 5 fruitbodies may be attached to a single pupa. The long, filamentous spores can be seen with a hand lens as they are discharged from the fertile head. The much rarer *Podostroma alutaceum* (not illustrated) is similar but has a pale yellow head and it grows on decaying wood, never parasitizing insects. *Militaris* means "soldier-like," referring to the likeness of the fruitbodies to the pendent ornamentations on some military uniforms.

Latin name: *Cordyceps ophioglossoides* (Ehrenb.) Link
Common name: Goldenthread Cordyceps
Order: Hypocreales
Family: Clavicipitaceae
Fruitbody: Up to 4 in. (10 cm) tall; club-like with a fertile head on a narrower, sterile stalk; **head** up to 1¼ in. (3 cm) long, cylindrical to spindle-shaped or oval; yellowish brown to dark reddish brown, becoming nearly black with age; surface roughened with raised dots (perithecia); **flesh** firm, white; **stalk** 1–3 in. (2.5–7.5 cm) tall, more or less equal; colored like the head or more yellow, with golden yellow rhizomorphs at the base.
Spores: 2–5 x 1.5–2 µm, elliptical, smooth, hyaline, non-amyloid.
Occurrence: Solitary or in small groups on the ground in woods, always attached with cord-like yellow rhizomorphs to buried false truffles (*Elaphomyces* spp.); mycoparasitic; summer-fall; uncommon.
Edibility: Unknown.

Comments: The golden yellow rhizomorphs are easily observed, but careful excavation is necessary to locate the attached false truffle. *Cordyceps capitata* (not illustrated) is similar but has a broadly conical to nearly globose head, and its stalk is directly attached to buried false truffles. *Trichoglossum hirsutum* (p.416) is also similar but has a jet-black flattened head, and is not attached to false truffles. *Ophioglossoides* means "snake-like," referring to the shape of the fruitbody that somewhat resembles the head of a snake.

Latin name: *Dictyophora duplicata* (Bosc) E. Fischer
Synonym: *Phallus duplicatus* Bosc
Common names: Netted Stinkhorn; Wood Witch
Order: Phallales
Family: Phallaceae
Fruitbody: Up to 8 in. (20 cm) tall; phallus-shaped with a bell-shaped to oval cap on a cylindrical stalk; **cap** up to 2 in. (5 cm) tall and 1½ in. (4 cm) wide; chambered with pits and ridges, with a white-rimmed perforation at the tip, covered at first with a fetid brownish olive slime; **stalk** tapering upward, hollow, emerging from a membranous saccate volva at the base and having a lacy, net-like skirt just below the cap; surface sponge-like, white to pale pinkish; **immature fruitbodies** more or less globose with thick whitish rhizomorphs at the base; whitish to pinkish; interior soft and gelatinous.
Spores: 3.5–4.5 x 1.5–2 μm, elliptical, smooth, hyaline.
Occurrence: Solitary or in small groups on the ground in rich woods, gardens, and landscaped areas, especially near well-decayed tree stumps and logs; saprobic; late summer-fall, occasional.
Edibility: Edible in the "egg stage."

Comments: As with other stinkhorns, this spectacular fungus can sometimes be located by its potent, fetid odor, which attracts flies, beetles, and other insects that feed on the sugary slime coating that contains the reproductive spores. When insects visit the fruitbodies, they inadvertently carry away and disseminate spores to new locations. After the greenish slime layer has been "cleaned off," the remaining pitted cap resembles a weathered morel. The immature "eggs" can be "hatched" if placed in a moist environment such as on damp moss within a container. Once the egg ruptures, the stalk expands quite rapidly. Slicing an immature egg lengthwise will reveal the structure of the undeveloped cap and stalk within. *Phallus impudicus* (not illustrated) is similar but lacks a net-like skirt below the cap. *Phallus ravenelii* (not illustrated) lacks a netted skirt on the stalk, and has a smooth cap beneath the slime layer. *Duplicata* means "growing in pairs."

Latin name: *Leotia viscosa* Fr.
Common name: Green-headed Jelly Club
Order: Helotiales
Family: Leotiaceae
Fruitbody: 1–2½ in. (2.5–6.5 cm) tall; club-like with a well-defined fertile head and sterile stalk; **head** nearly round to convex or irregularly cushion-shaped with an incurved margin, dark olive-green to nearly black, the sterile underside is paler; surface glabrous, viscid when wet; **flesh** gelatinous; **stalk** cylindrical to compressed, becoming hollow, rubbery; pale yellow to orange-yellow; surface smooth to somewhat granular, viscid when moist.
Spores: 17–26 x 4–6 µm, spindle-shaped, often curved, smooth, hyaline.
Occurrence: Gregarious, often in clusters on the ground in woods, especially in moss on sandy soil along streams and road banks, infrequently occurring on well-decayed wood; saprobic; summer-early winter; common.
Edibility: Non-poisonous.

Comments: When growing among mosses the Green-headed Jelly Club blends in and is easily overlooked. The stalks are slippery and difficult to grasp. *Leotia atrovirens* (not illustrated) is nearly identical but has a pale green stalk. *Leotia lubrica* (not illustrated) is also similar but has a dull yellowish to butterscotch-colored head and stalk. The color forms of these species overlap and DNA analysis may eventually alter their taxonomic status. *Viscosa* means "viscid."

Latin name: *Microglossum rufum* (Schwein.) Underw.
Common names: Orange Earth Tongue; Yellow Earth Tongue
Order: Helotiales
Family: Leotiaceae
Fruitbody: ¾–2½ in. (2–6.5 cm) tall; club-like with a well-defined fertile head and sterile stalk; **head** oval to spoon-shaped, usually compressed; orange-yellow to bright yellow; surface glabrous; **stalk** cylindrical; colored like the head or paler, surface granulose to scurfy, sometimes nearly glabrous.
Spores: 18–38 x 4–6 µm, sausage-shaped to spindle-shaped, segmented, smooth, hyaline.
Occurrence: In small groups or clusters on humus, well-decayed wood, or in moss; saprobic; summer; occasional.
Edibility: Unknown.

Comments: *Mitrula elegans* (not illustrated), which is somewhat similar, has an orange to pale pinkish orange fertile head. It grows in troops on decaying leaves and litter in wet areas in spring. *Neolecta irregularis* (not illustrated) has a more elongated, irregularly shaped, compressed and often twisted head with a short stalk-like base, and it grows on the ground or in moss in conifer woods. Compare with *Spathulariopsis velutipes* (p.415), which is ochre-yellow and has a broader, fan-shaped head. Compare also with *Cordyceps militaris* (p.409), which has a bright orange dotted head and is always attached to buried insect pupae. *Rufum* means "reddish" or "rusty brown," a name that doesn't apply well to this species, which is typically orange or yellow.

Latin name: *Mutinus elegans* (Mont.) E. Fischer
Common names: Elegant Stinkhorn; Witches' Eggs (immature stage)
Order: Phallales
Family: Phallaceae
Fruitbody: 4–7 in. (10–18 cm) tall; spike-shaped to phallus-shaped tapered cylinder with a blunt, perforated apex; **fertile surface** a fetid olive-brown slime layer (gleba) on the upper third of the stalk; **stalk** cylindrical, hollow, swollen in the mid portion or tapering upward, arising from a white saccate volva; deep pinkish red, paler to whitish near the base; surface chambered, sponge-like; **immature fruitbodies** somewhat globose at first, becoming ovoid to oblong before rupturing at the apex as the stalk expands from within, white to slightly pinkish, attached to the substrate with white rhizomorphs.
Spores: 4–7 x 2–3 μm, oblong-elliptical, smooth, embedded in the gleba.
Occurrence: Usually in groups or clusters on the ground in rich woods, urban and landscaped places, especially near well-decayed stumps and logs, and in wood chips; saprobic, summer-fall; common.
Edibility: Edible in the "egg" stage.

Comments: As with other stinkhorns, the spores are produced in a fetid, sugary slime layer that attracts various insects, which then become vectors for spore dispersal. The stalk beneath the slime layer is pinkish red. *Mutinus ravenelii* (not illustrated) has a less tapered stalk and a clearly differentiated swollen fertile head. The similar *Mutinus caninus* (not illustrated) has a swollen fertile head, a whitish stalk, and a relatively weak odor. *Elegans* means "graceful" or "elegant."

Latin name: *Spathulariopsis velutipes* (Cooke & Farl. ex Cooke) Maas Geest.
Synonym: *Spathularia velutipes* Cooke & Farlow
Common name: Velvet-foot Fairy Fan
Order: Helotiales
Family: Geoglossaceae
Fruitbody: 1–2½ in. (2.5–6.5 cm) tall; fan-like with well-defined fertile head and sterile stalk; **head** up to 1¼ in. (3 cm) wide; flattened or compressed, rounded to fan-shaped or spoon-shaped with a wavy margin, appearing to partially envelop the stalk; cream-colored to pale yellow-brown; surface smooth or wrinkled; **stalk** cylindrical, hollow, flexible; ochre to reddish brown, darker than the fertile head; surface minutely downy, often with orange hairs at the base.
Spore print: Yellow-brown.
Spores: 30–40 x 1.5–2 µm, thread-like, multiseptate, smooth, hyaline.
Occurrence: In groups or clusters on the ground, on wood mulch, or on well-decayed wood; saprobic; summer-early fall; common.
Edibility: Not edible.

Comments: *Spathularia flavida* (not illustrated) has a yellow to pale buff fertile head and a smooth stalk with whitish to yellow mycelium at the base. *Neolecta irregularis* (not illustrated) has a bright yellow contorted head and a white stalk. *Velutipes* means "velvet foot."

Latin name: *Trichoglossum hirsutum* (Pers.) Boud.
Common name: Velvety Black Earth Tongue
Order: Helotiales
Family: Geoglossaceae
Fruitbody: 1½–3 in. (4–7.5 cm) tall; club-like with an enlarged fertile head that tapers into a thin stalk; **head** variable from cylindrical to oval or spindle-shaped, often compressed or flattened with a lengthwise furrow; dark sooty brown to black; surface dry, minutely velvety; **flesh** brown, tough; **stalk** slender, about ⅛ in. (3 mm) thick, often curved, more or less equal to slightly compressed, tough; colored like the head; surface densely velvety.
Spores: 80–150 x 6–7 μm, cylindrical and thread-like, multiseptate, smooth, brown, non-amyloid.
Occurrence: Solitary or more often gregarious, sometimes clustered, on soil, well-decayed wood, or in moss, usually in woods but also in open wet areas; saprobic; summer-fall; common.
Edibility: Not edible.

Comments: These dark earth tongues are easily overlooked, but after one is spotted, others can often be found nearby. Similar species of *Trichoglossum* can only be separated microscopically. *Geoglossum* species (not illustrated) are very similar but have smooth fruitbodies, often with viscid heads. Compare with *Xylaria polymorpha* (p.418), which has a hard, finger-like fruitbody and a finely warted fertile surface. Compare also with *Cordyceps ophioglossoides* (p.410). *Hirsutum* means "hairy."

Latin name: *Vibrissea truncorum* (Alb. & Schwein.) Fr.
Common name: Water Clubs
Order: Helotiales
Family: Vibrisseaceae
Fruitbody: ½–1 in. (1.3–2.5 cm) tall; club-shaped with a well-defined fertile head and sterile stalk; **head** ⅛–¼ in. (3–6 mm) wide; hemispherical to convex or flattened cushion-shaped; yellow to yellow-orange; surface glabrous, shiny; **stalk** cylindrical, equal or tapering downward, flexible; whitish to grayish, or nearly black at the base; surface scurfy or dotted with dark tufts.
Spores: 150–250 x 1–1.5 μm, thread-like, multiseptate, smooth, hyaline.
Occurrence: Usually in clusters or rows on wet twigs, branches, or other woody debris in mountain streams and seeps; saprobic; late spring-summer; common in the mountains.
Edibility: Unknown.

Comments: Although Water Clubs are quite small, they are colorful and easily spotted. When mature, the long thread-like spores can be observed with a hand lens as they are released. Other small club-like species that are found in wet places include *Mitrula elegans* (not illustrated), which has an elongated, irregularly-shaped, pale pinkish orange to apricot-colored head, and *Cudoniella clavus* (not illustrated), which has a whitish to dull buff convex head. *Truncorum* means "blunt-ended."

Latin name: *Xylaria polymorpha* (Pers.) Grev.
Common name: Dead Man's Fingers
Order: Xylariales
Family: Xylariaceae
Fruitbody: 1–4 in. (2.5–10 cm) tall, up to 1½ in. (4 cm) thick; more or less club-shaped, often gnarled or knobby, usually somewhat cylindrical but sometimes branched or lobed; **fertile surface** crust-like and often cracked, white to grayish and powdery at first (asexual stage), then black and minutely warted; **stalk** short, stout, cylindrical, not always distinguishable from the fertile portion; **flesh** white, corky, tough.
Spore print: Black.
Spores: Ascospores 20–30 x 6–9 μm, spindle-shaped with one side flattened, smooth, dark brown; asexual spores (conidia) are elliptical, smooth, and hyaline.
Occurrence: Typically in clusters on decaying wood of broad-leaved trees, especially beech and maple, often on or near old stumps, or growing from buried wood; saprobic; year-round; common.
Edibility: Not edible.

Comments: When the fruitbodies first appear in spring, they are dingy white to grayish and soon covered with powdery asexual spores. Later they become black and have a crusty, charred appearance. At this stage the flasks (perithecia) that contain the asci and ascospores are visible as tiny warts or projections embedded in the surface. *Xylaria longipes* (not illustrated) is smaller, more slender, has a more distinct stalk, and is more likely to grow singly. *Xylaria hypoxylon* (not illustrated) has slender black branches with flattened, antler-like white tips. *Polymorpha* means "having many forms."

Clavaria zollingeri, p. 423

Coral Mushrooms and Look-Alikes

Section 6

The mushrooms in this section resemble marine corals. Some grow as clusters of simple tubular clubs (worm corals), while others are branched and shrub-like. Coral mushrooms are among the most beautiful mushrooms in our woods, but many are very difficult to identify without using a microscope. Those included here are relatively distinctive and can usually be recognized in the field. Remember, however, that other similar species are not featured in this guide. Although some coral mushrooms are good edibles, others are at least mildly poisonous, causing various degrees of gastrointestinal disorders. Because of the difficulty of identifying species, caution is advised when considering any coral mushrooms for the table.

Latin name: *Clavaria rubicundula* Leathers
Common name: Smoky Worm Coral
Order: Agaricales
Family: Clavariaceae
Fruitbody: Up to 5 in. (12.5 cm) high; composed of clustered slender cylinders with rounded or pointed tips, becoming hollow in age, sterile base not clearly defined; vinaceous buff to pale pinkish gray; **flesh** fragile; white, odor and taste not distinctive or faintly of iodine.
Spore print: White.
Spores: 5.5–8.5 x 3–4 μm, elliptical, smooth, hyaline, non-amyloid.
Occurrence: In dense caespitose clusters on the ground in woods; saprobic; summer-early fall; occasional.
Edibility: Unknown.

Comments: *Clavaria purpurea* (not illustrated) is similar but is purple to pale violet and more delicate in form. *Clavaria vermicularis* (not illustrated) is similar in form but is pure white overall. *Rubicundula* means "ruddy" or "reddish."

Latin name: *Clavaria zollingeri* Lév.
Common name: Magenta Coral
Order: Agaricales
Family: Clavariaceae
Fruitbody: Up to 3½ in. (9 cm) high and wide; shrub-like, multi-branched and repeatedly forked, arising from a common base; branches thick, round to somewhat compressed, tips rounded to blunt; amethyst to violet or reddish purple, base whitish, or grayish in age; surface smooth to slightly wrinkled; **flesh** pale violet, aging ochraceous, fragile, odor and taste not distinctive or somewhat radish-like.
Spore print: White.
Spores: 5.5–7 x 4.5–5.5 µm, broadly elliptical to subglobose, smooth, hyaline, non-amyloid.
Occurrence: Solitary or in groups or clusters on the ground in broad-leaved and mixed woods; saprobic; summer-fall; fairly common.
Edibility: Edible.

Comments: *Clavulina amethystina* (not illustrated) is similar but is more extensively branched, and has only 2 spores produced on a basidium. *Clavaria purpurea* (not illustrated) is more or less cylindrical, unbranched and usually paler in color. The specific epithet honors European mycologist Heinrich Zollinger (1818–1859).

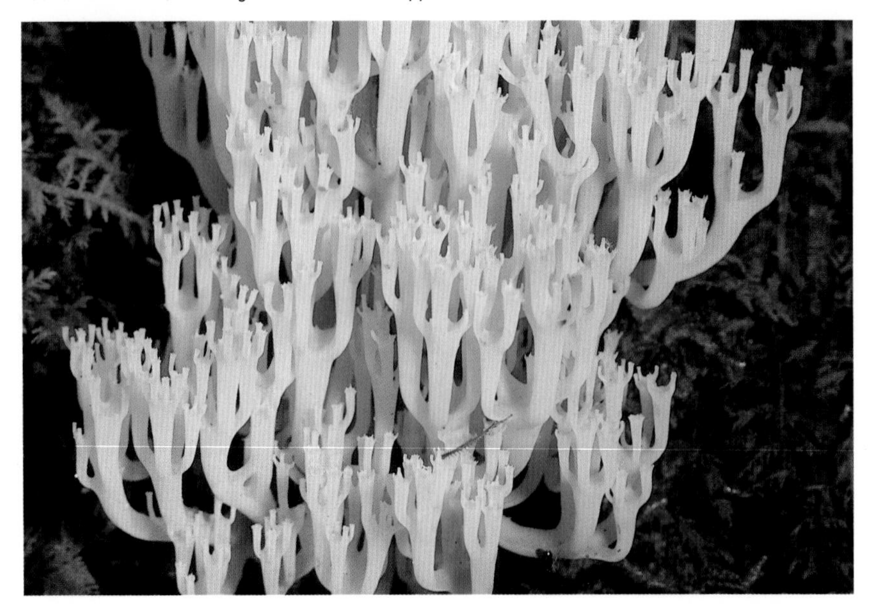

Latin name: *Clavicorona pyxidata* (Pers.) Donk
Synonym: *Clavaria pyxidata* Pers.
Common name: Crown-tipped Coral
Order: Russulales
Family: Auriscalpiaceae
Fruitbody: Up to 4 in. (10 cm) high and wide; multi-branched, candelabra-like coral arising from a common stalk-like base; forked branches arise in tiers from the tips of lower branches and culminate in "toothed" crown-like tips; white to pale yellow or pinkish buff, becoming ochraceous or darker in age; branches round to compressed, smooth; **flesh** whitish, elastic, odor not distinctive, taste mild to peppery.
Spore print: White.
Spores: 4–5 x 2–3 μm, elliptical, smooth, hyaline, amyloid.
Occurrence: Solitary or clustered on decaying logs, stumps, and fallen branches of broad-leaved trees; saprobic; late spring-early fall; common.
Edibility: Edible.

Comments: The Crown-tipped Coral is one of the few coral mushrooms that grows on wood. It is easily recognized by its box-like branches with crowned tips. *Ramaria stricta* (p.432) also grows on wood but has yellowish to pinkish buff parallel branches with yellowish pointed tips. *Pyxidata* is derived from the Greek word "pyxis," which refers to a small box.

Latin name: *Clavulina cristata* (Holmsk.) J. Schröt.
Synonyms: *Clavulina coralloides* (L.) J. Schröt.; *Clavaria cristata* Pers.
Common name: Crested Coral
Order: Cantharellales
Family: Clavulinaceae
Fruitbody: Up to 3 in. (7.5 cm) high and wide, shrub-like, branched, especially near the tips, arising from a stalk-like base; white to dingy white or ochraceous in age; branches rounded to compressed with flattened, fringed, "feathery" tips; surface smooth to finely wrinkled; **flesh** solid, somewhat brittle, odor and taste not distinctive.
Spore print: White.
Spores: 7–9 x 6–7.5 μm, subglobose, smooth, hyaline, non-amyloid.
Occurrence: Usually gregarious on the ground in woods; mycorrhizal; summer-fall; common.
Edibility: Edible.

Comments: This common coral mushroom is often infected with the mycoparasite *Helminthosphaeria clavariarum* (not illustrated), a pyrenomycete that covers the base and branches with blackish dots (perithecia). Parasitized specimens can be mistaken for *Clavulina cinerea* (not illustrated), which is pale gray to ash gray overall, but not dotted. *Ramariopsis kunzei* (not illustrated) is similar but has a pinkish tinge at the base and its branches lack "fringed" tips. *Tremellodendron pallidum* (p.435) has whitish, tough, cartilaginous branches with blunt tips. *Cristata* means "crested" or "tassel-like at the tip."

Latin name: *Clavulina ornatipes* (Peck) Corner
Synonym: *Clavaria ornatipes* Peck
Common name: Fuzzy-foot Coral
Order: Cantharellales
Family: Clavulinaceae
Fruitbody: Up to 2½ in. (6.5 cm) tall; simple or sparingly branched, more or less cylindrical to flattened, sometimes with irregularly shaped knobby branch tips; upper fertile surface smooth or wrinkled; gray to pinkish gray, becoming brownish in age; sterile base densely hairy, dark brown, tough; **flesh** brown, flexible, odor and taste not distinctive.
Spore print: White.
Spores: 8.5–11 x 6.5–8.5 µm, elliptical to subglobose, smooth, hyaline, non-amyloid.
Occurrence: Solitary or in small groups on the ground in wet woods; saprobic; summer-fall; rare.
Edibility: Unknown.

Comments: The dark brown hairy base is distinctive. *Clavulina cinerea* (not illustrated) is somewhat similar but grayish overall, more extensively branched, and lacks a hairy base. *Ornatipes* means "ornate foot."

Latin name: *Clavulinopsis aurantio-cinnabarina* (Schwein.) Corner
Synonym: *Clavaria aurantio-cinnabarina* Schwein.
Common name: Orange Spindle Coral
Order: Agaricales
Family: Clavariaceae
Fruitbody: Up to 5 in. (12.5 cm) tall; slender, spindle-shaped, unbranched or rarely branched hollow clubs, cylindrical to compressed or grooved, usually crooked, tips blunt or pointed; deep reddish orange to paler pinkish orange; base scarcely distinguishable from fertile portion, sometimes whitish; **flesh** brittle to somewhat flexible, odor not distinctive to slightly fetid, taste not distinctive.
Spore print: White to pale yellow.
Spores: 4.8–5.5 x 5.5–6.5 µm, subglobose, smooth, hyaline, non-amyloid.
Occurrence: In loose or compact clusters on the ground in woods; saprobic; summer-fall; common.
Edibility: Non-poisonous.

Comments: *Clavulinopsis fusiformis* (not illustrated) is nearly identical except it is bright yellow overall. *Clavulinopsis helveola* (not illustrated) has thinner, yellow to orange-yellow clubs with a distinct narrow stalk, and angular, warted spores. It typically grows singly or scattered in loose groups. *Aurantio-cinnabarina* means "orange and cinnabar-red."

Latin name: *Clavulinopsis corniculata* (Godey) Corner
Synonym: *Clavaria corniculata* Schaeff.
Common name: Yellow Antler Coral
Order: Agaricales
Family: Clavariaceae
Fruitbody: 1½–3 in. (4–7.5 cm) high; branched (sometimes sparingly), usually terminating with 2 incurved, antler-like tips; upper portion yellow to yellow-orange or yellow-brown, with a whitish, downy stalk-like rooting base; branches smooth, often compressed; **flesh** whitish to pale yellow, fairly firm and tough, odor not distinctive, taste mild or bitter.
Spore print: White.
Spores: 5.5–7 x 4–6 μm, subglobose with a prominent apiculus, smooth, hyaline, non-amyloid.
Occurrence: Solitary or in scattered groups on the ground in grassy places in woods and woodland margins; saprobic; early summer-fall; occasional.
Edibility: Edible.

Comments: *Calocera viscosa* (not illustrated) has smaller, golden yellow to orange-yellow cartilaginous branches, and grows on wood (sometimes buried). *Ramaria* species are more extensively branched, fleshier, and have a thick, stalk-like base. *Corniculata* means "having horn or spur-like appendages."

Latin name: *Macrotyphula juncea* (Fr.) Berthier
Synonym: *Clavariadelphus junceus* (Fr.) Corner
Common name: Fairy Thread Coral
Order: Agaricales
Family: Clavariaceae
Fruitbody: Up to 4 in. (10 cm) long; very thin, erect, or bent and partly reclining, cylindrical with pointed or blunt tips, becoming hollow; surface smooth; whitish to pale yellow-brown or pale reddish brown; slightly narrower sterile base barely distinct from fertile portion; **flesh** firm, tough, odor not distinctive, taste mild to somewhat sour.
Spore print: White.
Spores: 6–12 x 3.5–5.5 µm, elliptical to spindle-shaped, smooth, hyaline.
Occurrence: Gregarious on decaying twigs, rotting leaves and other debris in wet places in broad-leaved and mixed woods, sometimes embedded in the substrate with fine, hairy rhizomorphs; saprobic; fall; occasional.
Edibility: Unknown.

Comments: These thread-like fruitbodies might be mistaken for plant stems. They are inconspicuous and easily overlooked. The name *juncea* refers to "*Juncus*," which is a genus of rushes.

Latin name: *Ramaria botrytis* (Pers.) Ricken
Common names: Pink-tipped Coral; Clustered Coral
Order: Phallales
Family: Ramariaceae
Fruitbody: Up to 6 in. (15 cm) high and wide; repeatedly branched and cauliflower-like arising from a thick, rooting base; branches short, crowded, cylindrical or compressed with pointed or blunt tips; whitish at first, becoming ochraceous to tan in age, base whitish, tips pinkish to rose or purplish, fading with age; **flesh** white, firm, tips brittle, odor and taste not distinctive or slightly fruity.
Spore print: Pale yellow to ochraceous.
Spores: 11–16 x 3.5–5.5 µm, oblong-elliptical to nearly cylindrical, surface with longitudinal striations, yellowish, non-amyloid.
Occurrence: Solitary or in small groups, sometimes merging into larger clusters, on the ground in woods; mycorrhizal; summer-early fall; occasional.
Edibility: Edible and good, but produces a laxative effect in some individuals.

Comments: The pinkish to reddish purple branch tips and compact cauliflower-like form are distinctive. *Ramaria subbotrytis* (not illustrated) is bright coral pink overall at first, becoming cream-colored to ochraceous toward the base with age. Compare with *Ramaria formosa* (p.431), which has pinkish branches with yellowish tips. *Botrytis* means "clustered."

Latin name: *Ramaria formosa* (Pers.) Fr.
Synonym: *Clavaria formosa* Pers.
Common names: Yellow-tipped Coral; Beautiful Clavaria
Order: Phallales
Family: Ramariaceae
Fruitbody: Up to 5 in. (12.5 cm) tall and wide, shrub-like, densely branched from a common stalk-like base; branches pale salmon to orange-pink at first, fading and becoming yellowish to ochre with age, bruising brownish; branch tips often blunt, pale yellow to straw-yellow when young; surface smooth to wrinkled; base thick, pinkish to white; **flesh** colored like the branches, fibrous to moderately brittle, odor not distinctive, taste mild to bitter or astringent.
Spore print: Pale yellow-brown.
Spores: 10–14.5 x 4.5–6 μm, elliptical, minutely warted, buff.
Occurrence: Solitary to gregarious, sometimes growing in broad patches, rows, or arcs on the ground in broad-leaved and conifer woods; mycorrhizal; summer-fall; common.
Edibility: Poisonous, causing gastrointestinal disturbance.

Comments: The slender branches of this very beautiful coral often unite very low near the base and may appear to be clusters of independent fruitbodies. More compact forms may actually be *Ramaria formosa* while the one shown here could represent a closely related species. *Ramaria stricta* (p.432) has yellowish pointed branch tips, and grows on decaying wood. *Formosa* means "finely formed" or "beautiful."

Latin name: *Ramaria stricta* (Pers.) Quél.
Synonym: *Clavaria stricta* Pers.
Common name: Straight-branched Coral
Order: Phallales
Family: Ramariaceae
Fruitbody: Up to 4½ in. (11.5 cm) tall; repeatedly forked, erect, parallel branches arising from a short base; branches slender, cylindrical to compressed, pale yellow to buff-pink or pinkish tan, bruising brown; branch tips pointed, yellowish at first, later colored like the lower part; base white, with rhizomorphs often embedded in the substrate; **flesh** whitish, bruising brown, tough, elastic, odor sweet, spicy, taste bitter or metallic.
Spore print: Ochre-yellow.
Spores: 7.5–10 x 4–5 μm, elliptical, minutely warted, hyaline, non-amyloid.
Occurrence: Solitary or in clusters on decaying logs, stumps, and fallen branches of broad-leaved and conifer trees; saprobic; summer-fall; common.
Edibility: Not edible.

Comments: Compare with *Clavicorona pyxidata* (p.424), which also grows on decaying wood but has crown-like branch tips and lacks a distinctive spicy odor. *Stricta* means "straight."

Latin name: *Sparassis spathulata* (Schwein.) Fr.
Synonym: *Sparassis herbstii* Peck
Common name: Cauliflower Mushroom
Order: Polyporiales
Family: Sparassidaceae
Fruitbody: Up to 12 in. (30 cm) or more wide; a complex, more or less globose mass of wavy, flattened branches united into a thick stalk-like base; individual branches thin, wavy, broad at the tip, tapering toward the base; creamy white to pinkish buff or yellowish to pale ochraceous, often with horizontal bands near the edge; **flesh** whitish, flexible, becoming tough with age, odor and taste pleasant when fresh, becoming sourish or somewhat like latex when old.
Spore print: White.
Spores: 4–7 x 3–4 µm, oval, smooth, hyaline, non-amyloid.
Occurrence: Solitary or in groups on the ground, especially near the base of well-decayed oak stumps; saprobic; summer-fall; common.
Edibility: Edible and good when young.

Comments: Looking like a cluster of ribbon candy or a bath sponge, this large and distinctive fungus is not likely to be confused with any fungus other than *Sparassis crispa* (not illustrated), which has smaller, thinner, and more curly branches. Both species are prized edibles when young. *Spathulata* means "spoon-shaped," referring to the individual branches.

Latin name: *Thelephora palmata* (Scop.) Fr.
Common name: Fetid False Coral
Order: Thelephorales
Family: Thelephoraceae
Fruitbody: Up to 3 in. (7.5 cm) tall; coral-like tufts of erect forking branches originating from a stalk-like base; individual branches flattened; whitish at first, becoming dark gray to lilac brown or blackish brown with a blackish base; branch tips whitish to gray, fan-shaped, vertically grooved; **flesh** tough, flexible, odor unpleasant (like spoiled cabbage).
Spore print: Reddish brown.
Spores: 8–12 x 7–9 μm, elliptical to subglobose, spiny, brown.
Occurrence: Solitary or gregarious, sometimes in clusters on the ground in conifer and mixed woods; mycorrhizal; summer-fall; occasional.
Edibility: Not edible.

Comments: *Thelephora terrestris* (not illustrated) is somewhat similar but it has spreading fan-shaped to vase-shaped lobes that form rosette-like clusters. Some "true" coral mushrooms resemble the Fetid False Coral but have fleshy, brittle branches. *Palmata* means "having five or more divisions from one point."

Latin name: *Tremellodendron pallidum* (Schwein.) Burt
Synonym: *Tremellodendron schweinitzii* (Peck) G.F. Atk.
Common name: False Coral Mushroom
Order: Tremellales
Family: Exidiaceae
Fruitbody: Up to 6 in. (15 cm) wide; compact clusters of erect or spreading branches, arising from a central base; branches cylindrical or flattened, often fused together, tips blunt; white to dingy white or creamy white, sometimes greenish from algae; **flesh** cartilaginous, very tough and flexible, almost woody at the base, odor not distinctive, taste rancid.
Spore print: White.
Spores: 7.5–11 x 4–6 µm, elliptical to sausage-shaped, smooth, hyaline.
Occurrence: Solitary or in groups on the ground in broad-leaved and mixed woods, or in grassy places; saprobic; summer-fall; common.
Edibility: Non-poisonous.

Comments: The tough, almost indestructible fruitbody easily separates the False Coral from more fragile "true" coral mushrooms. Based on the microscopic feature of having basidia that are divided by partitions, this fungus is taxonomically aligned with the Jelly Fungi. *Pallidum* means "pale."

Geastrum fimbriatum, p. 442

Puffballs, Earthballs, Earthstars, and Other Puffball-Like Mushrooms

Section 7

The species in this section produce spores within an enclosed outer wall (peridium) or spore case. Unlike most other mushrooms, they do not forcibly discharge their spores. Puffballs are more or less spherical to pear-shaped, thin-skinned, and quite variable in size. Their flesh (gleba) is solid when young but becomes a dry powdery mass of spores when mature. Earthballs are similar to puffballs but have a hard, firm flesh when young and a thick outer rind. Earthstars are "glorified puffballs" that have a multi-layered peridium with an outer wall that splits into ray-like segments that open and recline to expose an inner spore case. This section also includes one hypogeous false truffle that resembles an earthball. Immature stinkhorn "eggs" look similar to puffballs, but at maturity they develop erect stalks and are grouped with the club fungi, in Section 5.

Latin name: *Astraeus hygrometricus* (Pers.) Morgan
Synonym: *Geastrum hygrometricum* Pers.
Common names: Hygrometer Earthstar; Barometer Earthstar
Order: Boletales
Family: Sclerodermataceae
Fruitbody: Up to 3½ in. (9 cm) across when fully expanded; at first more or less globose to onion-shaped and partially buried, emerging at maturity, at which time the outer wall splits at the apex into 6–15 star-like rays that spread and recline, revealing an inner spore case; **rays** thick and leathery, inner surface cracking, dingy white to yellowish, or grayish brown to dark brown with age, exterior surface fibrous roughened; tan to dark brown; **spore case** globose, up to 1¼ in. (3 cm) in diameter, sessile, thin and papery with an irregular perforation at the apex; white to gray or brown; surface felt-like or scurfy; **gleba** solid and white at first, becoming brown and powdery at maturity.
Spores: 7–11 µm, globose, spiny or finely warted, dark brown in mass, non-amyloid.
Occurrence: Solitary or more often in groups on the ground in woods, especially with oak and pine along streams and roadsides, often in sandy or loose soil; mycorrhizal; late summer-fall, sometimes overwinters; occasional to locally common.
Edibility: Not edible.

Comments: In wet weather the spreading star-like rays of this curious earthstar open and recline. In dry conditions the rays close back around the spore case. This hygroscopic property is easily demonstrated as the rays will repeatedly open and close if mature specimens are placed in shallow water, then removed and allowed to dry. *Hygrometricus* means "water measuring."

Latin name: *Calostoma cinnabarina* Corda
Common names: Gelatinous Stalked Puffball; Hot Lips
Order: Boletales
Family: Sclerodermataceae
Fruitbody: A stalked puffball, at first enclosed in a double-layered covering (exoperidium) consisting of a thick, gelatinous, transparent outer layer and a thin, red, cartilaginous inner layer. At maturity the exoperidium breaks away in fragments, exposing the spore case; **spore case** up to ¾ in. (2 cm) in diameter; globose, at first bright red from a powdery residue of the inner layer of the covering, but soon fading and eventually becoming buff or yellowish with a bright red suture-like opening at the top; **stalk** is a thick anastomosing bundle of translucent gelatinous strands forming a cylindrical spongy column up to 1½ in. (4 cm) tall, much of which is often below ground.
Spores: 14–20 x 6–9 μm, elliptical, pitted, hyaline.
Occurrence: In groups or clusters on the ground in woods, along forest roads and woodland margins; saprobic; spring-fall; locally common.
Edibility: Of no culinary interest.

Comments: This distinctive fungus is fairly common in the mountain counties of West Virginia. When the layered covering of the spore case first breaks away, it leaves an accumulation of gelatinous red-centered globules around the base. *Calostoma ravenelii* (not illustrated) is similar but much smaller and lacks a gelatinous layer on the spore case. *Calostoma lutescens* (not illustrated) has a well-defined collar at the base of the spore case, and except for a red "mouth," the spore case is dingy yellowish overall. *Cinnabarina* means "cinnabar-red."

Latin name: *Calvatia cyathiformis* (Bosc) Morg.
Common name: Purple-Spored Puffball
Order: Agaricales
Family: Lycoperdaceae
Fruitbody: 2½–7 in. (6.5–18 cm) broad; more or less globose or pear-shaped, often with folds or wrinkles at the base; **exterior surface** (peridium) composed of 2 layers that together form a thin, fragile rind; whitish to pale brown when young, becoming darker brown in age; surface smooth at first (like kid-leather), becoming areolate to coarsely scaly in the center, then cracking at the top and breaking away in flakes at maturity to expose the spore mass; **gleba** firm and white when young, becoming yellowish then finally purple-brown and powdery, eventually eroding until nothing remains except a ragged cup-like sterile base that may persist for months.

Spores: 4–7 μm, globose, spiny, purplish brown in mass.
Occurrence: Solitary or in groups on the ground in open grassy areas such as lawns, meadows, pastures, cemeteries, and ball fields; saprobic; summer-fall, common.
Edibility: Edible when young and completely white inside.

Comments: This common puffball is often found in company with *Agaricus campestris* (p.47) after rainfall in late summer. *Calvatia craniformis* (not illustrated) is similar but typically grows in woodlands, and the mature spore mass is yellowish brown. The Giant Puffball, *Calvatia gigantea* (not illustrated), is a much larger grassland species, often as large as 1½ feet (45 cm) or more in diameter. *Cyathiformis* means "cup-shaped," in reference to the form of the mature fruitbody and persistent sterile base.

Latin name: *Elaphomyces cyanosporus* Tul.
Synonym: *Elaphomyces persoonii* var. *minor* Tul. & Tul.
Common name: Deer Truffle
Order: Elaphomycetales
Family: Elaphomycetaceae
Fruitbody: ½–1¼ in. (1.3–3 cm) wide; more or less globose to ovoid, stalkless; outer wall (peridium) thick, leathery, dark brown to nearly black with a pale band inside the perimeter, surface densely warted; **gleba** firm, blackish blue, becoming powdery at maturity.
Spores: 23–26 µm, globose, reticulate, grayish blue.
Occurrence: Solitary or clustered in broad-leaved woods, partly embedded or entirely buried in the soil; mycorrhizal; summer-fall (year-round?); possibly rare.
Edibility: Unknown.

Comments: The first North American record for this species was from North Carolina in 1991. It may be rare, or perhaps it is only seldom reported because there are similar species with which it is easily confused. *Elaphomyces granulatus* (not illustrated) has a more brownish outer wall, and larger spores (28–45 µm). *Elaphomyces muricatus* (not illustrated) has a marbled inner peridium and a blackish brown gleba. Species of *Elaphomyces* are frequently parasitized by *Cordyceps capitata* (not illustrated) and *Cordyceps ophioglossoides* (p.410). These fungi produce small reddish brown or blackish club-like fruitbodies that are attached to the false truffles below ground.

Species of *Elaphomyces* are eagerly sought for food by deer and small mammals. Squirrels will only consume the outer rind of mature specimens, and they often leave the remaining powdered spores in mysterious clumps on rocks and logs. The genus name *Elaphomyces* means "deer fungus." *Cyanosporus* means "dark blue spores."

Latin name: *Geastrum fimbriatum* Fr.
Synonym: *Geastrum sessile* (Sowerby) Pouzar
Common name: Sessile Earthstar
Order: Phallales
Family: Geastraceae
Fruitbody: ¾–2 in. (2–5 cm) across when fully expanded; at first more or less globose and at least partially buried, emerging when mature, outer wall splitting at the apex into 5–9 star-like rays that spread and recurve beneath the exposed inner spore case; **rays** creamy white to pinkish on the upper surface; underside and base of outer wall whitish, encrusted with debris; **spore case** up to ¾ in. (2 cm) in diameter, sessile, globose with a conical, fringed apical pore, papery thin, smooth; buff to gray or grayish brown with a slightly paler zone around the pore opening; **gleba** solid and white at first, becoming brown and powdery at maturity.
Spores: 4–7 μm, globose, finely warted, light brown in mass.
Occurrence: Solitary or in small groups on the ground in conifer and broad-leaved woods; saprobic; summer-fall, persistent fruitbodies can be found year-round; occasional.
Edibility: Not edible.

Comments: As with many puffballs, the powder-like spores are ejected in bursts when raindrops hit the spore case wall, or they may be coaxed out more gradually by wind currents blowing across the pore opening. *Geastrum saccatum* (not illustrated) is very similar but has a more distinct halo with a well-defined margin encircling the pore opening. *Geastrum rufescens* (not illustrated) is larger with a pale pinkish buff to pinkish brown outer wall and rays. *Geastrum triplex* (p.443) is much larger and when expanded the peridium segments crack to form a concave saucer-like disc beneath the spore case. *Fimbriatum* means "fringed," referring to the pore opening of the spore case.

Latin name: *Geastrum triplex* Junghuhn
Common name: Collared Earthstar
Order: Phallales
Family: Geastraceae
Fruitbody: Expanded fruitbodies up to 4½ in. (11.5 cm) broad; more or less globose to onion-shaped at first, splitting at the apex into 4–8 pointed ray-like segments that open and expand radially then recurve beneath the fruitbody in order to elevate the exposed spore case within; **immature bulb** brownish, often with fissures on the exterior surface; **rays** pale tan to pale brown on the inner surface, which partially fractures crosswise about halfway as the rays recurve, leaving a prominent saucer-like disc beneath the spore case; **spore case** thin and papery, up to 1½ in. (4 cm) in diameter, subglobose, sessile, creamy white to grayish brown with a clearly defined halo surrounding a conical, fringed apical opening.
Spores: 4.5–5.5 µm, globose, warted, dark brown in mass.
Occurrence: Usually in groups, but occasionally solitary on the ground in woods, parklands, gardens, and on roadsides; saprobic; summer-fall; locally common.
Edibility: Not edible.

Comments: This showy earthstar is easily recognized by the saucer-like disc beneath the spore case. Immature bulbs can often be found in the vicinity of expanded fruitbodies. *Geastrum fimbriatum* Fr. (p.442) is similar but smaller, and lacks a saucer-like disc beneath the spore case. *Triplex* means "threefold," referring to the peridium that has three layers.

Latin name: *Lycoperdon americanum* Demoulin
Synonym: *Lycoperdon echinatum* Pers.
Common name: Spiny Puffball
Order: Agaricales
Family: Lycoperdaceae
Fruitbody: ¾–2 in. (2–5 cm) in diameter; subglobose to pear-shaped, **exterior surface** at first covered with spines up to ¼ in. (6 mm) long; spines pointed and often convergent at the tips, white at first, becoming dark brown in age, easily detachable and typically falling away with age; **spore case** papery, developing an apical pore at maturity; pale brown to dark brown; surface shiny, pitted or reticulate; **gleba** white at first, becoming brown and powdery as spores mature; **base** sterile, chambered, sometimes poorly developed.
Spores: 4–6 μm, globose, warted, chocolate-brown in mass, sometimes tinged purple.
Occurrence: Usually in small groups but sometimes solitary on the ground in woods, woodland borders, and grassy places; saprobic; late spring-fall; occasional.
Edibility: Edible when young and white inside.

Comments: *Lycoperdon pulcherrimum* (not illustrated) is nearly identical, but its spines do not turn brown in age, and the spore case has a smooth rather than pitted surface beneath the spines. *Vascellum curtisii* (p.449) is generally smaller (less than 1 in. [2.5 cm] in diameter), has almost no sterile base, and usually grows in tight clusters in lawns, pastures, and other grassy areas. *Lycoperdon marginatum* (not illustrated) is a grassland puffball with spines that slough off the mature spore case in compound flakes or sheet-like fragments. Compare with *Lycoperdon perlatum* (p.445). The specific epithet *americanum* refers to the American distribution of this species. Some authors consider *Lycoperdon echinatum* to be a separate, closely related European species rather than a synonym.

Latin name: *Lycoperdon perlatum* Pers.
Synonym: *Lycoperdon gemmatum* Batsch
Common name: Gem-studded Puffball
Order: Agaricales
Family: Lycoperdaceae
Fruitbody: Up to 3½ in. (9 cm) tall; pear-shaped to sometimes nearly globose with a tapered base; **outer surface** composed of conical spines and irregular-shaped whitish to cream-colored warts that become brownish with age and eventually fall away, leaving pits or net-like scars on the spore case beneath; **spore case** divided into an upper fertile portion (gleba) and a sterile, stalk-like base, tan to pale brown or grayish brown, darker in age, thin-walled, papery; **gleba** white and fleshy at first, becoming olive-brown and powdery as the spores mature; **base** sterile, chambered, occupying about a third to one half of the fruitbody.
Spores: 3.5–4.5 μm, globose, minutely warted, pale yellow (yellow-brown to olive-brown in mass).
Occurrence: Usually in groups, often clustered on the ground or sometimes on well-decayed woody debris in woods and open areas; saprobic; summer-fall, spore cases persistent; common.
Edibility: Edible when immature and entirely white inside.

Comments: *Lycoperdon molle* (not illustrated) has a cream-colored to pale yellow-brown spore case, which at first is lightly covered with short, delicate (almost granular) grayish brown spines, and its spore case is smooth after the spines fall away. *Calvatia excipuliformis* (not illustrated) is more pestle-shaped with an elongated sterile base, and has fugacious fine spines or granules on the exterior. Unlike species of *Lycoperdon,* its entire spore case wall disintegrates at maturity. *Lycoperdon pyriforme* (not illustrated) has a minutely roughened outer surface and grows in dense clusters on decaying logs and stumps. See *Lycoperdon americanum* (p.444) for additional comments about other spiny puffballs. *Perlatum* means "widespread," referring to the broad distribution of this species.

Latin name: *Morganella subincarnata* (Peck) Kreisel & Dring
Synonym: *Lycoperdon subincarnata* Peck
Common name: Ruddy Puffball
Order: Agaricales
Family: Lycoperdaceae
Fruitbody: Up to 1¼ in. (3 cm) wide; globose to depressed globose or pear-shaped, attached to the substrate by white rhizomorphs; **outer surface** made up of short purplish brown to reddish brown warts and spines with convergent tips, which fall away leaving distinct pits or reticulations on the spore case beneath; sterile base present but usually scant; **spore case** pale brown to grayish brown, somewhat leathery, developing an irregular-shaped apical pore at maturity; **gleba** white and fleshy at first, becoming brownish or tinged purple and powdery as the spores mature.
Spores: 3.5–4 μm, globose, finely warted, purplish brown in mass.
Occurrence: Usually in small groups on decaying, often moss-covered, logs and stumps of broad-leaved trees, especially beech, birch, and maple; saprobic; late summer-fall; common.
Edibility: Unrecorded but presumed to be edible.

Comments: *Lycoperdon pyriforme* (not illustrated), which also grows on decaying logs and stumps, frequently in massive clusters, has a rounded to pear-shaped brownish spore case with a scurfy to granulose surface that becomes smooth and darker brown at maturity. *Subincarnata* means "somewhat flesh-colored."

Latin name: *Phallogaster saccatus* Morgan
Common names: Stink Poke; Club-shaped Stinkhorn
Order: Phallales
Family: Hysterangiaceae
Fruitbody: Up to 2 in. (5 cm) high x 1¼ in. (3 cm) wide; club-shaped to pear-shaped, or somewhat globose; **surface** glabrous, often lumpy and developing depressions; creamy white to pinkish tan or lilac, whitish toward the base, eventually separating into lobes at the apex and disintegrating to expose the mature gleba; **gleba** chambered, light green when young, permeated with translucent gelatinous material, becoming dark green and slimy with a strong fetid odor when mature; **sterile base** stalk-like, attached to the substrate with white to pinkish branching rhizomorphs.
Spores: 4–5.5 x 1.7–2.3 µm, cylindrical-elliptical, smooth, hyaline (greenish in mass).
Occurrence: Solitary or more often in clusters on well-decayed logs, stumps, and woody debris, also in wood chip piles and mulched landscape beds; saprobic; late spring-summer; occasional.
Edibility: Not edible.

Comments: Before the outer wall ruptures, the young fruitbodies lack the fetid odor of mature specimens. *Saccatus* means "saccate" or "bag-shaped."

Latin name: *Scleroderma citrinum* Pers.
Synonyms: *Scleroderma vulgare* Hornem.; *Scleroderma aurantium* (L.) Pers.
Common names: Common Earthball; Poison Pigskin Puffball
Order: Boletales

Family: Sclerodermataceae
Fruitbody: 1–4 in. (2.5–10 cm) wide; globose to subglobose, often flattened on top, stalkless, attached to the substrate by dense strands of mycelium; **outer wall** (peridium) up to ⅜ in. (1 cm) thick, tough, leathery, whitish in cross section, tearing irregularly at the apex to expose the mature gleba, eventually eroding away except for a persistent basal disc; **outer wall surface** with large, coarse scales or flattened pyramidal warts, yellow to orange-yellow or pale brown to yellow-brown or golden brown; **gleba** firm and hard when young (like a potato); white at first, soon turning grayish to purplish black with whitish marbled veining, eventually becoming powdery olive-brown to blackish brown, odor somewhat waxy (like crayons) when young.
Spores: 10–13 μm, globose, surface with spines and ridges forming a partial reticulum, blackish brown in mass.
Occurrence: Solitary or gregarious, sometimes clustered on the ground or on well-decayed mossy logs and stumps in woods or woodland margins; mycorrhizal; summer-fall; common.
Edibility: Poisonous.

Comments: *Scleroderma areolatum* (not illustrated) is similar but has a yellowish brown to bronze-colored outer wall that is dotted with flattened dark brown scales, and spores that are not reticulated. *Scleroderma citrinum* is a frequent host for *Boletus parasiticus* (p.330). *Citrinum* means "citron yellow."

Latin name: *Vascellum curtisii* (Berk.) Kreisel
Synonyms: *Lycoperdon curtisii* Berk.; *Lycoperdon wrightii* Berk. & M.A. Curtis
Common names: Clustered Spiny Puffball; Curtis' Puffball
Order: Agaricales
Family: Lycoperdaceae
Fruitbody: ⅜–¾ in. (1–2 cm) wide; more or less globose, or compressed and misshapened when tightly clustered, sessile, with a fibrous pad of mycelium and soil at the base; **exterior** consisting of crowded clusters of white to buff-colored spines that converge at the tip, and gradually fall away in age; **spore case** cream to pale buff when first exposed, becoming pale brown in age; surface minutely velvety to smooth after the spines fall away, developing an apical pore at maturity; **gleba** firm and white when young, becoming greenish yellow then olive-brown and powdery; **sterile base** scant, chambered.
Spores: 3.5–4.5 μm, globose, minutely spiny, olive-brown in mass.
Occurrence: Typically in groups, often densely clustered on the ground in lawns, pastures, along grassy footpaths, or in other grassy areas, especially on hard-packed soil or bare ground; saprobic; summer-fall; fairly common.
Edibility: Edible when young.

Comments: *Lycoperdon marginatum* (not illustrated) is a similar but larger grassland puffball up to 2 in. (5 cm) wide, with a spine layer that detaches from the spore case in compound flakes. *Lycoperdon americanum* (p.444) is similar but grows in woods or grassy woodland borders, and has a pitted spore case after the spines fall away. The epithet *curtisii* honors American mycologist Moses Ashley Curtis (1808–1872).

Tremella mesenterica, p.454

Jelly Fungi

The mushrooms grouped here have a gelatinous to rubbery consistency when moist. In dry weather these fungi shrivel and become relatively inconspicuous, then revive when wet conditions return. The Jelly fungi (Tremellales) are microscopically distinct from other basidiomycetes by having a septate (divided by partitions) basidium. Some members of in this taxonomic group, such as *Pseudohydnum gelatinosum* (p.402) and *Tremellodendron pallidum* (p.435), are superficially more similar to the mushrooms in other sections and are placed elsewhere in this guide.

Latin name: *Auricularia auricula* (Hook) Underw.
Synonyms: *Auricularia auricula-judae* (Fr.) J. Schröt.; *Hirneola auricula-judae* (L.) Berk.
Common names: Wood Ear; Tree Ear
Order: Auriculariales
Family: Auriculariaceae
Fruitbody: 1–4 in. (2.5–10 cm) wide, sometimes larger; circular to irregularly cup-like or ear-shaped, margin lobed and wavy; **outer surface** (usually upper side) light brown to reddish brown or grayish brown, finely tomentose, sometimes appearing frosted, often with anastomosing ribs or veins, especially at the point of attachment; **inner fertile surface** (usually under side) smooth, often wrinkled or veined, colored like the outer surface; **flesh** thin, rubbery, and somewhat translucent when moist, hard when dry, odor and taste not distinctive; **stalk** rudimentary or absent.

Spores: 13–16 x 5–6 μm, cylindrical to sausage-shaped, smooth, hyaline (white in mass), non-amyloid.
Occurrence: Usually gregarious, often in overlapping clusters on logs, stumps, and fallen branches of broad-leaved trees, also reported to occur on conifer wood; saprobic; spring-fall, occasionally in winter; common.
Edibility: Edible (even raw); dries well for future consumption.

Comments: The similar *Tremella foliacea* (not illustrated) is pale brown to orangish brown or vinaceous brown and grows in a somewhat globose cluster of flattened, wavy branches arising from a common base. *Exidia recisa* (not illustrated) is golden brown to cinnamon-brown and grows as irregular-shaped, contorted clusters that become a flattened crust when dry. A few brown species of *Peziza* grow on wood, but they are fragile and brittle. The closely related *Auricularia polytricha* (not illustrated) is widely used in Asian cooking, and is commonly available from specialty food shops under the name "Cloud Ears." *Auricula* means "with ear-like lobes."

Latin name: *Exidia glandulosa* (Bull.) Fr.
Common name: Black Witches' Butter
Order: Tremellales
Family: Exidiaceae
Fruitbody: Irregular disc-shaped to convoluted pillow-like, usually fused into clusters or rows, or forming a spreading mass up to 10 in. (25 cm) or more wide; grayish to blackish brown or black; **surface** shiny, wrinkled, and glandular dotted; **flesh** gelatinous, deliquescing in prolonged wet weather, becoming a thin, inconspicuous black dotted crust when dry.
Spores: 12–14 x 4.5–5 μm, sausage-shaped, smooth, hyaline (white in mass), non-amyloid.
Occurrence: Clustered on stumps, logs, and fallen branches of broad-leaved trees, especially oak; saprobic; year-round; fairly common but easily overlooked when dried.
Edibility: Unknown, of no culinary interest.

Comments: *Bulgaria inquinans* (not illustrated) also grows in clusters on deadwood of oak and other broad-leaved trees, but it has dark brown, rubbery, turban-shaped fruitbodies with a flat or sunken black disc at the top and a scurfy exterior. *Exidia recisa* (not illustrated) is similar but golden brown to cinnamon-brown. *Glandulosa* means "full of glands."

Latin name: *Tremella mesenterica* Retzius
Synonym: *Tremella lutescens* Pers.
Common name: Witches' Butter.
Order: Tremellales
Family: Tremellaceae
Fruitbody: A complex gelatinous mass up to 4 in. (10 cm) or more across; convoluted at first, becoming lobed and irregularly folded; pale yellow to golden yellow or orange-yellow, becoming dark orange on drying; surface viscid when moist; **flesh** soft, translucent, and jelly-like when fresh, becoming shriveled and hard when dry, odor and taste not distinctive.

Spores: Basidiospores 10–18 x 8–12 µm, broadly elliptical, smooth, hyaline (white to yellowish in mass), non-amyloid; also producing asexual spores (conidia) 3–4.5 x 2.5–3.5 µm, subglobose, smooth, hyaline.
Occurrence: Solitary or in clusters on fallen branches, stumps and logs of broad-leaved trees, especially oak and beech; parasitic on wood-inhabiting fungi; spring-fall (occasionally in winter), common.
Edibility: Non-poisonous.

Comments: Witches' Butter fruits in rainy periods and will repeatedly rehydrate and dry according to prevailing moisture conditions. It has a conidial stage in which asexual spores are released from the surface of the fruitbody before it produces basidiospores. Witches' Butter is parasitic on crust fungi (*Peniophora* spp.), but the connection may not be apparent since the host fungus is often present only as mycelium. *Dacrymyces palmatus* (not illustrated) is very similar but grows on conifer wood and has a white base where it is attached to the substrate. *Mesenterica* means "like the middle intestine." Egads, such imagination!

Latin name: *Tremella reticulata* (Berk.) Farlow
Common names: White Coral Jelly; Dead Man's Rubber Glove
Order: Tremellales
Family: Tremellaceae
Fruitbody: Up to 6 in. (15 cm) tall and wide; forming a more or less hemispherical compound mass of upright, gelatinous, hollow, branched lobes, often fused and sometimes finger-like with blunt tips; white when young, becoming creamy to yellowish brown with age, somewhat translucent; flesh gelatinous to cartilaginous, odor and taste not distinctive.
Spores: 9–11 x 5–6 μm, broadly ovoid, depressed on one side, smooth, hyaline.
Occurrence: Solitary or gregarious on the ground or well-decayed wood in broad-leaved and mixed woods; saprobic and perhaps mycoparasitic; summer-fall; occasional to locally common in some years.
Edibility: Unknown.

Comments: During prolonged wet weather this odd fungus can fruit prolifically. *Reticulata* means "forming a network," alluding to the complex form of the fruitbodies.

Humaria hemisphaerica, p.465

Cup Fungi and Bird's Nest Fungi

Although superficially similar, the cup-shaped fungi in this section represent two very different taxonomic groups. The true cup fungi are ascomycetes, in which the reproductive spores are produced inside a sac-like structure called an ascus (pl. asci). At maturity the ascus ruptures and forcibly ejects the spores. The fertile inner surface of a true cup fungus (apothecium) is composed of densely packed asci that act like microscopic cannons as they "shoot" their spores into the air, often simultaneously in astronomical numbers. This amazing display of fungal fireworks is not only visible but sometimes audible as a soft "hiss." Spore dispersal of this type is triggered by rapid changes in light, air pressure, or physical contact with the fruitbody. It can often be initiated simply by touching the cups or blowing air on the inner surface.

The Bird's Nest fungi are related to puffballs. Their spores are produced within the "eggs" (peridioles). The peridioles are dislodged from the "nest" by raindrops that hit the cup and spatter them outward. Each peridiole has a sticky tail-like appendage that on contact attaches to nearby objects and vegetation. At this point the outer wall of the peridioles may disintegrate and release the spores, or the peridioles may be incidentally consumed by grazing animals, which facilitate the dispersal of spores that pass unharmed through their digestive system.

The species included here are saprobes. Few have any value as edibles, but they are interesting and attractive fungi that add to the beauty and diversity of our woods.

Latin name: *Aleuria aurantia* Peck
Synonym: *Peziza aurantia* Pers.
Common name: Orange Peel Fungus
Order: Pezizales
Family: Pyronemataceae
Fruitbody: Up to 3 in. (7.5 cm) broad, more or less cup-shaped and stalkless, spreading and becoming shallow or flattened with age; **inner fertile surface** brilliant orange to reddish orange, smooth; **exterior surface** pale orange, downy, sometimes appearing frosted when young, becoming smooth; **flesh** thin, whitish, brittle, odor and taste not distinctive.
Spores: 17–24 x 9–11 μm, elliptical, coarsely reticulated, hyaline (white in mass).
Occurrence: Gregarious, often in compressed clusters on the ground, frequently on bare soil in open areas, on disturbed ground, along woodland trails, and along roadcuts; saprobic; spring-fall; common.
Edibility: Edible but flavorless.

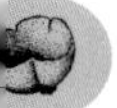

Comments: At first glance this fungus might be dismissed as a discarded orange peel. *Caloscypha fulgens* (not illustrated) fruits in conifer woods in early spring and is somewhat similar, but the cups are orange-yellow with bluish green stains on the exterior. *Aurantia* means "orange colored."

Latin name: *Chlorociboria aeruginascens* (Nyl.) Kanouse ex Ramamurthi, Korf & Batra
Synonym: *Chlorosplenium aeruginascens* (Nyl.) P. Karst.
Common name: Blue-green Wood Stain
Order: Helotiales
Family: Helotiaceae
Fruitbody: Small cups up to ½ in. (1.3 cm) wide, soon becoming flattened and disc-like; **inner fertile surface** smooth, bright turquoise or developing dull yellowish tones with age; **exterior surface** somewhat scurfy, blue to bluish green; **flesh** thin, blue green, odor not distinctive.
Stalk: Up to ¼ in. (6 mm) long, slender, central or eccentric.
Spores: 6–10 x 1.5–2 μm, spindle-shaped, smooth, hyaline.
Occurrence: In small clusters on decaying logs, branches, and stumps of broad-leaved trees, especially oak; saprobic; early summer-fall; occasional.
Edibility: Not edible.

Comments: The mycelium of this uniquely colored fungus stains the woody substrate bluish green. Although the fruitbodies are infrequently encountered, decaying twigs and branches with telltale bluish green staining is commonplace. *Chlorociboria aeruginosa* (not illustrated) is nearly identical but has larger spores (9–14 x 2–4 μm). *Aeruginascens* relates to "verdigris," which is a greenish blue substance that forms on copper or brass.

Latin name: *Crucibulum laeve* (Huds.) Kambly
Synonym: *Crucibulum vulgare* Tul. & C. Tul.
Common names: White-egg Bird's Nest; Common Bird's Nest
Order: Agaricales
Family: Nidulariaceae
Fruitbody: ¼–⅜ in. (6–9 mm) high and wide; at first nearly globose to somewhat barrel-shaped, tapering toward the base, becoming cup-like at maturity when the lid disintegrates to reveal numerous spore bearing "eggs" (peridioles) within; **inner surface** smooth, satiny, creamy white, with up to 15 or more peridioles nearly filling the cup at first; peridioles 1.5–2 mm in diameter, white, lens-shaped and smooth, each attached to the inner cup wall by a cord-like strand; **exterior surface** velvety, coarsely scurfy on top, ochraceous to tawny, becoming grayish with age; stalk absent.

Spores: 7–10 x 4–6 μm, elliptical, smooth, hyaline, non-amyloid.
Occurrence: Typically in clusters on a variety of organic substrates such as twigs and branches, compost, straw, leaf mold, decaying weeds, and miscellaneous woody debris; saprobic; summer-fall; common.
Edibility: Of no culinary value.

Comments: This little fungus is always a pleasure to find. It rarely occurs on larger stumps and logs, but almost anything else containing cellulose is fair game. The photo shows specimens growing on a discarded cotton work glove. *Cyathus stercoreus* (not illustrated) grows on dung or manure-enriched soil, and has dark gray to black peridioles. *Cyathus striatus* (p.461) has a densely hairy outer surface and lengthwise grooves on the inner wall. *Laeve* means "smooth," referring to the inner cup wall.

Latin name: *Cyathus striatus* (Huds.) Hoffm.
Common names: Fluted Bird's Nest; Splash Cups
Order: Nidulariales
Family: Nidulariaceae
Fruitbody: Up to ¾ in. (2 cm) high and ⅜ in. (1 cm) wide; vase-shaped with a narrow stalk-like base, covered at first with a thin membrane that disappears at maturity to expose 12–16 spore-bearing "eggs" (peridioles) deep in the bottom of the cup; **inner surface** shiny, pale gray to grayish brown, with lengthwise grooves or furrows; peridioles lens-shaped, 1.5–2 mm in diameter, pale gray to blackish, each attached to the inner cup wall by an elastic, thread-like strand; **exterior surface** tawny to dark brown or grayish brown, densely covered with coarse, shaggy hairs.
Spores: 12–21 x 7–12 μm, elliptical, smooth, hyaline.
Occurrence: Typically in clusters on fallen branches, twigs, bark, and other decaying wood, often attached to the substrate by a dark brown mycelial mat; saprobic; summer-fall, empty cups persist through winter; fairly common.
Edibility: Of no culinary value.

Comments: The similar *Cyathus olla* (not illustrated) has a flaring margin, a smooth inner surface, and larger peridioles (up to 3.5 mm wide). *Crucibulum laeve* (p.460) has a yellow-brown felted exterior surface and white peridioles. *Striatus* means "having grooves or ridges."

Latin name: *Disciotis venosa* (Pers.) Arnould
Synonym: *Peziza venosa* Pers.
Common name: Veined Cup; Cup Morel
Order: Pezizales
Family: Morchellaceae
Fruitbody: Up to 8 in. (20 cm) across; medium to large irregular-shaped cups with a short stalk-like base; **inner fertile surface** yellowish brown to reddish brown or dark brown, smooth at first, becoming wrinkled to vein-like in the center; **exterior surface** whitish to buff or ochraceous, somewhat velvety, roughened with scaly patches, especially near the margin; **flesh** thin, fragile, whitish to watery brown, odor not distinctive or mildly of chlorine; **base** short, broad, usually less than ¾ in. (2 cm) long, folded or wrinkled at the point of attachment.

Spores: 22.5–25 x 12–15 µm, oval, smooth, pale yellow (brownish in mass).
Occurrence: Solitary or more often in scattered groups or compressed clusters on the ground, especially on bare soil and along trails in rich broad-leaved woods and woodland margins; saprobic; spring; occasional to fairly common in some years.
Edibility: Edible with caution. Poisonous raw, and not easily distinguished macroscopically from similar species of unknown edibility.

Comments: The Veined Cup appears in early spring at about the same time as the first morels. When young, the cups are moderately deep and partially closed from the inrolled margin. Later the cups expand and become shallow to nearly flattened. *Peziza badioconfusa* (p.466) is also found in spring in similar habitats but has a smooth, dark brown inner surface and a dark brown exterior surface. *Peziza vesiculosa* (not illustrated) is smaller, pale ochraceous overall, and grows on dung or dung-enriched soil. *Peziza domiciliana* (not illustrated) grows on walls, basement carpets, or elsewhere in damp buildings or cellars, and has a pale yellow-ochre to buff inner surface and a short, often ribbed stalk. *Venosa* means "veined."

Latin name: *Galiella rufa* (Schwein.) Nannf. & Korf
Synonym: *Bulgaria rufa* Schwein.
Common name: Hairy Rubber Cup
Order: Pezizales
Family: Sarcosomataceae
Fruitbody: Up to 1½ in. (4 cm) wide; more or less globose to top-shaped at first, then opening at the top to form a shallow cup with a toothed margin; **inner fertile surface** smooth, tan to pale orange-brown or reddish brown; **exterior surface** wrinkled, blackish brown, densely hairy; **flesh** thick, gelatinous, rubbery, odor and taste not distinctive.
Stalk: Up to ⅜ in. (1 cm) long, or absent.
Spores: 10–20 x 8–10 μm, elliptical; finely warted, hyaline.
Occurrence: Usually in small groups or clusters on fallen branches, and buried wood of broad-leaved trees; saprobic; spring-summer; common.
Edibility: Not edible.

Comments: These dull-colored rubbery cups can blend in with the leaf litter and be difficult to spot. *Bulgaria inquinans* (not illustrated), which grows in dense clusters on logs and stumps, is similar in form but has a jet-black inner surface. Compare with *Wolfina aurantiopsis* (p.473), which has a shallow cup-like, almost woody fruitbody with a yellowish inner surface. *Rufa* means "rusty" or "reddish brown."

Latin name: *Helvella macropus* (Pers.) P. Karst.
Synonym: *Macroscypha macropus* (Pers.) Gray
Common name: Scurfy Elfin Cup
Order: Pezizales
Family: Helvellaceae
Fruitbody: ¾–1½ in. (2–4 cm) wide, shallow to nearly flattened cup on a slender stalk, margin incurved at first; **inner fertile surface** gray to grayish brown or yellowish brown, smooth, dull; **exterior surface** pale gray, granular to scurfy; **stalk** up to 2½ in. (6.5 cm) tall, cylindrical to somewhat compressed, equal or enlarged downward; colored like the underside of the cup, hairy-scurfy; **flesh** white, thin, brittle, odor and taste not distinctive.

Spores: 20–24 x 10.5–12.5 µm, elliptical to spindle-shaped, smooth to finely warted, with 1 large and 2 smaller oil droplets within, hyaline.
Occurrence: Solitary or in small groups on the ground, especially on sandy soil in broad-leaved or conifer woods; saprobic; early summer-fall; occasional.
Edibility: Not edible.

Comments: *Macropus* means "large foot," referring to the proportionately long stalk.

Latin name: *Humaria hemisphaerica* (Wigg.) Fuckel
Synonym: *Peziza hemisphaerica* Wigg.
Common name: Brown-haired Fairy Cup
Order: Pezizales
Family: Pyronemataceae
Fruitbody: Stalkless cups up to 1¼ in. (3 cm) wide; subglobose at first, soon expanding and cup-like; **inner fertile surface** white to grayish white, smooth; **exterior surface** densely covered with stiff brown hairs that form a fringe on the cup margin, whitish beneath; **flesh** thin, whitish, brittle, odor and taste not distinctive.
Spores: 20–24 x 10–13 µm, elliptical, finely warted, hyaline.
Occurrence: In small groups or clusters on the ground or on well-decayed wood; saprobic; summer-fall; fairly common.
Edibility: Not edible.

Comments: This small cup fungus is particularly attractive when found growing on old moss-covered stumps and logs. *Hemispherica* means "half round."

Latin name: *Peziza badioconfusa* Korf
Synonym: *Peziza phyllogena* Cooke
Common name: Common Brown Cup
Order: Pezizales
Family: Pezizaceae
Fruitbody: 1½–4 in. (4–10 cm) wide; stalkless deep or shallow cups that are compressed and misshapened when growing in clusters, becoming more or less flattened with age; **inner fertile surface** reddish brown, often with purplish tones, smooth; **exterior surface** reddish brown to purplish brown, somewhat dull, scurfy; **flesh** thin, brittle, fragile, odor and taste not distinctive.
Spores: 16–21 x 8–10 μm, elliptical, finely warted, 1 or 2 oil drops within, hyaline.
Occurrence: Solitary or in clusters on the ground or on well-decayed wood, especially on or near old moss-covered stumps; saprobic; spring-early summer; common.
Edibility: Reported to be edible if cooked.

Comments: This cup fungus is frequently encountered during morel season in April and May. The epithet *badioconfusa* indicates that this species can be confused with *Peziza badia* (not illustrated), a very similar fall fruiting species that has partially reticulated spores. Compare with *Auricularia auricula* (p.452), which always grows on wood and has rubbery, flexible flesh. Compare also with *Disciotis venosa* (p.462), which has a wrinkled or vein-like interior surface and a pale buff to ochraceous exterior.

Latin name: *Peziza succosa* Berk.
Common name: Yellow Sap Cup
Order: Pezizales
Family: Pezizaceae
Fruitbody: ¾–2 in. (2–5 cm) wide; more or less cup-shaped at first, expanding and becoming shallow to nearly flat with age, stalkless; **inner fertile surface** pale grayish brown, often with an olivaceous tint, smooth; **exterior surface** whitish to pale grayish or yellowish, staining yellow when bruised, somewhat scurfy; **flesh** thin, brittle, whitish, staining yellow from a sap that exudes when the flesh is broken, odor and taste not distinctive.
Spores: 16–22 x 8–12 µm, elliptical, coarsely ornamented with warts and ridges, hyaline (white in mass).
Occurrence: Solitary or in small groups on the ground, especially on bare soil in woods, parklands, along trails, streams, and forest roads; saprobic; summer; occasional.
Edibility: Unknown.

Comments: The similar *Peziza vesiculosa* (not illustrated), which grows on manure-enriched ground or dung, is pale ochraceous overall, has an incurved margin, and thick flesh that does not exude yellow-staining sap when broken. *Succosa* means "full of sap."

Latin name: *Rhizina undulata* Fr.
Synonym: *Rhizina inflata* (Schaeff.) Quél.
Common names: Pine-fire Cushion; Crust-like Cup
Order: Pezizales
Family: Rhizinaceae
Fruitbody: Up to 4 in. (10 cm) wide (or larger when fused in compound clusters); convex, crust-like to nearly flat, irregular in outline, margin wavy; **upper fertile surface** smooth, undulating, reddish brown to blackish brown, margin white; **underside** hollowed, dingy yellowish to ochraceous, attached to the substrate by thick whitish to brownish branching strands (rhizoids); **flesh** firm, becoming hard and leathery; stalk absent.

Spores: 30–40 x 7–10 µm, spindle-shaped with small appendages at each end, smooth or finely roughened, hyaline (white in mass).
Occurrence: Solitary or gregarious, at times in fused clusters on previously burned ground in conifer woods; saprobic and parasitic; summer-fall; uncommon.
Edibility: Not edible.

Comments: This unique fungus is not likely to be mistaken for any other. Its spores lie dormant in the soil until heat from fire initiates germination. The fruitbodies appear 1 or 2 years later. This fungus spreads as a saprobe on dead conifer roots but also invades the roots of living trees, causing a root rot that can eventually kill those that become infected. *Undulata* means "undulating" or "wavy."

Latin name: *Sarcoscypha austriaca* (O. Beck ex Sacc.) Boud.
Common name: Scarlet Cup
Order: Pezizales
Family: Sarcoscyphaceae
Fruitbody: ¾–2 in. (2–5 cm) wide; shallow cup-shaped, margin often lobed and irregular in outline; **inner fertile surface** scarlet red, smooth, shiny; **exterior surface** whitish to pale ochraceous or pinkish red, granulose to scurfy; **flesh** thin but not fragile, pinkish, odor and taste not distinctive; **stalk** usually short and thick, colored like the undersurface of the cup, sometimes lacking.
Spores: 20–38 x 9–16 μm, elongated-elliptical, smooth, with several oil droplets at each end; hyaline (white in mass).
Occurrence: Solitary or more often in groups or compressed clusters on fallen branches and twigs (often buried) in damp broad-leaved and mixed woods, especially along streams, near seeps, and in low wetland borders; saprobic; early to late spring; common.
Edibility: Non-poisonous.

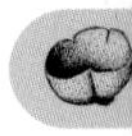

Comments: Although frequently concealed among leaf litter, even partially exposed cups will attract attention with their brilliant splash of color in early spring. *Sarcoscypha occidentalis* (p.470) has smaller cups and relatively long, slender stalks. *Austriaca* means "from Austria," where this widely distributed species also occurs.

Latin name: *Sarcoscypha occidentalis* (Schwein.) Cooke
Common name: Stalked Scarlet Cup
Order: Pezizales
Family: Sarcoscyphaceae
Fruitbody: Stalked shallow cups up to ½ in. (1.3 cm) wide; **inner fertile surface** scarlet red, smooth; **exterior surface** whitish or tinged pinkish red, glabrous; **stalk** up to ¼ in. (6 mm) long, nearly equal, whitish or tinged pinkish red; **flesh** thin, odor and taste not distinctive.
Spores: 18–22 x 10–12 µm, elliptical, smooth, hyaline.
Occurrence: Solitary or in clusters on decaying fallen branches and twigs (sometimes buried) in wet broad-leaved woods; saprobic; spring-summer; common.
Edibility: Of no culinary value.

Comments: These small cups appear in spring but usually later than the equally brilliant red *Sarcoscypha austriaca* (p.469), which is larger and sometimes stalkless. *Microstoma floccosum* (not illustrated) is another springtime small red cup that grows on fallen branches. Its exterior surface is densely covered with long whitish hairs. Compare with *Scutellinia scutellata* (p.471), a stalkless shallow red cup with black hairs fringing the margin. *Occidentalis* means "western," presumably referring to the broad distribution of this species in the western hemisphere.

Latin name: *Scutellinia scutellata* (L.) Lambotte
Synonym: *Peziza scutellata* L.:Fr.
Common names: Eyelash Cup; Molly Eye-winker
Order: Pezizales
Family: Pyronemataceae
Fruitbody: Up to ¾ in. (2 cm) wide; somewhat rounded at first, soon expanding into a stalkless shallow cup or nearly flattened disc; **inner fertile surface** cherry red to bright orange-red, smooth; **exterior surface** pale brown, more or less covered with stiff, dark brown to black hairs that form a fringe on the cup margin; **flesh** thin, brittle, reddish, odor and taste not distinctive.
Spores: 18–20 x 10–12 μm, elliptical, minutely warted, several oil droplets within; hyaline (white in mass).
Occurrence: Solitary or more often in clusters on damp soil or well-decayed logs, stumps, branches, and other woody debris in wet woods; saprobic; late spring-fall; common.
Edibility: Of no culinary value.

Comments: Use a hand lens to fully appreciate the delicate beauty of these small cups. *Scutellinia erinaceus* (not illustrated) is similar but has a dull orange to orange-yellow inner surface, and smooth spores. *Scutellata* means "shield-shaped."

Latin name: *Urnula craterium* (Schwein.) Fr.
Common name: Devil's Urn
Order: Pezizales
Family: Sarcosomataceae
Fruitbody: 1¼–2½ in. (3–6.5 cm) wide, 1½–4½ in. (3–11.5 cm) tall; club-shaped when very young, soon expanding into a deep, rounded cup that tapers to a slender, stalk-like base (often buried); **inner fertile surface** dark brown to jet black, smooth; **exterior surface** blackish to grayish brown or pinkish brown, scurfy; margin toothed or ragged; **flesh** thin, soon leathery, odor and taste not distinctive.
Spores: 22–37 x 10–15 μm, elliptical, smooth, hyaline.
Occurrence: Usually in groups or clusters on the ground, but associated with decaying branches and buried wood of broad-leaved trees; saprobic; spring; common.
Edibility: Generally considered inedible though apparently non-poisonous.

Comments: The Devil's Urn is easily overlooked due to its dark color that blends in with the forest floor. It often occurs in rows alongside fallen branches. The cups are sometimes buried to the rim in the litter, thus appearing like holes in the ground. *Craterium* means "goblet-shaped."

Latin name: *Wolfina aurantiopsis* Eckblad
Common name: None
Order: Pezizales
Family: Sarcosomataceae
Fruitbody: Up to 2½ in. (6.5 cm) wide; stalkless shallow cups; **inner fertile surface** pale yellow to ochraceous or somewhat orangish; smooth; **exterior surface** with projecting folds, covered with brownish to black mycelium; **flesh** thick, tough, becoming corky on drying; whitish; odor not distinctive.
Spores: 16–18 x 27–33 µm, broadly elliptical, granular within, hyaline to slightly yellowish.
Occurrence: Solitary or in small groups or clusters broadly attached to decaying wood in mixed woods, especially with pine and hemlock, also reported to grow on soil; saprobic; summer-fall; uncommon.
Edibility: Unknown.

Comments: Until the firm, corky texture is noted, this uncommon cup fungus is likely to be mistaken for a pale form of *Galiella rufa* (p.463). *Aurantiopsis* means "appearing orange-colored."

Latin name: *Wynnea americana* Thaxter
Common names: Rabbit Ears; Moose Antlers
Order: Pezizales
Family: Sarcoscyphaceae
Fruitbody: Compound clusters up to 5 in. (12.5 cm) tall and wide consisting of elongated, almost spindle-shaped cups that are open on one side and united at the base; **inner fertile surface** smooth, pinkish to reddish brown at first, becoming dark brown to blackish with age; **exterior surface** blackish brown to dark reddish brown, coarsely granular to warty; **flesh** brown, firm, odor and taste not distinctive; **stalk** rudimentary, attached to a woody underground sclerotium.
Spores: 32–40 x 15–16 µm, elliptical with pointed tips, surface banded lengthwise, several oil drops within, hyaline.
Occurrence: Solitary or in scattered groups on the ground in broad-leaved and mixed woods; saprobic; summer-fall; uncommon.
Edibility: Unknown.

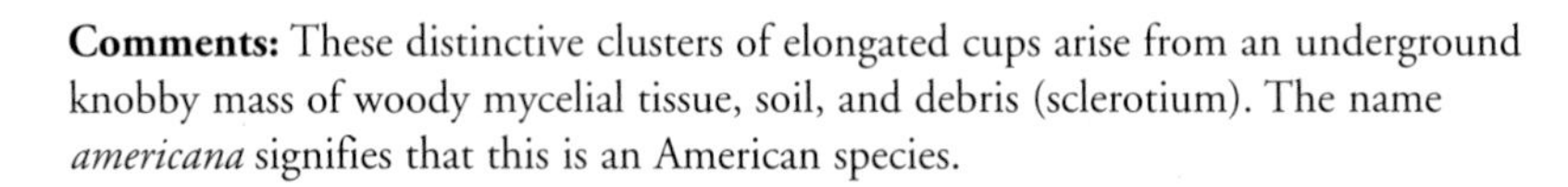

Comments: These distinctive clusters of elongated cups arise from an underground knobby mass of woody mycelial tissue, soil, and debris (sclerotium). The name *americana* signifies that this is an American species.

Helvella crispa, p.482

Morels, False Morels, and Elfin Saddles

The species contained in this section are of great importance to mushroom hunters since they include the true morels (genus *Morchella*), which are perhaps the most highly prized of all edible wild mushrooms. Here too are the false morels (genus *Gyromitra*) and elfin saddles (genus *Helvella*), which are generally regarded—or suspected—to be poisonous when consumed, especially if eaten raw.

True morels and elfin saddles contain toxic hemolysins. These substances destroy red blood corpuscles, but are rendered harmless in cooking. False morels have a different and potentially more dangerous toxic property. They contain gyromitrin, a carcinogenic toxin that breaks down to monomethylhydrazine, a potentially lethal compound that is similar in structure to solid rocket fuel.

Most species in this section fruit in the spring, usually from late March to early May. In some years true morels can be plentiful if one knows the places to hunt for them. False morels, on the other hand, are never very abundant. They usually grow singly here and there, or in small groups. Elfin saddles bear a resemblance to certain false morels, but they can be found anytime throughout the growing season. All of these mushrooms are ascomycetes and are closely related to the cup fungi.

Experienced collectors are not likely to confuse edible true morels with poisonous false morels, but they resemble each other superficially, and often share the same habitat. It is well to remember that true morels have obvious pits (cups) with ridge-like borders. False morels have convoluted, brain-like, or folded caps that lack clearly defined individual pits.

Latin name: *Gyromitra brunnea* Underw.
Synonym: *Gyromitra fastigiata* (Krombh.) Rehm
Common names: Elephant Ear; Brown False Morel; Gabled False Morel
Order: Pezizales
Family: Discinaceae
Fruitbody: 2½–6 in. (6.5–15 cm) tall; **cap** irregularly lobed and folded to multifaceted saddle-shaped, often compressed against and variously attached to the stalk; outer fertile surface yellow-brown to orange-brown or reddish brown, wrinkled, glabrous; underside white to cream or pale brown; **flesh** thin, brittle, odor not distinctive; **stalk** massive, more or less equal or broader at the base, often branching near the apex, interior stuffed or chambered; surface smooth or velvety, with irregular folds and indentations, white.
Spores: 24–28 x 11–15 μm, elliptical, some with multiple appendages at the end, surface faintly reticulate, 1–3 oil drops within, hyaline.
Occurrence: Solitary or in small groups on the ground, especially near rotting stumps in broad-leaved woods; saprobic; spring; uncommon.
Edibility: Possibly poisonous. Reports vary, but several closely related species are known to be toxic.

Comments: This spectacular false morel appears in the spring at the same time as the true morels. *Gyromitra gigas* (p.480) is smaller and has a more convoluted, brain-like cap. *Gyromitra infula* (not illustrated) has a saddle-shaped cap, and fruits in late summer and fall. *Brunnea* means "russet-brown."

Latin name: *Gyromitra esculenta* (Pers.) Fr.
Synonym: *Helvella esculenta* Pers.
Common names: Beefsteak Morel; False Morel; Lorchel; Turban Fungus
Order: Pezizales
Family: Discinaceae
Fruitbody: 1½–4 in. (4–10 cm) tall; **cap** more or less globose to irregular; outer fertile surface wrinkled to folded or convoluted, reddish brown to blackish brown, shiny; underside whitish; **flesh** thin, brittle, odor mild to somewhat fruity; **stalk** cylindrical to compressed, base often enlarged, hollow or nearly so; dingy white to tan or pinkish tan; surface smooth or granular.
Spores: 18–26 x 9–12 μm, elliptical, smooth, 2 oil drops within, hyaline (pale yellowish in mass).
Occurrence: Solitary or more often in groups on the ground under pines and other conifers, rarely fruiting on decaying wood; saprobic; early spring; occasional.
Edibility: Dangerous, potentially lethal.

Comments: This is one of the first fleshy mushrooms to appear in the springtime, usually a little earlier than true morels, although their fruiting periods overlap. Despite the name *esculenta,* which means "edible," this mushroom has caused fatalities when eaten. It contains gyromitrin, a carcinogenic compound that is converted to more acutely dangerous monomethylhydrazine. Although some persons continue to eat this mushroom with impunity, others have suffered tragic consequences. Whether this is due to individual thresholds for toxic doses or to varying ecological factors is unclear. Parboiling and thorough cooking apparently destroys some of the toxin. It also has been suggested that the age of the fruitbodies may influence toxicity. In any event, the Beefsteak Morel is far too dangerous to experiment with. *Gyromitra gigas* (p.480) is similar but has a tan to tawny cap, and usually grows in broad-leaved woods.

Latin name: *Gyromitra gigas* (Kromb.) Cooke
Synonym: *Gyromitra korfii* (Raitv.) Harmaja
Common name: Bull-nose False Morel
Order: Pezizales
Family: Discinaceae
Fruitbody: 1½–5 in. (4–12.5 cm) tall; **cap** more or less globose, convoluted and brain-like with a wavy margin; outer fertile surface ochre to yellow-brown or somewhat reddish brown, glabrous, lustrous; interior complex folded and chambered; underside white to buff, granular or scurfy; **flesh** whitish, brittle, odor and taste not distinctive; **stalk** massive, somewhat cylindrical with lengthwise folds and channels, base often enlarged, hollow to stuffed and chambered, surface smooth or grainy, white.
Spores: 26–34 x 11–13.5 μm, broadly elliptical with multiple blunt appendages at the ends, smooth to wrinkled, several oil droplets within, hyaline (pale yellow in mass).
Occurrence: Solitary or loosely scattered in small groups on the ground, in broad-leaved and mixed woods; saprobic; spring; occasional.
Edibility: Not recommended, almost certainly poisonous when raw, although according to some authorities it is edible when cooked.

Comments: *Gyromitra esculenta* (p.479) is similar but grows under conifers and has a dark reddish brown cap. *Gigas* means "gigantic," which is a bit of a stretch for most collections.

Latin name: *Helvella atra* J. König
Synonym: *Leptopodia atra* (J. König) Boud.
Common name: Black Helvella
Order: Pezizales
Family: Helvellaceae
Fruitbody: Up to 3 in. (7.5 cm) tall; composed of a saddle-shaped to folded and nearly flattened saucer-shaped cap with a free margin, and a slender stalk; **cap** ½–1¼ in. (1.3–3 cm) wide, **outer fertile surface** smooth, blackish brown to black; **sterile undersurface** somewhat scurfy; smoky brown to blackish, usually paler than fertile surface; **flesh** thin, odor and taste not distinctive; **stalk** cylindrical to compressed, tough, equal or tapering upward; colored like the cap, base white; surface scurfy to velvety.
Spores: 15–18.5 x 10–12 µm, broadly elliptical, smooth, with 1 large oil drop within, hyaline.
Occurrence: Solitary or in small groups on the ground in mixed woods, and woodland clearings and margins; saprobic; summer-fall; uncommon.
Edibility: Unknown.

Comments: *Helvella elastica* (not illustrated) is very similar but has a yellow-brown to grayish brown cap, and a whitish to cream stalk. *Helvella lacunosa* (not illustrated) has a grayish to blackish cap and a thick, deeply furrowed and pitted stalk. *Atra* means "very dark."

Latin name: *Helvella crispa* (Scop) Fr.
Common names: White Elfin Saddle; Fluted White Helvella
Order: Pezizales
Family: Helvellaceae
Fruitbody: Up to 4½ in. (11.5 cm) tall; **cap** irregularly lobed or folded to more or less saddle-shaped with a wavy margin, white to cream or pale ochre; outer fertile surface wrinkled to lumpy, glabrous; underside pale buff, downy; **flesh** thin, brittle, odor and taste not distinctive; **stalk** variable from short and squat to elongated, equal or tapering upward, deeply furrowed and ribbed, colored like the cap.
Spores: 17–12 x 10–14 µm, broadly elliptical, smooth, 1 or more oil droplets within, hyaline (white in mass).
Occurrence: Solitary or scattered on the ground in broad-leaved and conifer woods, at woodland margins, and on roadsides; saprobic; summer-fall; fairly common.
Edibility: Poisonous raw. Young specimens edible after parboiling and prolonged cooking.

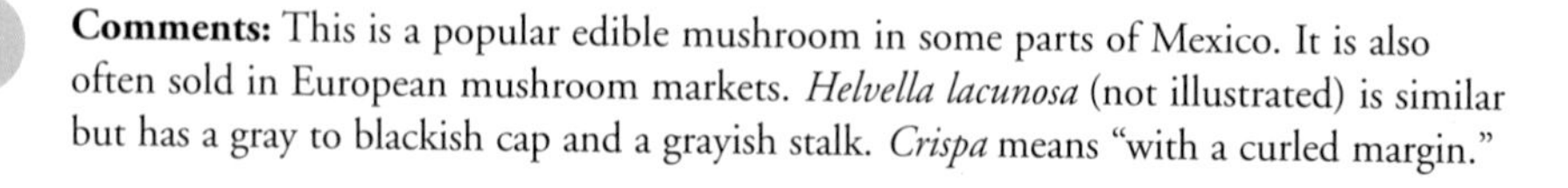

Comments: This is a popular edible mushroom in some parts of Mexico. It is also often sold in European mushroom markets. *Helvella lacunosa* (not illustrated) is similar but has a gray to blackish cap and a grayish stalk. *Crispa* means "with a curled margin."

Latin name: *Morchella angusticeps* Peck
Common names: Peck's Morel; Narrow-headed Morel; Black Morel; Snake Head
Order: Pezizales
Family: Morchellaceae
Fruitbody: Up to 6 in. (15 cm) tall; consisting of a conical pitted head on a long stalk; **head** honeycombed with dark brown pits, separated by blackish ridges and more or less arranged in lengthwise rows, fused to the stalk at the base, hollow inside; **flesh** thin, odor mild, earthy; **stalk** ¾–1½ in. (2–4 cm) in diameter, cylindrical, more or less equal or flared at the apex, smooth to slightly furrowed, hollow, brittle, whitish to pale brownish yellow; surface scurfy to granulose.
Spores: 24–28 x 12–14 μm, elliptical, smooth, hyaline (yellowish ochre in mass).
Occurrence: Solitary or loosely scattered on the ground in broad-leaved woods and woodland margins; saprobic, but also capable of forming mycorrhiza; spring; occasional.
Edibility: Edible when well cooked.

Comments: According to some mycologists, *Morchella augusticeps* is synonymous with *Morchella elata* (p.484). Whether they are two separate species or merely forms of one species, both are included in this guide since they are distinctive in the field. Intermediate specimens are sometimes found, which adds to the confusion. *Angusticeps* means "narrow head."

Latin name: *Morchella elata* Fr.
Common names: Black Morel; Snakehead
Order: Pezizales
Family: Morchellaceae
Fruitbody: Up to 6 in. (15 cm) tall; consisting of a conical to elongated, bluntly triangular pitted head on a cylindrical, hollow stalk; **head** hollow inside, fused to the stalk at the base, honeycombed with dark brown pits separated by blackish ridges; pits arranged more or less in parallel lengthwise rows; **flesh** thin, odor mild, "earthy"; **stalk** up to 1 in. (2.5 cm) in diameter, more or less equal, or swollen at the base, often vertically wrinkled or grooved, off-white to pale yellowish; surface scurfy to granulose.
Spores: 18–25 x 11–15 μm, broadly elliptical, smooth, hyaline.
Occurrence: Usually gregarious on the ground in woods and near trees in more open areas, such as woodland clearings and margins, especially near tulip poplar and ash, also on burnsites; saprobic and mycorrhizal; spring; common.
Edibility: Edible and delicious when cooked (see comments below).

Comments: The "blacks" are usually the first of the true morels to appear in spring and are much sought after for their fine flavor. As with other morels, they are poisonous if eaten raw. Some individuals suffer stomach upsets from eating black morels even when they are well-cooked, especially if consumed with alcoholic beverages. Compare with *Morchella angusticeps* (p.483), which may or may not be a distinct species. The latter often can be found at the same time in the same habitat, but it has a much longer stalk-to-head ratio. *Morchella semilibera* (p.487) has a lighter brown head with a flaring bottom margin that is free from the stalk. Other morels have yellow-brown or grayish heads. *Elata* means "tall."

Latin name: *Morchella esculenta* (L.) Pers.
Common names: Common Morel; Yellow Morel; Molly Moocher; Sponge Mushroom; Haystack; Blond Morel; Dryland Fish
Order: Pezizales
Family: Morchellaceae
Fruitbody: 2½–7 in. (6.5–18 cm) or more tall; consisting of a cylindrical stalk with an enlarged sponge-like head; **head** variable in shape, often more or less rounded to oval, or somewhat oblong-conical, fused to the stalk at the base, honeycombed with tan to yellow-brown or grayish pits separated by distinct paler ridges, hollow; **flesh** brittle, odor "mushroomy"; **stalk** cylindrical but usually enlarged at the base, variously ribbed or furrowed, hollow at maturity; whitish to yellowish, sometimes with brownish stains toward the base in age; surface roughened to granulose.
Spores: 20–26 x 12–16 μm, broadly elliptical, smooth, hyaline (yellowish ochre in mass).
Occurrence: Solitary or more often gregarious, sometimes clustered on the ground in a variety of habitats, especially on soil with a limestone base. Favorable sites include tulip poplar and ash woods, apple orchards or anywhere there are very old apple trees, near dying elm trees, burned areas, and along wooded streams and riverbanks; saprobic and mycorrhizal; spring; common (plentiful in some years).
Edibility: Edible and delicious when cooked, poisonous raw.

Comments: There are several forms of *Morchella esculenta,* including the "Big Foot Morel," which some authors call *Morchella crassipes* (illustrated). The Big Foot, which appears late in the morel season, can sometimes be up to 12 in. (30 cm) or more tall. The "White Morel," which some authors call *Morchella deliciosus* (illustrated), has gray pits with white ridges. Another distinct form, which is sometimes called the "Tulip Morel," is a slender and much smaller version of the yellow morel, rarely reaching more than 3 in. (7.5 cm) tall. The Tulip Morel is found in woods beneath tulip poplar trees (illustrated). *Esculenta* means "esculent" (edible). *Continued on next page*

Tulip Morel

White Morel

Latin name: *Morchella semilibera* DC.
Synonym: *Mitrophora semilibera* (DC.) Lév.
Common names: Half-free Morel; Dog Pecker
Order: Pezizales
Family: Morchellaceae
Fruitbody: 2½–6 in. (6.5–15 cm) tall, consisting of a flaring, skirt-like, conical, pitted head on a cylindrical stalk: **head** up to 1½ in. (4 cm) broad; conical or broadly bell-shaped and pointed, with longitudinal ridges and cross ridges forming pits more or less arranged in rows; yellow-brown to dark brown, sometimes with olivaceous tones, ridges darker than the pits; margin at the base flaring and free from the stalk; **flesh** thin, whitish, brittle, odor "earthy"; **stalk** cylindrical, equal or more often tapering upward from a broader base, often with shallow lengthwise furrows or ridges, hollow, brittle; creamy white to yellowish white; surface roughened to granulose.
Spores: 22–30 x 12–18 μm, broadly elliptical, smooth, hyaline (pale ochre in mass).
Occurrence: Usually gregarious but sometimes solitary on the ground in broad-leaved woods and grassy woodland margins, especially near tulip poplars; saprobic; spring; common.
Edibility: Edible when cooked but inferior to other morels.

Comments: This is the only true morel with a margin not attached to the stalk at the base of the head. Slicing the morel lengthwise will show that the stalk attachment is about halfway up from the margin of the head. The somewhat similar *Verpa conica* (not illustrated) has a domed, bell-shaped head that is attached only at the top of the stalk, and although the surface of the head can be somewhat wrinkled, it lacks pits and ridges. *Verpa bohemica* (not illustrated) has a deeply wrinkled head that is thinly attached to the stalk at the apex. *Semilibera* means "half free."

Xylaria magnoliae, p.497

Mycoparasites and Miscellaneous

This is a catch-all group of odd fungi that do not fit into any of the other sections. All are distinctive and easily recognized by their overall form and growth habitat.

Latin name: *Asterophora lycoperdoides* Ditmar
Synonym: *Nyctalis asterophora* Fr.
Common name: Powder-cap Asterophora
Order: Agaricales
Family: Tricholomataceae
Width of cap: ½–1 in. (1.3–2.5 cm)
Length of the stalk: ¾–2 in. (2–5 cm)
Cap: Hemispheric with an inrolled margin, becoming convex; white, surface dull, dry, fibrillose-roughened, soon disintegrating into a mass of brown powder (asexual spores); **flesh** whitish, odor and taste farinaceous.
Gills: Adnate; well-spaced, thick, blunt, at times reduced to little more than veins, often sterile; whitish to pale gray or pale pinkish buff.
Stalk: More or less equal or tapering at the base; white at first, staining brown with age; surface smooth to slightly velvety.
Spore print: White from basidiospores.
Spores: Basidiospores (when produced) 5–6 x 3.5–4 μm, broadly elliptical, smooth, non-amyloid. Asexual chlamydospores 14–17 x 12–16 μm, subglobose, spiny, brown.
Occurrence: Clustered in small groups on decaying russulas, frequently on *Russula dissimulans;* mycoparasitic; summer-fall; occasional to common in some years.
Edibility: Not edible.

Comments: Although this mushroom can produce two types of spores, the basidiospores that are formed on gills are often lacking. The powdery brown, chlamydospores are produced on the upper surface of the cap. These asexual spores have spiny "star-like" projections, which accounts for the genus name *Asterophora,* meaning "star-bearer." The specific epithet *lycoperdoides* means "resembling *Lycoperdon,*" which is a genus of puffballs. Other small, whitish mushrooms that grow on the remains of decomposed mushrooms include *Collybia tuberosa* (not illustrated) and *Collybia cookei* (not illustrated). Both of these have thin caps, close gills, and small tuber or seed-like nodules at the stalk base. Compare with *Asterophora parasitica* (p.491).

Latin name: *Asterophora parasiticus* (Bull.) Singer
Synonym: *Nyctalis parasitica* (Bull) Fr.
Common name: Parasitic Asterophora
Order: Agaricales
Family: Tricholomataceae
Width of cap: ⅜–¾ in. (1–2 cm)
Length of the stalk: ½–2 in. (1.3–5 cm)
Cap: Bell-shaped to convex with an incurved margin when young, later becoming broadly convex with an uplifted margin; whitish to brownish gray, sometimes tinged violaceous; surface smooth, silky-fibrillose; **flesh** thin, pale brown, odor and taste somewhat farinaceous, unpleasant.
Gills: Adnate to notched with a decurrent tooth, thick, widely spaced, occasionally forked; whitish to grayish brown.
Stalk: Slender, equal, sometimes curved, becoming hollow in age; colored like the cap, often with brownish stains near the base; surface fibrillose to somewhat scurfy.
Spore print: White.
Spores: Basidiospores 5–6 x 3–4 μm, elliptical, smooth, hyaline, non-amyloid; chlamydospores 12–16 x 9–11 μm, elliptical to spindle-shaped, smooth, hyaline, non-amyloid.
Occurrence: In small groups clustered on decaying fruitbodies of russulas, frequently on *Russula dissimulans* and its close relatives; mycoparasitic; summer-fall; uncommon.
Edibility: Not edible.

Comments: Reduced numbers of basidiospores are produced on the gills of this unusual mushroom, then the gill tissue itself becomes a source of asexual chlamydospores. Compare with and see notes under *Asterophora lycoperdoides* (p.490). *Parasitica* means "parasitic."

Latin name: *Cryptoporus volvatus* (Peck) Shear
Synonym: *Polyporus volvatus* Peck
Common name: Veiled Polypore
Order: Polyporales
Family: Polyporaceae
Fruitbody: ½–2 in. (1.3–5 cm) wide; a more or less globose to domed hoof-shaped, stalkless, developing an opening on the underside at maturity; **exterior surface** tough, leathery, smooth; whitish to yellowish at first, becoming tan to brownish; **interior** with white, solid flesh up to ½ in. (1.3 cm) thick, and a layer of tubes beneath the flesh; pore surface white to brown, pores small (3–4 per mm).
Spore print: Pinkish buff.
Spores: 8–12 x 3–5 µm, elliptical to cylindrical, smooth, hyaline.
Occurrence: Solitary or gregarious on bark of standing dead conifers, also reported to occur on living trees; saprobic; spring-summer; uncommon.
Edibility: Not edible.

Comments: This fungus might be mistaken for a puffball, but slicing it open lengthwise will reveal a hidden cavity and pore layer within. Beetles and other small insects which investigate the small opening that eventually develops on the underside may be partially involved with the dispersal of spores. *Cryptoporus* means "hidden pores." *Volvatus* means "having a volva," referring to the outer wall that encloses the tube layer.

Latin name: *Daldinia concentrica* (Bolton) Ces. & De Not.
Common names: Carbon Balls; Cramp Balls; King Alfred's Cakes
Order: Xylariales
Family: Xylariaceae
Fruitbody: ½–2 in. (1.3–5 cm) in diameter; hemispherical to more or less globose or pear-shaped woody balls, occasionally with a short stalk-like base; pinkish brown to reddish brown at first (asexual stage), becoming grayish brown to black (sexual stage); **exterior surface** somewhat roughened, obscurely dotted with tiny flask openings (perithecia); **interior** hard, brittle, with a silvery sheen and clearly visible concentric bands of purple-brown, grayish brown, and black when cut vertically, odor not distinctive, taste nutty.
Spore Print: Black.
Spores: 12–17 x 6–9 μm, elliptical with 1 flat side, smooth, dark brown; asexual spores (conidia) are colorless under the microscope.
Occurrence: Usually in clusters on logs, stumps, and larger fallen branches of broad-leaved trees; saprobic; year-round; common.
Edibility: Not edible.

Comments: In the asexual stage the fruitbodies are covered with whitish to pale fawn powdery conidiospores. The sexual stage fruitbodies look like charred blackened balls. These release prodigious numbers of black ascospores that often accumulate like coal dust on the surrounding substrate. The name *concentrica* refers to the concentric zones in the interior.

Latin name: *Hypomyces chrysospermus* Tul. & C. Tul.
Common name: Golden Hypomyces
Order: Hypocreales
Family: Hypocreaceae
Fruitbody: A mold that infects various boletes in three phases. The asexual stage (*Verticillium*) appears first, enclosing the host with a white mold. Asexual spores (conidia) are produced on the mold surface. The white mold is replaced by the second stage (*Sepedonium chrysospermum*), which is yellow or yellow-brown and powdery. This stage also produces asexual spores (aleuriospores). The final stage, which doesn't appear until the bolete host is well-decayed, consists of orange to reddish brown, round or flask-shaped pimples (perithecia) embedded in the remains of the bolete. The sexual stage ascospores are produced within these vessels.
Spores: Conidia 10–30 x 5–12 µm, elliptical, single-celled, smooth, hyaline; aleuriospores 10–25 µm, globose, single-celled, warted, yellow to yellow-brown; ascospores 15–30 x 4–6 µm, spindle-shaped, two-celled, hyaline.
Occurrence: Infects several species of boletes, especially *Boletus chrysenteron* (p.317) and *Boletus parasiticus* (p.330). Also reported to occur on the gilled mushroom *Paxillus involutus* (p.149); parasitic; summer-fall; common.
Edibility: Not edible, possibly poisonous.

Comments: This mycoparasite is especially common in wet summers. Any boletes infected with this or any other mold should not be consumed. *Chrysospermus* means "having golden-yellow seeds."

Latin name: *Hypomyces hyalinus* (Schwein.) Tul. & C. Tul.
Common name: Amanita Mold
Order: Hypocreales
Family: Hypocreaceae
Fruitbody: A whitish to pale pinkish mold that forms on species of *Amanita,* causing them to abort normal development and become club-like or phallic-shaped structures up to 10 in. (25 cm) tall; surface roughened with dots (perithecia).
Spores: Ascospores 13–22 x 4.5–6.5 µm, spindle-shaped, divided into 2 cells, warted, hyaline.
Occurrence: Host mushrooms occur singly or in groups on the ground in woods; summer-fall; common.
Edibility: Not recommended, potentially dangerous. The parasite infects both edible and poisonous species.

Comments: Although the most common host for this parasitic fungus is the edible *Amanita rubescens* (p.60), the mold has also been reported to infect some other Amanitas, including members of the deadly *Amanita virosa* (p.62) group. *Hyalinus* means "colorless."

Latin name: *Hypomyces lactifluorum* (Schwein.) Tul. & C. Tul.
Common name: Lobster Mushroom
Order: Hypocreales
Family: Hypocreaceae
Fruitbody: A fungus forming a parasitic crust-like surface on certain gilled mushrooms; size and shape of the host is more or less funnel-shaped, and up to 6 in. (15 cm) or more wide; infected mushrooms become bright orange to deep, rich orange-red, sometimes mottled with white areas; surface of the crust roughened with tiny embedded flasks (perithecia) in which ascospores are produced; **flesh** (of the host) is firm and white, frequently exuding a white latex when broken, odor not distinctive, taste mild to slightly peppery.
Spores: 35–50 x 4.5–7 *µ*m, elongated spindle-shaped, divided into 2 cells, prominently warted, hyaline.
Occurrence: The parasite infects various white species of *Lactarius* and *Russula,* which grow on the ground, sometimes partially buried, in woods; summer-fall; occasional to locally common in some years.
Edibility: Edible by most accounts (see comments below).

Comments: This striking fungus is unmistakable. When *Hypomyces lactifluorum* infects a mushroom, normal development and function of the fruitbody ceases and the gills are transformed into encrusted blunt folds or ridges. At this stage the host is impossible to identify. Although it goes against all common sense to consume any wild mushroom that cannot be identified with certainty, the Lobster Mushroom is by most accounts regarded as a safe edible. However, there are reports of gastrointestinal-type poisonings associated with the Lobster mushroom and caution is advised. *Lactifluorum* means "milk flowing," in reference to the latex-exuding flesh of host species of *Lactarius.*

Latin name: *Xylaria magnoliae* J.D. Rogers
Common names: Magnolia Cone Xylaria; Cone Flickers
Order: Xylariales
Family: Xylariaceae
Fruitbody: 1½–4 in. (4–10 cm) long; thin, wavy, spike-like projections emanating from decaying Magnolia cones; surface somewhat tomentose or powdery and white at first (asexual stage), later becoming black and roughened with protrusions of minute flasks (containing ascospores).
Spores: 11–15 x 3–5 μm, crescent to boat-shaped, smooth, light yellow.
Occurrence: Clustered on fallen, blackened Magnolia cones; saprobic; summer-fall; fairly common.
Edibility: Not edible.

Comments: This fungus is easily recognized since it only occurs on the cones of magnolia trees. The specific epithet refers to this association. *Xylaria hypoxylon* (not illustrated) is similar but grows on decaying wood.

Glossary

acrid (taste): intensely sharp, burning
adnate (gill or tube attachment): broadly attached to the stalk at a right angle
adnexed (gill attachment): narrowly attached to the stalk where only the upper portion of the gill edge is connected
amatoxins (= amanitins): cyclic octopeptide compounds that are highly toxic
amyloid: exhibiting a blue or blue-black color reaction in Melzer's reagent
anastomosing: having interconnections between adjacent gills, or vertical lines that intersect on a stalk
annular zone: a thin ring zone of fibrils or gluten on a stalk; less substantial than an annulus
annulus: remains of a partial veil leaving a ring of tissue on the stalk of some mushrooms
apex (of the stalk): the upper portion just beneath where the cap is attached
apiculus: a projection on a basidiospore where it was attached to the basidium
apothecium (apothecia pl.): cup-shaped fruitbody containing a layer of asci lining the inner surface
appendiculate (cap margin): with overlapping or adhering flaps or fragments of tissue
appressed-fibrillose: with fibrils that lie flat on the surface
appressed-scaly: having flattened scales
areolate: surface cracked or divided into angular block-like units
ascending-adnate (gill attachment): gills curving upward and attached to the stalk at less than a right angle
ascospore: sexual spore of an ascomycete
ascomycete: a fungus that produces spores in an ascus, including morels, false morels, cup fungi, true truffles, and others
ascus (asci pl.): fertile cell of an ascomycete that contains spores, usually 8 in number
basidiomycete: a fungus that produces spores on a basidium, including gilled mushrooms, boletes, coral mushrooms, and others
basidium (basidia pl.): microscopic club-like spore-bearing cell of basidiomycetes
bloom: with a delicate powdery appearance
bolete: a general name for a stalked fleshy mushroom having a removable layer of tubes on the underside of the cap
bracket fungus: a general term for polypores and tooth fungi having a shelf-like, often woody or leathery fruitbody, commonly called a "conk"
buff (color): pale brownish yellow or grayish yellow
bulbous (stalk): having a swollen base
caespitose: term used to describe a cluster of fruitbodies that have stalks growing close together but not fused

cap: variously shaped part of the fruitbody that supports the spore-bearing tissue, also called the pileus
cartilaginous: firm and tough but pliable
chlamydospore: thick-walled asexual spore arising from hyphae
close (gill attachment): a relative term for gills that nearly touch each other but with visible space between
compressed (stalk shape): flattened in cross section
concolorous: with 2 or more parts of a fruitbody having matching colors
concrescent: grown together
conididum (conida pl.): asexual spore, also called conidiospore
convex (cap shape): rounded
corrugated: wrinkled
cortex (of stalk): rind, outer layer
cortina: a partial veil constructed of web-like filaments
crenulate: having very small rounded teeth
crossveins: interconnections between parallel gills or lines of a hymenium
crowded (gill spacing): crammed together with little space between
cuticle (of the cap or stalk): outermost layer of tissue, also called pileipellis as applied to the cap
decurrent (gill attachment): attached gills that extend down the stalk
decurrent tooth (gill attachment): with a tooth-like prolongation of a gill that extends down the stalk
deliquescing: being reduced to liquid
depressed (cap shape): with a sunken center
dextrinoid: exhibiting a brick red or purplish brown color reaction in Melzer's reagent
disc: center of a mushroom cap, usually the area directly above the attachment of the stalk
eccentric (stalk attachment): off-center but not on the cap edge
endoperidium: inside layer of a peridium
equal (stalk shape): with a stalk diameter unchanging for the entire length
eroded (gill edge): uneven, irregularly degraded
evanescent (= fugacious): briefly present, then vanishing
exoperidium: outer layer of a gasteromycete when more than 1 layer is present
fairy ring: circular or arched zone of multiple fruitbodies
farinaceous (odor or taste): of freshly ground flour; (**of a surface**) mealy, covered with fine particles
fertile surface (= hymenium): layer of tissue on which the spore-bearing structures are produced
fibril: minute thread-like fiber
fibrillose: covered with delicate fibrils or hairs

fibrillose-matted: surface texture of interwoven fibrils, felted
fibrillose-scaly (of cap or stalk): having scales made up of fibrils
flesh: sterile tissue of a fungus fruitbody, also called the context
floccose: covered with loose, cotony scales
fruitbody (= carpophore, sporocarp): the stage of a fungus that produces and disperses reproductive spores
fugacious: lasting a short time (see evanescent)
gasteromycete: a fungus with a hymenium enclosed within 1 or more layers, and which does not forcibly discharge its spores (ex. puffballs and earthstars)
germ pore: a zone of reduced wall thickness on a spore through which the spore may germinate
gills: radiating plate-like structures on the underside of the cap of many mushrooms, also called lamella (lamellae pl.)
glabrous: smooth, unadorned, bald
gleba: interior fertile part of a gasteromycete or truffle
globose: spherical, round
gluten: a viscous substance resulting from the dissolution of gelatinous tissue
glutinous: covered with a gluten-like substance
gregarious: growing in groups in a small area but separated
hemispheric: half-round
homogeneous (of the flesh): undifferentiated
hyaline: colorless and clear
hygrophanous: changing color or transparency with the influence of moisture
hygroscopic: readily absorbing moisture and altering the tissue structure as a result
hymenium: spore bearing layer
hypha (hyphae pl.): individual microscopic strand of a mycelium or fruitbody
hypogeous: developing below ground
inrolled (cap margin): rolled inward
KOH (potassium hydroxide): caustic soda used as a chemical reagent
labyrinthine: maze-like, convoluted
lamellula (lamellulae pl.): a short gill between normal gills (lamellae) that does not reach all the way to the stalk
latex: a fluid that exudes from cut or broken tissue of some mushrooms, especially in the genus *Lactarius*
lignicolus: growing on or in wood
lignin: the component of wood that gives it structural strength and durability
long-decurrent (gill attachment): gills that extend down the stalk for much of their length
lubricous: slippery, greasy to the touch
luminescent (= bioluminescent): emitting light, glowing in the dark

macrofungi (= macromycetes): any fungus that produces a fruitbody that is visible to the unaided eye
macroscopic: feature large enough to be seen without the aid of a microscope
Melzer's reagent: a chemical compound made up of chloral hydrate, potassium iodide, and iodine used to test for amyloid or dextrinoid reactions
membranous: resembling a thin skin
mycelium (mycelia pl.): mass of nutrient absorbing hyphae that forms the vegetative stage of a fungus
mycoparasite: a fungus that parasitizes another fungus
mycophagy: consuming fungi as food
mycorrhiza: mutually beneficial association between a fungus mycelium and the root system of a vascular plant
non-amyloid (spores or tissue): remaining colorless or yellowish in Melzer's reagent
notched (gill attachment): narrowly attached gills that have a notch or wedge-like space where the gill turns upward before contacting the stalk
ochraceous (color): dull brownish yellow
oil drops (within spores): what appears as oily droplets under a microscope
olivaceous (color): with olive green tones
ovate: an egg-shaped two-dimensional surface such as spores as seen under a microscope
ovoid: egg-shaped
pallid: barely colored (can be various hues)
parasitic: a fungus that obtains its nutrients from a living organism
partial veil: a protective layer of tissue that in some mushrooms extends from the cap margin to the stalk, enclosing the immature hymenium
pathogen: a parasitic fungus that causes disease or death of a host
pellicle (= pellis): a viscid, peelable outermost layer of a mushroom cap
perennial: a fruitbody that persists and produces spores for more than 1 year
peridiole: a lens-shaped spore case of a bird's nest fungus
peridium: outer layer enclosing the spore case of a gasteromycete
perithecium (perithecia pl.): flask-shaped structure containing asci
polypore: a general term used to describe a fruitbody with a poroid fertile surface that cannot easily be separated from the cap, and is often of a leathery or woody texture
pore: variously shaped mouth of a tube on boletes and polypores
poroid: a fertile surface having circular or almost circular openings in which spores are produced
potassium hydroxide: see KOH
pruinose: appearing to be covered with a fine powder
recurved (cap margin or scales): bent backward

resupinate: lying flat on a surface
reticulate: having a net-like pattern
rhizoid: branching strands of hyphae at the base of a stalk that help anchor a fruit-body and absorb nutrients
rhizomorph: aggregation of hyphae with an inner core of conductive tissue, often penetrating a substrate
rudimentary: under developed
russet (color): reddish brown
saprobe (= saprotroph): a fungus that obtains nutrients from dead and decaying organic matter
scabers: roughened projections on the stalk of some boletes, especially in the genus *Leccinum*
sclerotium (sclerotia pl.): a hard, compact mass of hyphae that is a resting stage for some fungi, and from which a fruitbody may arise
scurfy (= furfuraceous): coarsely granular or bran-like
serrulate (gill edge): notched or irregularly toothed
sessile: lacking a stalk
short-decurrent (gill attachment): only slightly extending down the stalk
sinuate (gill attachment): having a concave indentation (notch) near the stalk
spathulate: shaped like a spoon or spatula, oval with a narrowed base
spermatic (odor): resembling the odor of human semen
sphaerocyst: spherical cell
spore: a microscopic reproductive unit of fungi
spore case (= spore sac): structure containing the spore mass of a gasteromycete
spore print: a mass deposition of spores
stalk: the structure that supports and elevates the cap or head of a fruitbody, also called a stipe
sterile: tissue incapable of producing spores
striate: marked with lines or grooves
stroma (stromata pl.): a hard mass of vegetative hyphae in which perithecia are embedded in certain ascomycetes
subdecurrent (gill attachment): tending toward being decurrent
subdistant (gill spacing): intermediate between close and widely spaced
subglobose: nearly round
substrate: substance on or within which a fungus grows and obtains nourishment
sulcate: with lines more deeply grooved than striate
tawny (color): yellowish brown like a lion's fur
teeth: spike-like or spiny spore-bearing structure on some fungi, especially Hydnaceae
terrestrial: growing on the ground
tomentose: covered with fine, densely matted hairs, woolly

tomentum: an aggregate of soft, matted hairs resembling wool
translucent: a substance through which light can pass
translucent-striate (cap margin): having visible gill lines on the upper side of a cap
truncate: ending abruptly, appearing chopped off
tuberculate-striate (cap margin): having small bumps and striations
umbilicate: depressed, like a navel
umbo: a raised knob or protrusion at the center of a cap
universal veil: an outer covering that encloses the entire immature fruitbody of some mushrooms
vinaceous (color): dull pinkish brown to grayish purple
violaceous (color): having purplish tones
viscid: sticky or slimy
volva: the remains of a universal veil, situated at the stalk base
warts (cap surface): small patches of universal veil distributed on the cap of some mushrooms
warts and ridges (spore ornamentation): having raised projections and connecting lines
zonate: having concentric bands of different color or texture

References and Recommended Reading

Additional Photo-illustrated Guide Books Useful for the Central Appalachians

Barron, George. 1999. *Mushrooms of Northeast North America.* Renton, Wash.: Lone Pine Publishing. 336 pp. A well-illustrated compact field guide that covers northeastern Canada and the United States as far south as West Virginia. Includes section on slime moulds.

Bessette, Alan E., Arleen R. Bessette, and David W. Fischer. 1997. *Mushrooms of Northeastern North America.* Syracuse, N.Y.: Syracuse Univ. Press. 582 pp. Well-illustrated and comprehensive, covering a large area of the northeast and upper midwest and extending as far south as South Carolina. Includes field keys.

Bessette, Alan E., William C. Roody, and Arleen R. Bessette. 2000. *North American Boletes.* Syracuse, N.Y.: Syracuse Univ. Press. 400 pp. Well-illustrated. Comprehensive coverage of boletes, including nearly all species known to occur in the Central Appalachians. Includes field keys.

Lincoff, G.H. 1981. *National Audubon Society Field Guide to North American Mushrooms.* New York, N.Y.: Knopf. 926 pp. The most compact field guide that covers all of North America. Well-illustrated and solid information.

Miller, O.K. 1973. *Mushrooms of North America.* New York, N.Y.: E.P. Dutton. 360 pp. A classic. Out of print and in need of revision but still useful. Good field keys. Photos mostly very good.

Phillips, R. 1991. *Mushrooms of North America.* Boston, Mass.: Little, Brown and Company. 319 pp. Out of print but worth searching for. Approximately 1000 species illustrated with studio photographs. Covers all of North America.

Recommended Books on Selected Topics

Benjamin, Dennis R. 1995. *Mushrooms: Poisons and Panaceas.* New York, N.Y.: W.H. Freeman and Company. 422 pp. An overview of mushroom toxins and related topics.

Bessette, Arleen R., and Alan E. Bessette. 2001. *The Rainbow Beneath My Feet.* Syracuse, N.Y.: Syracuse Univ. Press. 176 pp. Comprehensive guide to using mushroom pigments for dying textiles. Well-illustrated.

Both, E.E. 1993. *The Boletes of North America—A Compendium.* Buffalo, N.Y.: Buffalo Museum of Science. 436 pp. No illustrations but an invaluable reference for the study of boletes.

Kendrick, Bryce. 1992. *The Fifth Kingdom,* 2nd ed. Newburyport, Mass.: Focus Information Group. 406 pp. An excellent overview of the kingdom of fungi, covering biology, ecology, and other related topics. Illustrated. Somewhat technical.

Largent, D.L. 1977. *How to Identify Mushrooms to Genus I: Macroscopic Features.* Eureka, Calif.: Mad River Press. 86 pp. Very useful line-illustrated guide to the features and stature types used in the field identification of gilled mushrooms.

Largent, D.L., and T.J. Baroni. 1988. *How to Identify Mushrooms to Genus VI: Modern Genera.* Eureka, Calif.: Mad River Press. 277 pp. A valuable reference to the genera of gilled mushrooms. Includes keys to genera, charts, a glossary of terms, and genus profiles. No illustrations.

Schaecter, Elio. 1997. *In the Company of Mushrooms: A Biologist's Tale.* Cambridge, Mass.: Harvard Univ. Press. 280 pp. An engaging narrative sampling of many fascinating aspects of mushroom biology, ecology, and folklore. Illustrated.

Stamets, P. 2000. *Growing Gourmet and Medicinal Mushrooms,* 3rd ed. Berkeley, Calif.: Ten Speed Press. 574 pp. An in-depth guide for growers with information on medicinal properties of mushrooms. Well-illustrated. Comprehensive.

Stephenson, Steven L., and Henry Stempeh. 1994. *Myxomycetes, A Handbook of Slime Molds.* Portland, Oreg.: Timber Press. 183 pp. An excellent account of the slime molds. Well-illustrated.

Weber, Nancy Smith. 1988. *A Morel Hunter's Companion: A Guide to the True and False Morels of Michigan.* Lansing, Mich.: Two Peninsula Press. 209 pp. Informative guide to the morels and their relatives, includes recipes. Well-illustrated.

Resources

Field Mycology. Cambridge Univ. Press, 40 West 20th St., New York, NY 10011-4211. A quarterly publication from the British Mycological Society with emphasis on field identification of mushrooms plus other interesting articles and features. Its focus is mostly on mushrooms and fungi of Britain and Europe, but it is also surprisingly relevant for American enthusiasts.

Mushroom: The Journal of Wild Mushrooming, Box 3156, Moscow, ID 83843. A quarterly magazine devoted to all aspects relating to the enjoyment of mushrooms, including mushroom hunting, cooking, growing, arts & crafts, book reviews, events, and more.

North American Mycological Association, 6615 Tudor Ct., Gladstone, OR 97027-1032, website: <www.namyco.org>. A nationwide organization that promotes amateur mycology. NAMA provides many educational services and conducts annual mushroom forays, photo contests, and more. Membership includes a bimonthly newsletter and a yearly journal.

Index to Scientific and Common Names

Page numbers in bold indicate detailed description and photo

About the Author

William C. Roody is a seasonal biologist with the Wildlife Diversity Program at the West Virginia Division of Natural Resources. He is co-author of *North American Boletes: A Color Guide to the Fleshy Pored Mushrooms* and *Preliminary Checklist of Macrofungi and Myxomycetes of West Virginia.* He has won numerous awards for his mushroom photography, and his photos have appeared in several mushroom guidebooks. In the year 2000 Bill was the recipient of the North American Mycological Association Award for Contributions to Amateur Mycology. He and his wife, Donna Mitchell, live in Barbour County, West Virginia, where together they study and collect mushrooms for science and gastronomy.